Praise for *Capture*

"Kessler proposes an original theory of the mind. His cogent argument is that a great deal of the apparently inexplicable behavior of human beings is the result of impulses, drives, and obsessions that may share fundamental neural and psychodynamic mechanisms. This carefully researched book is both startling and engaging, and is written with brio."

—Andrew Solomon, National Book Award–winning author of *The Noonday Demon*

"In this richly documented, beautifully written, and original work, David Kessler has given us an idea that explains one of the most strange and most powerful processes in the human brain."

—E. O. Wilson, University Professor Emeritus, Harvard University

"*Capture* is a breakthrough book. In a world of increasingly specialized knowledge, it takes a particular gift and some stubbornness to cut across the fields of neuroscience, psychiatry, philosophy, and psychology, and to ask the fundamental question: Why is it that we allow our best selves to be captured and torpedoed by thoughts and actions that sink us? Kessler's exploration of the question makes for a compelling read. His ultimate answer is profound and one that could be life changing and life saving. I know I will be handing this book out for just that reason."

—Abraham Verghese, MD, author of *Cutting for Stone*

"This book offers a bold, overarching explanation for many of the great problems of the mind, problems that are often merely named. Dr. Kessler writes persuasively and with unusual clarity. *Capture* is an engrossing book, impressive in its cultural as well as its scientific reach."

—Tracy Kidder, Pulitzer Prize–winning author of *The Soul of a New Machine* and *Mountains Beyond Mountains*

"*Capture* defines a shape of human experience about which I'm pretty sure no one has ever written before, a shape that seems to inform everything from the smallest action to the largest life aim, a unified-field theory of human activity that draws in how we form thoughts, manage trauma, and even try to reconcile will and cause."   —Chris Ware, author of *Building Stories*

"Kessler writes about the concept of capture, or forces that strongly influence the mind, overriding reason and will. Invoking novelists, Freudian drives, and current neuroscience, Kessler explains how capture motivates, clarifies thoughts, and provides insight. A challenging and rewarding book for both scholars and lay readers."
    —*Library Journal*

"Kessler is an excellent storyteller, and *Capture* is bursting with human drama drawn from real lives rather than the bland, composite case studies that clinicians tend to favor."
    —*Washington Post*

"[A] Big Idea about how to conceptualize the mind and the brain. . . . *Capture* offers a new lens through which to understand human behavior."   —*New York Times Book Review*

CAPTURE

ALSO BY DAVID A. KESSLER

The End of Overeating

A Question of Intent

CAPTURE

UNRAVELING THE MYSTERY OF

MENTAL SUFFERING

DAVID A. KESSLER, MD

HARPER PERENNIAL

NEW YORK • LONDON • TORONTO • SYDNEY • NEW DELHI • AUCKLAND

For Ben

A hardcover edition of this book was published in 2016 by Harper Wave, an imprint of HarperCollins Publishers.

The names and identifying characteristics of some individuals discussed in this book have been changed to protect their privacy. This book is for informational purposes only. You should seek advice from your medical professional for any specific health concerns you may have.

HarperCollins books may be purchased for educational, business, or sales promotional use. For information, please e-mail the Special Markets Department at SPsales@harpercollins.com.

Grateful acknowledgment is made for permission to reprint from the following:

The Collected Poems of Sylvia Plath, edited by Ted Hughes. Copyright © 1960, 1965, 1971, 1981 by the Estate of Sylvia Plath. Editorial material copyright © 1981 by Ted Hughes. Reprinted by permission of HarperCollins Publishers and Faber & Faber Ltd.

Sexton, Anne. *Live or Die*. Copyright © 1966 by Anne Sexton, renewed 1994 by Linda G. Sexton. Reprinted by permission of Houghton Mifflin Harcourt Publishing Company. All rights reserved.

Stein, Edith. *The Hidden Life: Hagiographic Essays, Meditations, Spiritual Texts*, edited by Dr. L. Gelber and Michael Linssen, OCD. Translated by Waltraut Stein, PhD. English translation copyright © 1992 Washington Province of Discalced Carmelites. ICS Publications, 2131 Lincoln Road N.E., Washington, DC 20002-1199, www.icspublications.org.

Robert Graves: The Centenary Selected Poems, edited by Patrick Quinn. Copyright © 1995 by the Trustees of the Robert Graves Copyright Trust. Reprinted by permission of Carcanet Press Ltd.

FIRST HARPER PERENNIAL EDITION PUBLISHED 2017.

Designed by Jo Anne Metsch

Library of Congress Cataloging-in-Publication Data

Names: Kessler, David A., 1951– author.
Title: Capture : unraveling the mystery of mental suffering / David A. Kessler, MD.
Description: First edition. | New York, NY : Harper Wave,
    [2016] | Includes bibliographical references.
Identifiers: LCCN 2015030362| ISBN 9780062388513
    (hardcover) | ISBN 9780062388537 (ebook)
Subjects: LCSH: Mental illness. | Mental illness—Etiology. | Mental illness—
    Psychological aspects. | Psychology, Pathological. | Psychiatry.
Classification: LCC RC454 .K47 2016 | DDC 616.89—dc23 LC record
    available at http://lccn.loc.gov/2015030362

ISBN 978-0-06-238852-0 (pbk.)

17 18 19 20 21   LSC   10 9 8 7 6 5 4 3 2 1

CONTENTS

An unfamiliar light—
and suddenly it's your whole life.
Black out the windows and board them up
and it's still there in the bowl of the spoon
and in the soup
and in the lamp and in the air after the lamp goes out.
The others can't see it. They creep up, pitying you.
All their love won't make it go away.
Look at this instead, they say, offering up some old argument.
The light never goes out. It is so beautiful.
Someday you might learn to ignore it . . .

—SARAH MANGUSO

PART I

I

A HUMAN MYSTERY

THE TERRIBLE MASTER

He left more than a dozen lamps burning in his workroom. They shone upon the desk, and on the unfinished manuscript neatly stacked on top of it. Next to the manuscript was a two-page letter. This was the scene on the evening David Foster Wallace hanged himself.

Wallace's suicide at the age of forty-six devastated the literary community. He was, at the time, acclaimed as the boldest, most innovative writer of his generation. His novel *Infinite Jest* was widely lauded by critics and thought to have redefined postmodern American fiction. The manuscript on the desk, which he despaired of ever completing, would be published posthumously as *The Pale King*. Though it was fragmentary, the novel, many would later argue, contained some of his best work. Despite Wallace's frustration with his inability to complete the book, in some ways his life had never been better. He had married four years earlier and was comfortably settled in California, with a teaching job he loved. Why, then, did he take his own life?

It's not that Wallace's suicide came as a surprise to anyone who knew him well. He had been troubled since adolescence: brilliant, yet stricken with self-doubt and, at times, a paralyzing self-awareness. As a young man, he depended on alcohol and marijuana to dampen his constant anxiety. He had come close to suicide before, and in his fiction he had written of the state of mind that drives one to that abyss. Yet he had also tried to save himself. He had been prescribed antidepressants in college and continued taking them throughout the rest of his life. At the time of his death, he was a dedicated member of Alcoholics Anonymous, having successfully left behind both alcohol and marijuana years before. Though prone to fits of anger and reclusive behavior, he fought his self-destructive impulses and sought community—with his fellow AA members, devoted students, and literary colleagues.

The questions remain: Why would someone who struggled so long not just to stay alive but also to stay vital and connected end his life so violently? Why would someone with such recognized talent choose not to go on? What was it that was beyond the grasp of his clearly formidable intellect and will? What was the underlying cause of the depression that governed Wallace's deep unhappiness? "Depression" is a label used to describe a group of symptoms. It is not a cause.

From an early age, Wallace wanted to be exempt from the ordinary. He wanted to excel—first as a student, and later as a writer—and he wanted others to recognize his genius. Yet, as soon as he succeeded—if he earned an A-plus or received critical acclaim—he grew uneasy, and then despairing. He wanted to be a good person, but suspected something crooked about the way in which he'd achieved success, something false in himself.

Those closest to Wallace recall that this inner conflict developed in boyhood. Contradictory impulses—yearning for greatness yet feeling like a fake with every new achievement—pushed him further into himself. He wrote about the phenomenon in his short story "Good Old Neon," in which an advertising executive describes his own suicide posthumously: "The more time and effort you put into trying to appear impressive or attractive to other people, the less impressive or attractive you felt inside—you were a fraud. And the more of a fraud you felt like, the harder you tried to convey an impressive or likable image of yourself so that other people wouldn't find out what a hollow, fraudulent person you really were."

Occasionally Wallace worked double time in order to bend reality to meet his fears, employing a kind of meta-logic reminiscent of his fiction. "A lot of his criticism wasn't that he was stupid in the whole bell curve of the population, but that what he was really good at was pretending to be *really* smart," the novelist Mark Costello told me about his old friend and roommate. "The way Dave kept self-pessimism alive was by creating this other narrative where he said, 'Well, actually, what I am is . . . a false sort of smart.' "

Wallace was haunted by the "fraudulence paradox," as he called it in "Good Old Neon." As an adult, he was always on high alert, always sensitive to signs of the beguiling impostor that, though he must have known it exists in all of us, he could not allow for in himself. He once scribbled in the margin of a book, "Grandiosity, the constant need to be and be seen as superstar." Something about this notion stuck and became a reflexive thought—one that made him feel very bad—when he encountered something that threatened his sense of credibility.

And any number of things could threaten his sense of credibility: critical praise, academic success, romantic attention, somebody laughing at his jokes. These were all land mines and prompted by them Wallace felt an immediate split between how he was perceived and who he really was. In these moments, his life became a lonely performance. Everything else receded into the background. This feeling encompassed him more strongly each time he experienced it, gaining traction in his mind.

Depression involves a continual focus on negative thoughts, experiences, memories, and feelings to the exclusion of all else—a kind of inverse learning. As a person narrows his or her attention, focusing on only the most negative stimuli, the mind slowly devours itself. This process seemed to be particularly true for Wallace. It would be impossible to know just how, and in how many ways, he was gripped by self-doubt, but it seems fair to say that he was seized by his self-destructive refrain. He knew it, but felt powerless to change it.

"What goes on inside," he wrote in "Good Old Neon," "is just too fast and huge and all interconnected for words to do more than barely sketch the outlines of at most one tiny little part of it at any given instant."

In a 2005 commencement address he delivered at Kenyon College, Wallace advised the graduating class to "think of the old cliché about 'the mind being an excellent servant but a terrible master.' This, like many clichés, so lame and unexciting on the surface, actually expresses a great and terrible truth," he said. "It is not the least bit coincidental that adults who commit suicide with firearms almost always shoot themselves in the head. They shoot the terrible master."

CAPTURE

What happens when our rational minds feel as though they've been hijacked by something we cannot control? I hope to begin to unravel the mystery of mental suffering, to decipher the underpinnings of a range of intense mental afflictions—including addiction, depression, anxiety, mania, obsessive thoughts, and violent anger.

For more than two decades, I have researched the ways in which certain substances—specifically, tobacco and food—come to influence and, in some cases, control our actions. What most fascinates me is how these substances seem to override both reason and will, directing our thoughts, feelings, and behavior, apparently without our consent. Nobody ever decides he's going to smoke 780,000 cigarettes over the course of a lifetime—people think they're going to try smoking once, or perhaps a few times, to see what it is like. We don't want to eat until we feel sick, but many of us do it anyway. What else, I wondered, could exert such control over our thoughts and actions? Is it possible that the same biological mechanism that selectively controls our attention and drives us to chain-smoke or overeat—in other words, to behave in ways that are not beneficial to our well-being—is also responsible for a range of emotional suffering?

After years of research, I have come to the conclusion that there is, in fact, a common mechanism underlying many of our emotional struggles and mental illnesses. Simply put: a stimulus—a place, a thought, a memory, a person—takes hold of our attention and shifts our perception. Once our attention becomes increasingly focused on this stimulus, the way we think and feel, and often what we do, may not be what we consciously want. I have termed this mechanism "capture." Capture underlies many forms of human behavior, though its effects may be detrimental or beneficial. By viewing our behavior through this lens, I hope to help explain the power that capture has over us when it drives us to destructive impulses.

The theory of capture is composed of three basic elements: narrowing of attention, perceived lack of control, and change in affect, or emotional state. Sometimes these elements are accompanied by an urge to act. When something commands our attention in a way that feels uncontrollable and, in turn, influences our behavior, we experience capture.

Things command our attention all the time. We notice anomalies in our surroundings; a yellow dot on a screen of black dots will attract us, as will a sudden movement or the sound of someone calling our name. But capture entails more than just that initial marshaling of our attention. Capture changes our mood by evoking memory or imagination, desire or fear. When I am captured, I think that the person calling my name in a certain tone is not simply saying my name, but offering an implicit criticism or personal slight that makes me feel insecure or anxious; soon, it is all I can think about.

There are thousands of examples of mini-captures in our day-to-day lives. Here is one example from my life. I am sitting on a plane planning to write about the process of "being captured." I have settled in for the long flight ahead, one that I view as an opportunity for uninterrupted writing. I have a plan. I open my laptop, but as soon as I do, I begin to hear the voices of the two men sitting next to me rapt in a loud conversation. To me they are talking at the decibel level equivalent to truck traffic. As I try to concentrate on my work, I keep on hearing their conversation. The vrooming noise of the jet engines nearby, with an output of twenty-five thousand pounds of force thrust, does not enter into my consciousness. The more I try to focus on my work, the more I hear the conversation. I become more and more annoyed. The more fervently I don't want to hear the conversation, the more I do. If I could shut it out, I certainly would.

While the exact nature of such everyday experiences varies, the experiences themselves are not uncommon—yet the exact object or thought that has the power to capture is highly personal. Interestingly, I had not even begun to write when I noted the men's voices in conversation. Did I have a background expectation that they would disturb me based on prior experience? Otherwise, why would I have focused on them in the first place? Or were they just so loud that the sensory quality of the noise captured my attention? I have always hated noisy distractions when working, but why had I not acquired the capacity to strategically shift my attention? Whenever I tried, my attention was pulled back to the voices. It would be fair to conclude that I "attended" excessively to those voices.

We routinely scan our environments, allocating and shifting attention based on characteristics of stimuli we encounter and the goals we intend to pursue.

Capture seizes our attention quietly. We may sense a mental shift, but we do not understand where it comes from. The experience occurs outside our conscious control, and we surrender to it before we perceive it. It is the result of something—an idea, sensation, person—that comes to dominate our minds, thrusting aside all else and occupying the center of our consciousness. Cues that are associated with the source of capture can become as significant as the source itself. Whether a perfume scent or a song, the sound of a helicopter or the sight of a fire truck, a good grade on a test or the prospect of making a speech—association with any and all stimuli affects everyone differently (or not at all), and the power it asserts will be relevant only to those primed for it. Even the briefest encounter with one of these cues can narrow our attention and affect the way we feel.

When we are drawn to a particular stimulus, we act in response to a feeling or need aroused by it. Every time we respond, we strengthen the neural circuitry that prompts us to repeat these actions. As we continue to react in the same ways to the same stimulus over time—thereby sensitizing the learning, memory, and motivational circuitry of our brains—we create emotional and behavioral patterns. Our thoughts, feelings, and actions begin to arise automatically. What started as a pleasure becomes a need; what was once a bad mood becomes continuous self-indictment; what was once an annoyance becomes persecution. This process of neural sensitization occurs, and grows stronger, over the course of a lifetime. It becomes increasingly difficult for us to resist its pull. Eventually, what captures us can become so concentrated and overwhelming that, in its most drastic forms, it feels as if we are being driven by something outside our control.

While capture is often the source of great pain and suffering, it can also grip us in positive ways. The joy of hearing a beloved song, a visit to the quiet interior of a church, the pursuit of a worthy cause in which we believe—all these, too, can be sources of capture.

The genesis of capture is profoundly individual. Our life histories and narratives result from the singular totality of our actions and experiences. What captures me affects who I am—and who I am affects what can capture me.

An ideal way to understand capture is to see it at work—to experience it, albeit indirectly, through the voices of those who have struggled with it. The narratives in this book express a range of thoughts, feelings, and actions—some transient, but many life-altering. These narratives of mental distress may seem unrelated, but the central role of capture is evident in each of them.

THE NATURE OF MENTAL DISTRESS

David Foster Wallace had come closer than any psychologist, psychiatrist, or other human being to narrating how our minds can seize us, compelling us to act against our will and reason. Though I didn't say so explicitly, I hoped to enlist the help of his parents, James and Sally Wallace, in understanding the "terrible master" their son had described. They agreed to speak with me about David.

I visited James and Sally in their home in Urbana, the house in which David grew up, located near the campus of the University of Illinois Urbana-Champaign, where James once taught as a professor. Built in the 1950s, the modest one-and-a-half-story house sits on a quiet tree-lined street, is graced with a southern exposure, and is surrounded by a deep-green lawn. It is a home like so many in America and it produced one of the most prodigious minds of a generation.

Underneath David's fierce talent and preternatural intelligence lay the seeds of what ultimately led to his suicide. The medical model has long been that depression is the manifestation of a "broken" mind—a biological error—that causes people to experience what feels like unbearable pain. This oversimplification does nothing to help us understand the debilitating force of what we have come to call depression.

Sally tried to explain what precipitated David's pain. "The first time David manifested any behavior that we thought was troublesome, he was about sixteen," she said. "He was our first child, and we didn't know much about adolescents. He was mean to his sister; he was grouchy, moody. He was smoking some marijuana, but we knew about it. We didn't know exactly how much. We assumed that if we said it was okay for him to smoke it in the storage room right off his bedroom, then at least he wouldn't be out driving. Then Jim took him around to colleges, and David was terrified."

"He had anxiety attacks," James recalled, "really bad ones, ashen gray, vomiting, trembling . . ."

"He was glad Amherst accepted him early so he wouldn't have to go through that again," Sally said. "When it came time to think about going off to college, though, he got terribly upset, saying that he didn't want to go, could he please just stay home. We said, 'Yes, if you want to stay home. But know that you can always come home, too.' So off he went. So there was that anxiety. But then he took two separate semesters off. He would start the semester and then have to come home; he just couldn't function. Then he took himself to a psychiatrist—because he'd outgrown the pediatrician we all loved— and was given some medication. I think he started off with Nardil."

We talked for hours about what it might have felt like for David. At one point, without prompting, Sally whispered the word "fire."

"Kate Gompert?" I said, invoking David's fictional character.

In *Infinite Jest*, David wrote as Kate Gompert, who struggles with a "predator-grade depression" referred to simply as "It": "Make no mistake about people who leap from burning windows. Their terror of falling from a great height is still just as great as it would be for you or me standing speculatively at the same window just checking out the view. . . . The variable here is the other terror, the fire's flames: when the flames get close enough, falling to death becomes the slightly less terrible of two terrors."

"Yes, jumping out of a burning building," Sally remarked, reflecting on this passage. "That's the way I view David's suicide. Having to end the pain no matter what," she said. "David was going to be free from the pain. He needed to do it. It wasn't that he wanted to do it. He needed to do it."

She began to cry. "I'm sorry," she said, raising her hands to her face.

My conversations with the Wallaces, always candid and often difficult, kept coming back to this idea: David was driven, in the end, by something that he could not control. His father, a professor of moral philosophy, examined this notion—that there are passions that arise within us that we cannot resist—through the lens of philosophy and history. We discussed Socrates and the Socratic doctrine, which posits that nobody willingly does what he or she regards as irrational.

"Socrates would say, if a person goes on taking drugs, for example, then the person is ignorant. The person really doesn't think that the thing to do is to stop," James Wallace offered. "Or if the person isn't ignorant but keeps on doing it, it must be involuntary, there must be something forcing the person to do it."

But how, then, I asked him, do we allow for the seemingly conscious component of extreme emotion and behavior?

"Plato seemed to get something like this," James said. "Plato realized that there were certain kinds of desires that sometimes couldn't be reasoned away. You look at the ocean water from your raft, and you're terribly thirsty; you want to drink it. You know damn well if you drink it, it won't do any good; it will only make things worse. But the desire doesn't go away with those reasons. You crave it."

And that craving for salt water cannot be tamed by reason?

"Plato thought, by an act of will," James replied, "that you could train yourself, but it wasn't enough to say, 'I'm going to do this now.' You had to work at it until you got strong enough to master these nonrational thoughts. But I'm sure he would agree that there are people who are just out of control completely; their passions have become coercive."

"It's how depression takes hold," I suggested. "It's also how thoughts of suicide can take hold."

The month before he ended his life, his parents recalled, David had seemed a little better. He was more engaged, in fact, than they had seen him for some time. There had been an extended period of crisis prior to this: he'd tried going off Nardil, the antidepressant he'd been taking continually for nearly two decades. "It was, I think, the fatal—the tragic—misstep of deciding to find another medication," Sally suggested. But David had been eager for a life without Nardil. It belongs to an older class of antidepressants, and despite stabilizing his mood, it caused a number of unpleasant side effects. Once he tapered off, however, he discovered that no other drug seemed to work as effectively. He resumed taking Nardil, hoping that it would return him to his steadier self, but to no avail. "The last resort," Sally explained, "wasn't working."

David panicked. He felt that there was now nothing left that could help him get better. He was convinced that there was no way out.

When his parents visited their son for nearly two weeks in August 2008—they had agreed to stay with him while his wife, Karen, was out of town—they noticed a shift in his demeanor. When they first arrived, he looked drawn and exhausted, like somebody who'd been suffering from a serious illness for a long time. But toward the end of their stay, he seemed to have pulled himself together. On the second-to-last day of their visit, they accompanied David to the airport to pick up his wife. "Karen skipped down the escalator, and he ran over and hugged her, and it was just joyous," Sally said. Riding home, she recalled, "we sat in the backseat going back to the house, and although the car was a manual transmission, he drove mostly with his hand on Karen's knee. And I thought, 'Oh, thank God. Thank God. Things are going to be okay.' "

"He fooled everybody," James said. "He fooled his psychiatrist. He fooled his wife."

I wondered aloud whether he had known he was fooling anybody. Perhaps he had made a decision—one that simultaneously offered an answer and an end.

"But everyone was fooled anyway," James countered. "Maybe not intentionally by David, but we were. We all thought he'd turned a corner."

He went on: "What struck me as so strange was that the week or two weeks before he ended his life, he felt better. I guess he found a solution to his problem, right? The problem was this monster, which was himself."

Understandably, James and Sally Wallace grappled with their son's pain and his ultimate resolution. It may well be beyond comprehension for any parents to accept that their children want to end their lives. What is clear from my conversations with them, however, is their penetrating awareness of the negative loop in which their son was caught. No matter what his success, either personal or professional, David filtered out everything that reflected well on him and took in everything that could be construed badly. This kind of detrimental filtering can only lead to crushing self-doubt. Sally and James were witness to the "terrible master" at work, and felt helpless to release David from it.

In this heartbreaking life, a self-perpetuating spiral led to suicide. In other instances, it can lead to self-harm of different kinds, or to violence against others, even homicide. It is my hope that by delving deeper into our understanding of capture and by examining the various experiences of it, we might begin to learn how to release those caught in its viselike grip.

THE SEARCH FOR A
COMMON MECHANISM

Psychiatry was supposed to figure out the terrible master. Yale's Department of Psychiatry is generally regarded as one of the best in the world, one that excels in both research and clinical work. Steve Bunney, once the department chair and now a professor emeritus, is an old friend and colleague.

Steve knew only the bare outlines of David Foster Wallace's story: a gifted writer, in the prime of his life, critically acclaimed, and recently married, decides to kill himself. I filled in some of the details, telling him about Wallace's battle with intense self-doubt, even as he entertained grand ambitions. Wallace hoped to be read in one hundred years, and yet, at the same time, he felt utterly incapable and unworthy of success. He worried that *Infinite Jest*, the novel that catapulted him to fame, had used up all his literary energy, that he might never be able to complete another book again.

I asked Steve about the nature of Wallace's pain. Why does this feeling become crippling for some people but not for others? What is emotional pain? How might he begin to describe it?

"For someone like Wallace, in the simplest form, it is about a disconnect between the person you want to be and the person you perceive yourself to be," Steve posited. "There is a feeling of losing control; this is one of the biggest issues in psychiatry. If you don't have control, then that's when you can get into trouble—whether you have anxiety or depression or whatever—because that can be a threat to your very existence. I can't fully imagine the anguish that makes an otherwise healthy person want to end his life. But I do know that suicidal people feel there is no other way to escape from the negative thoughts and feelings. One of the paradoxes of suicide is that it becomes the last and only way that a person can exert control."

Steve pointed out that it's not uncommon for successful people from all walks of life to feel a similar kind of distress. Research has shown that highly accomplished people often report the feeling of being a "fraud." Many successful people grow concerned, as Wallace did, that if people really knew them, they'd realize they weren't deserving of their achievements. "So it's not uncommon—and in some ways it's useful, I suppose, this self-doubt," Steve said. "But it can also become a pathology, dangerous, even fatal if it gets out of control."

Over the course of his career, Steve has studied multiple ways of evaluating psychiatric distress. He began as an internist with an interest in psychoanalysis, but he devoted much of his time to studying the biological causes of mental disorders. Throughout, however, he remained attuned to the impact of life experiences on brain function and their role in the genesis of mental illness.

I asked Steve who, if anyone, in his field was thinking about these issues in a comprehensive way. "Well," he said, "psychiatry today is mostly about prescribing. It's about finding the right medication."

There was something in his voice that sounded a little wistful. "Don't get me wrong: You have to be happy for the great progress we have made in pharmacology. We can bring real relief, often lifesaving relief, to a lot of people. Maybe medication, in the end, didn't work for David Foster Wallace, but from what you have told me, it bought him a lot of time and helped him to accomplish a great deal.

"The problem is not medication, but our perception of it. I'm afraid it tempts people, especially those outside of the profession, but even some within, to consider mental illness as a simple malfunctioning of the brain," he continued. "Whereas mental illness almost always has to do with the interplay between a predisposition and things that happen outside, in the real world. It is often the interaction between the physical brain and the life experience that the mind is trying to deal with—that's what we have to keep our eyes on. You want to know not just what these disorders respond to but also where these things come from, in the hope that you may find a better treatment."

Steve knew I had become interested in addiction during my tenure as commissioner of the Food and Drug Administration. As part of our investigation of the tobacco industry, I had to learn everything I could about the science of addiction. Steve was also aware of my later research into obesity, in which I examined the ways people grapple with their desire to eat more than they need. I explained to Steve that I was seeing similarities among addiction, overeating, and mental illnesses—anxiety, obsession, impulse disorder, phobia, panic, depression, mania, hypochondria, and even some aspects of psychosis.

"Well, this is interesting," Steve replied. "On the one hand, the effort of the profession has been to make finer and finer diagnostic distinctions. We have the *DSM-V*, which we use to diagnose and classify hundreds of mental disorders. And yet the biology is suggesting that these disorders are related. It's no longer that schizophrenia is completely different from depression—it looks very much as if there's this spectrum that's interrelated rather than separate neurotransmitter system diseases. And the drugs, wonderful as they may be, are not so precisely targeted as is often claimed. We see one drug working for a variety of problems—or not working. It seems to be often largely a matter of individual response. There is also the growing evidence of the huge placebo effect in psychoactive drugs, which is, or ought to be, humbling.

"Physics and chemistry are very complicated," he went on. "Astronomy is very complicated. But if we keep working on a problem, we arrive at laws that govern these very complex systems. The laws of thermodynamics, for example, explain a lot of really complicated things we couldn't understand before. I see that you are on a similar path."

Steve's analogy articulated the goal of my work over the last two decades. By looking for commonalities across the intricate array of human emotional experience and behaviors, I have been working to unlock our understanding of the complex and potentially destructive constructs that govern our actions.

The past one hundred years of insights, brought about by advances in psychology and psychiatry, have left us without a firm grasp of what causes mental illness. Major depressive disorder ranks second in disabilities globally. The pharmacological remedies to which Steve refers are of limited value and in certain instances may be doing more harm than good. Perhaps most important, they separate us from our feelings and do nothing to deepen our understanding about why we do the things we do. Of course, this is not a new puzzle, but rather an age-old one. In order to see how we might make progress in this formidable quest, it helps to see how others have grappled with these questions.

2

THE HISTORICAL AND SCIENTIFIC CONTEXT OF CAPTURE

"We pass through terrible trials on account of not understanding our own nature," St. Teresa of Avila wrote in the sixteenth century. "This is the cause of suffering for many people."

For millennia, gods, priests, and philosophers were the doctors of the soul. Faced with the mysteries of the human psyche, the Greeks found in mythology a set of explanations for the otherwise inexplicable. If anguish or insanity was indeed the result of evil deities, then it could be tamed by the intervention of priests. Likewise, emotional outbursts could be ascribed to the vengeful and fickle gods, whose wrath could be assuaged through sacrifice and prayer. By placing the forces responsible for madness outside the individual subject, the Greeks created a rich system for diagnosing, treating, and understanding mental illness and bizarre behavior—a way of naming the ineffable.

In ancient Greece, religious belief mingled freely with the knowledge of anatomy; the recognizable physicality of the Greek gods, their involvement in the everyday fabric of living, blurred the line between the mythological and biological. When gods were offended, they sent invisible fluids into the offending mortal's soul, thereby taking possession, quite literally, of the human mind.

Indeed, vestiges of the demonic—the belief in outside forces that can take up residence in the mind—persisted well into the modern period. Arguing that modern psychotherapy has become a system of mere gimmicks, the American psychologist Rollo May claimed that all human motivation was ultimately "daemonic," or ascribable to Eros, anger, or the desire for power. The demonic, however, need not be destructive; it can be a profoundly creative force. It is most readily apparent in the irresistible attractions of sexual love, or the seemingly superhuman creativity of the poet, artist, or religious leader. According to May, the demonic pervades even our more humble thoughts and actions; it forms the basis of our relationships and many of our creative endeavors. For Goethe, it was "the power of nature"; for Yeats, an "eternal battle" between "the daemons and the gods . . . in my heart."

Only in the eighteenth century did natural philosophy begin to challenge the authority of previously accepted explanations. A new medicine of the mind was in its infancy, one that sought to elucidate the causes of mental distress through logical deduction and scientific rationality. One of the most influential thinkers of the Scottish Enlightenment, David Hume, asserted that our rational faculties always work in the service of the passions. "Reason," Hume writes, "is utterly impotent in this particular. The rules of morality therefore are not conclusions of our reason." Cool-headed reasoning and logical argumentation were, in fact, anything but dispassionate: at best, our rational faculties could excite the passions, which would prompt us to action.

The Enlightenment marked a watershed in the history of Western psychology, as scientists began to articulate the first primarily cognitive model of human behavior. Feelings, however, represented slippery ground for early modern science: they are, by definition, ephemeral and diffuse mental states, with no known anatomical or structural basis. Nonetheless, the Scottish physician Alexander

Crichton insisted that the passions be given due attention in the evolving scientific understanding of human physiology:

> The passions are to be considered, in a medical point of view, as a part of our constitution, which is to be examined with the eye of a natural historian, and the spirit and impartiality of a philosopher. It is of no concern in this work whether passions be esteemed natural or unnatural, or moral or immoral affections. They are mere phenomena, the natural causes of which are to be inquired into; they produce constant effects on our corporeal frame, and change the state of our health.

For Dr. Crichton, all mental suffering could be traced to an "irritability" of the nervous system. Anxiety, grief, anger, and elation all had a physiological basis, even if the precise nature of that basis remained uncertain.

For Crichton, all emotion began with the mental impression of an object—a blooming tree, a human face, an abstract idea. These objects, whether internal or external, excited our passions and, by extension, directed our behavior. As the scientific community gradually assimilated these ideas, consciousness itself lost its aura of divine mystery: once the work of unknown and unknowable forces, our emotional life became the result of a physiological process.

WILLIAM JAMES AND ATTENTION

William James, the philosopher, psychologist, and author of such enduring works as *Varieties of Religious Experience*, identified an essential question at the heart of capture: "Why certain ideas are so strong," he once jotted in his notes for a lecture, as to coerce attention.

James was intellectually driven to make sense of the world for not only himself but also an ever-widening audience. As with David Foster Wallace, James's need to understand human experience was propelled by his torturous emotional life. He repeatedly fell into spirals of self-loathing or terror. "I was on the continual verge of sui-cide," he wrote to a friend in his early twenties, after several months of unrelenting mental and physical pain. (The two were often inter-twined for James.) "My own condition, I am sorry to say, goes on pretty steadily deteriorating in all respects," he wrote to his brother, the novelist Henry James, several years later. Even after James achieved career success, settled into a marriage he described as al-lowing him to be "born again," and spent decades as a beloved pro-fessor at Harvard, he found himself intermittently battling his demons. Yet he managed always to forge ahead, redeeming his strug-gle by using it as the basis for numerous personal and philosophical breakthroughs.

Early in his life, James showed a penchant for the arts; he was a talented painter. While an apprentice to the artist and teacher Wil-liam Morris Hunt in Newport, Rhode Island, he discovered that painting offered him not only a creative outlet but also a window onto nature and its cycles. This interest in landscape painting gradu-ally led to a curiosity about the underlying laws of the natural world and an affinity for science.

James ultimately received his MD degree in 1869, but he did not practice medicine. "I embraced the medical profession a couple of months ago. My first impressions are that there is much humbug therein, and that with the exception of surgery, in which something positive is sometimes accomplished, a doctor does more by the moral effect of his presence on the patient and family, than by anything else," he wrote in a letter to his mother—adding wryly, "He also extracts money from them." He told a friend, "I am about as little fitted by nature to be a worker in science of any sort as anyone can be and yet . . . my only ideal life is a scientific life."

James became interested in physiology as well as psychology, which at that time was considered a branch of philosophy. In 1870 two unexpected developments—James's back went out, leaving him nearly incapacitated, and his beloved cousin died—plunged his mood precipitously downward.

Thereafter, he experienced a crisis. "Whilst in this state of philo-sophical pessimism and general depression of spirits about my prospects," he wrote, "I went one evening into a dressing-room in the twilight to procure some article that was there; when suddenly there fell upon me without any warning, just as if it had come out of the darkness, a horrible fear of my own existence."

He was suddenly faced with the image of an epileptic patient he'd encountered as a medical student—"a black haired youth with greenish skin." The boy would "sit all day on one of the benches, or rather shelves against the wall, with his knees drawn up against his chin, and the coarse grey undershirt, which was his only garment, drawn over them enclosing his entire figure."

James would recount this story, more than three decades later, as the experience of a fictional Frenchman in his book *Varieties of Religious Experience*—he later confessed it was based on his own life. It was the embodiment of his mental deterioration.

The image of this mute, unhappy boy left a searing impression on James. "There was such a horror of him, and such a perception of my own merely momentary discrepancy from him, that it was as if something hitherto solid within my breast gave way entirely, and I became a mass of quivering fear. . . . The experience has made me sympathetic with the morbid feelings of others ever since."

While seeking a revelation strong enough to pull him up from his deep-rooted misery, James turned to the writings of Charles Renouvier, a French philosopher who would, in the years to come, become a lasting influence. But at first, it was by opposing Renouvier that James was able to hoist himself back onto his feet.

"I finished the first part of Renouvier's second *Essais* and saw no reason why his definition of free will—the sustaining of a thought *because I choose to* when I might have other thoughts—need be the definition of an illusion. At any rate, I will assume for the present—until next year—that it is no illusion. My first act of free will shall be to believe in free will."

James came to one of the cornerstones of his intellectual and emotional life by taking issue with Renouvier's idea that we deceive ourselves when we believe in our power over the mind. He pushed against the destructive current of his thoughts and would reclaim a measure of control over his mood.

Regardless of whether his belief was credible, it galvanized James. He soon realized it was not enough to adopt a strong point of view; he needed to act on it. "Now I will go a step further with my will, not only act with it, but believe as well," he wrote in his journal; "believe in my individual reality and creative power."

Once James's spirit had begun to improve, he ventured still further, declaring that he'd given up the belief that mental disorders had a physical basis. When his father asked him how he'd come to make this change in himself—from sadness to "great effusion"—James replied in part that "he saw that the mind did act irrespectively of material coercion, and could be dealt with therefore at first-hand, and this was health to his bones." He reasoned that disorders such as melancholy or hypochondria—two afflictions that James himself battled—were not simply irreversible physiological inheritances, but in fact, could be deliberately refused.

In taking this up as a cause, in his own life and as an organizing pursuit within the study of psychology, James granted himself both strength of conviction and the positive emotional charge that accompanies it. He had stepped out of the jaws of misery—even if only for a time. This triumph would set the tone for the course of his career. "Despair lames most people," he would proffer in a lecture more than thirty years later, "but it wakes others fully up."

James was not the first, of course, to pose questions of how we surrender authority over, or regain control of, our thoughts and actions. As we've learned, the ancient Greeks sought to explain the mysteries of human behavior in supernatural or cosmological terms. In Homer's epic poem *The Iliad*, King Agamemnon, leader of the Greek army, blames the gods when he is compelled to apologize for having acted unreasonably toward his most powerful soldier. "It was not that I did it: Zeus and Fate and Erinys that walk in darkness struck me mad when we were assembled on that day that I took from Achilles the prize that had been awarded to him," he pleads. "What could I do?"

In the century that followed, the influential triumvirate of Greek philosophers, Socrates, Plato, and Aristotle, shifted the discourse from such magical thinking to a more discerning view of human conduct. By examining human impulses, they sought to uncover the ways in which we are directed from within. Of course, these Greek philosophers were not speculating within a psychological context; rather, they were striving to understand the human spirit through the lens of ethics and the desire for a just society.

Socrates's argument that humans are motivated by reason above all else essentially suggests we do wrong only when we don't know it to be wrong. His protégé, Plato, put forth a more complex view. He argued that the soul is governed by three distinct forces: reason, appetite, and spirit—each with its own needs. Reason loves learning and wisdom; spirit yearns for fame and distinction; and appetite seeks bodily pleasure. In the virtuous person, these three elements exist in harmony, with reason holding sway over both appetite and spirit. For others, however, persistent conflict among these urges results in psychological turmoil, a civil war within the soul.

With this concept, Plato not only moved the seat of the struggle to the mind itself but also presented the mind as a dynamic system of opposing forces.

James used Plato's concept to build a psychological framework for understanding human behavior. Among the Greek philosophers, however, it is Aristotle who was James's closest intellectual forebear. Perception, Aristotle suggested, is not a passive state through which we take in the world around us, but a form of active selection. The virtuous person, he argued, directs his attention toward the most worthy subjects, in a process of sustained and deliberate contemplation. Consequently, behavior is more a matter of focus than reason: careful attention to suffering will inspire generosity, for example, and contemplation of the beautiful will initiate a desire to create.

James came to a similar conclusion, but from a different angle. He defined consciousness as a rushing stream of thoughts and insights— fleeting, idiosyncratic, often irrational and unrelated—that nonetheless cohere into the impression of a continuous life. He framed his investigation of consciousness in evolutionary terms: why might human consciousness be necessary in the first place?

Expanding on Aristotle's notion, he, too, concluded that the answer lay in attention. Consciousness allows us to attend judiciously to one—or two or ten—of the infinite array of things available to the mind at any given moment and, just as important, to ignore the rest. What distinguishes the human mind, he explained, is this very particularity: "Millions of items of the outward order are present to my senses which never properly enter into my experience. Why? Because they have no *interest* for me. Only those items which I *notice* shape my mind—without selective interest, experience is an utter chaos."

What James called "interest"—the process by which each of us chooses what to attend to—could likewise be termed investment or care. James offered further explanation (harking back to his days as a painter) for how this process creates an individual world: "Interest alone gives accent and emphasis, light and shade, background and foreground—intelligible perspective, in a word."

After publishing, in 1890, his first major work, *Principles of Psychology*, which explores his theories of consciousness, James shifted his focus toward understanding mental illness. In a lecture he gave in 1896, he described a wide range of extreme conditions (kleptomania, alcoholism, sexual impulsivity, pyromania, suicide, and even homicide) and proposed that each one held within it the seed of everyday anxieties and impulses. When we dutifully pay the cashier for our items, for example, thoughts of petty theft might dart through our minds. Likewise, a degree of paranoia and fear accompanies other common activities: making sure that the door is locked, the gas is off, no one is hiding under the bed. Those who cannot resist such inclinations and, in turn, amplify them, according to James, suffer from a loss of control over the faculty of attention. James saw in the extreme behaviors of the mentally ill the key to understanding the dynamics of attention in everyday life. Building on a concept developed by nineteenth-century French psychologist Théodule-Armand Ribot known as *idée fixe* (literally translated as "fixed idea"), James formulated a theory of mental illness.

For Ribot, the answer lay in the passions, states of mind suspended somewhere between the fleeting world of emotions and the permanence of complete insanity. On one side of Ribot's spectrum lay the everyday emotional states of surprise, envy, boredom; on the other were the ongoing numbness of melancholia, the wandering detachment of psychosis. When emotions outlasted their natural duration, when they became "fixed" in the mind, they slid along this spectrum, away from recognizable forms of human experience and toward mental illness.

Ribot believed that a single mechanism lay behind this process. How and why might a momentary feeling gradually take root in the mind, losing any recognizable connection to the stimulus (a blooming flower, the death of a lover, a joke) that gave rise to it?

Like many other psychologists and physicians of the nineteenth century, Ribot thought that our mental life was governed not by God, nor by any other nonmaterial force; rather, the mind operated according to strictly mechanistic principles, which could be understood through rational inquiry. The human mind was, in short, a complex machine, akin to the increasingly complex industrial machinery that rapidly transformed the European landscape over the course of the nineteenth century. Ribot therefore sought to make sense of the passions in mechanical terms: an abnormality in the brain became his explanation for behaviors that would otherwise elude explanation.

In the *idée fixe*, or monomania, Ribot found a model for inherently irrational states of mind. The *idée fixe* took root during a period of incubation, when patients began to display symptoms of generalized agitation and psychological suffering. Monomania gradually infected the workings of the healthy mind, slowing and eventually stopping the gears and levers of rational thought. Once in the thrall of an *idée fixe*, the impassioned brain redirected all its energies away from goal-driven projects and toward impulsive behavior. As the monomaniacal patient deteriorated, the physician became a powerless witness to a violent struggle between passion and reason, between "the overmastering impulse and the arrestive powers of the will."

Though he could not prove his hypothesis with medical imaging or brainwave monitoring, Ribot believed that all impassioned states, or *idées fixes*, resulted from excessive irritations of an isolated group of brain cells. The psychologist Pierre Janet later refined this idea: Janet used the term to describe not a rare form of pathology but, rather, obsession as a "response to the assault of the everyday"— anxiety created as a diversion from life.

James later expanded his study of extreme preoccupation to include public figures fervently committed to a social or political cause: Dorothy Dix, an advice columnist who questioned the sexist mores of her day; Henry Bergh, who founded the Society for the Prevention of Cruelty to Animals; and Henry Parkhurst, who single-handedly fought organized crime and pork-barrel politics in New York City. These people relinquished themselves, albeit in a more successful direction, to their own kind of single-mindedness; their passions, too, as James put it, "show under a microscope the play of human nature."

James recognized that in all people there was a need for transforming personal change, drawn from the "hot place" of consciousness. He saw this as much in patients overcome by obsession as in those consumed by an external cause; he even identified it as an urge behind religious ardor.

"The higher and lower feelings, the useful and erring impulses, begin by being a comparative chaos within us," James writes in *Varieties of Religious Experience*, which comprised lectures he gave throughout 1901 and 1902 on the nature of religion. Religion, to James, was defined as "the feelings, acts, and experiences of individual men in their solitude, so far as they apprehend themselves to stand in relation to whatever they may consider divine." He is referring not to institutional dogma or to those for whom religious observance is merely a "dull habit," but rather to the experiences of the faithful for whom belief in God is the "very inner citadel." "Remember," he writes, "that the whole point lies in really believing." It was the belief in belief that mattered.

Yet James also understood that the torment of the mind is not easily settled by religion. Belief alone is not enough; the emotional charge that accompanies such belief is equally important. In order to achieve transcendence, he posited, belief has to be compelling enough to subsume lesser tendencies; it has to have, in James's words, "that touch of explosive intensity."

Meaningful religious experience, James adds, involves the "push of the subconscious, the irrational, instinctive part . . . vital needs and mystical overbeliefs." He was the first scientist to implicate the mind in religious experience—to assert that hidden faculties controlled "the sense of union with the power beyond us."

James offers the words of St. Augustine, describing a moment when Augustine discovers his overpowering weakness, as an example of a higher desire failing to conquer: "The new will which I began to have was not yet strong enough to overcome that other will, strengthened by long indulgence. So these two wills, one old, one new, one carnal, the other spiritual, contended with each other and disturbed my soul. . . . It was myself indeed in both the wills, yet more myself in that which I approved in myself than in that which I disapproved in myself. Yet it was through myself that habit had attained so fierce a mastery over me. . . . I made another effort and almost succeeded, yet I did not reach it, and did not grasp it." Conversion, James claims, is one of the ways in which people find their way to happiness—as St. Augustine eventually did.

Yet religion, James reassuringly continues, "is only one out of many ways of reaching unity; and the process of remedying inner incompleteness and reducing inner discord is a general psychological process, which may take place with any sort of mental material. . . . For example, the new birth may be away from religion into incredulity; or it may be from moral scrupulosity into freedom and license; or it may be produced by the irruption into the individual's life of some new stimulus or passion, such as love, ambition, cupidity, revenge, or patriotic devotion."

A radical shift in perspective was, for James, the only way to release the ever-tightening grip of capture on our attention. Like Aristotle and Plato, however, James could not explain the pathophysiology of capture. Why—and how—can a single thought seize our attention without our awareness and become fixed in our minds?

FREUD AND DRIVE

Any historical discussion of why we do, think, and feel what we wish we did not do, think, and feel must include an examination of Sigmund Freud's work. At the center of Freud's theory of the mind is his concept of drive: a source of stimulation or pressure, a "demand made upon the mind for work." Freud understood drive as though it were a physical force, animated by a certain quantity of energy and balanced by counterforces. Approaching the mind as a neurologist, Freud postulated that there were two physiological components underlying all drives: neurons in a state of excitation and "flow"—a nonspecific measure of energy. This rather nebulous concept reflected the difficulty of explaining mental distress using only the science of his day. As Freud himself said, he sought his answer to the problem of human behavior "between medicine and philosophy," "between the mental and physical."

Freud needed a theory to explain why people did, thought, and felt as they did. A student of the Viennese psychologist Josef Breuer, Freud was exposed at an early age to one of Breuer's patients, whose bizarre condition would have a substantial influence on the young physician's theory of mind. The patient was unable to drink any liquids whatsoever. She recalled during her treatment that in the home of her governess, she once saw a dog drink out of a glass and had felt a profound wave of disgust. Freud hypothesized that the patient's memory of this disturbing image, albeit unavailable to the conscious mind, persisted throughout adolescence and adulthood. The emotional valence of the experience survived, hidden in the depths of the unconscious.

For Freud, repressed memories such as these lay at the root of almost all human suffering. Even after they drifted beyond the boundaries of conscious life, mental representations of certain objects or events continued to cause mental distress. The repressed memory, and the emotions attached to it, could thus serve as an unconscious motive for certain patterns of behavior. Freud was particularly interested in the inevitable conflicts between the demands of the present moment and these early memories.

Freud's journey into the unconscious led him to conclude that, beginning in early childhood, human drives were systematically repressed by social norms. The realm of the unconscious thus broadened to include not only ideas and memories but drives themselves—fundamental biological needs and urges. Chief among these were the libido, or the sexual drive, and the death drive, the primary source of human aggression. This latter drive aimed at the violent undoing of the world, including the destruction of the self. For Freud, all mental activity occurred as an attempt to let off mental steam, to reduce the accumulated tension of these neglected drives. This need was expressed in wishes and urges; any reduction in mental pressure was experienced as pleasure. Painful memories, then, were not the only thoughts that were shunted into the realm of the unconscious. Stifled at an early age, our most basic drives also occupied this largely invisible terrain.

Central to Freud's theory was the belief that the unconscious fashioned out of experience a complex symbolic language that disguised the socially unacceptable content of our drives. This "psychic alchemy" involved transforming repressed or unfulfilled drives into phobias, obsessions, and other neuroses. The surreal landscape of dreams became the primary site of these transformations, as the unconscious mind refashioned the elements of daily living into a coded representation of desire. Freud's theory borrowed from the functionalist and associationist schools of the English psychologists the idea that, through experiences, particular phenomena become associated with pleasure or pain. For Freud, it was repressed drives that were ultimately responsible for these associations, which collectively gave rise to affect, or the emotional lens through which a given person sees the world. Moreover, these associations were all embedded in a complex web of meaning. Over time, certain stimuli became linked to other stimuli; affect attached to one object or idea could be "displaced" onto another.

Perhaps Freud's most enduring observation was that motivations have both biological and psychological correlates. He was right that stimuli, both external and internal, can elicit emotional reactions, that the motivational forces behind our actions are not necessarily apparent, and that emotional processing can cause intense distress. In short, he recognized that the boundary separating wellness and distress is fluid. Freud's unconscious was populated not by submerged ideas or memories that floated, from time to time, to the surface of consciousness. (That, for Freud, was the "preconscious.") Rather, the unconscious had its own structure, rules, and dynamics that walled its contents off from the conscious mind. Freud postulated that drives remained in the unconscious because their content was antithetical to the very foundations of culture. He compared the forces of repressed desire to the "legendary Titans, weighed down since primeval ages by the massive bulk of the mountains which were once hurled upon them by the victorious gods and which are still shaken."

Only in his later years did Freud fully articulate his structural model of the mind. On this model, the id was the origin of the aggressive and libidinal drives; "a cauldron full of seething excitations," the id has "no organization, produces no collective will, but only a striving to bring about the satisfaction of the instinctual needs subject to the observance of the pleasure principle." These diffuse energies could be sublimated into the ego, a more mature, integrated conception of the self.

There was little empirical evidence to support Freud's hypotheses. The only tangential evidence for the unconscious that Freud could muster was memory: if we can know something at two distinct points in time, then such memories must reside somewhere in the interim. Moreover, any study of the clinical efficacy of psychoanalysis was necessarily uncontrolled. As several writers have noted, Freud's clinical strategy was largely fitting traces of evidence into an organized, comprehensible, and compelling whole. "When we invent an unconscious mind to give coherence and continuity to the conscious, we voluntarily leave the sphere of fact for the sphere of fiction," wrote the psychologist E. B. Titchener in 1910.

Nonetheless, by shifting the study of the mind away from morality and rationality and toward the unstable ground of desire, Freud moved science toward a clearer understanding of human thought and behavior.

At the same time, even Freud's own disciples lamented that Freud did not pay enough attention to the problem of affect. As one scholar of Freud concluded, "We do not possess a systematic statement of the psychoanalytic theory of affects." Another scholar described Freud's theory of affect as "fitful" and "largely unfocused." Defining the unconscious as the source of all drives ultimately proved futile: by focusing on altered states of consciousness such as hypnosis and dreams, Freud failed to prove the relevance of his theory to everyday mental life. By the end of the twentieth century, orthodox Freudianism had become the province of an increasingly small group of clinicians and theorists.

Freud's downfall reflected not only the fantastical nature of his theories, which reduced all drive to sex and aggression, but also the clinical failure of psychoanalysis, or of the talking cure, to eradicate psychopathology. As scholars adopted an increasingly critical perspective on Freudian analysis, the content of analysis came to be seen as the mental projection of the treating physician. After all, Freud rejected the possibility that insight could result from conscious mental activity on the part of the patient. By subtly planting suggestions, the therapist could convince the patient that a fantastical narrative was not only plausible but irrefutably true. The results of this process were more akin to "theatrical scripts or stage directions" than to scientific findings, according to a scholar. This is not to say that therapy in itself had no value. In addition to revealing the inadequacies of a given narrative, the therapist could work with the patient to develop a new theatrical script. The sum total of our past experiences and anxieties about the future would still have a role in determining what and how we think, feel, and behave, but these individual constellations of meaning could always be reorganized, or viewed from a more flattering angle.

One of the most important, albeit unsung, inheritors of the Freudian tradition was an aspiring playwright turned psychologist named Silvan Tomkins. Like Freud, Tomkins valued intellectual originality and refused to make his theories conform to accepted scientific dogma. His influence, however, has barely been recognized by the psychiatric establishment.

Though Tomkins agreed with Freud that mental illness is ultimately emotional in nature—a disorder of feelings—he thought that Freud was unduly wedded to drive theory. Freud saw in the obstinate recurrence of traumatic memories a hidden desire to rewrite one's personal history, to be given a second chance at redemption. Patients enthralled by trauma were acting out the haunting "if only" of remorse: *if only I had dodged the bullet; if only I had rescued the child.*

For Tomkins, however, more important than drives or wish fulfillment was how such thoughts became dominant, or immune to our feeble attempts at controlling them. In the case of a soldier tormented by traumatic memories of war, for instance, Tomkins saw no wish fulfillment at work: what end could such repetition of trauma possibly serve? Rather than posit entire narratives of submerged or hidden desire, Tomkins turned his attention to the forces that reactivated these traumatic memories and invested them with so much power.

Tomkins found the germ of an answer in awareness, or attention. Over the past six decades, the cognitive sciences have gradually returned to these basic questions: How do our brains represent the world around us (sounds, smells, images, and thoughts)? How is information processed and transformed as it travels through neurons and across synapses? Rejecting the dogma that only observable actions can be studied, cognitive science has slowly but surely rekindled the study of the human experience. As scientists shifted their focus from behavior to consciousness, the acquisition, processing, storage, and transformation of information became important subjects of study.

Yet missing from this approach is any attempt to understand the phenomenology of subjective experience—what it is like to feel, think, and be. For Freud, drives, not experience, were central to understanding the mind. Affect was simply the conscious perception of underlying desire, the sex drive or the death drive. Tomkins saw affect as an object of study in its own right, akin to drive, perception, memory, and cognition. Whereas drives represented biological needs, affects were innate reactions to the environment that could either amplify or inhibit drives: in the absence of an excited affect, Tomkins argued, the sex drive becomes rather impotent. Over the course of his life, he identified nine basic categories of affect, including interest, fear, distress, rage, and humiliation. Though these affects triggered thoughts and desires, the triggering process itself was automatic and virtually undetectable.

Tomkins's contemporaries largely ignored subjective experience in favor of objective, observable action. The mind was an information processing system: science made little provision for what it feels like to be a thinking being. Yet for Tomkins, affect was the all-important lens through which we came to know the world. It was not the content of perception (what we hear, touch, taste, and think) that defined our mental life; rather, it was the affective state we brought to bear on those perceptions. Over the course of the twentieth century, Tomkins's claims gradually took hold in the scientific community: experimentalists turned to the role of feelings in psychotherapy, and cognitivists such as Aaron Beck and Albert Ellis included affect in their models of psychopathology.

Applying the ideas of Stoicism to clinical practice, Ellis argued that "faulty" ideas were the ultimate cause of distressing thoughts. Suffering was the by-product of dysfunctional beliefs or a distorted worldview, which generated what an IT consultant might call systemic errors in processing information. Elaborating on Ellis's theory, Beck noted that distressing thoughts cannot be simply turned off or ignored; they are functionally automatic. His solution was to bring to the surface the faulty reasoning, the irrational beliefs, that lay behind them. Cognitive behavioral therapy, or CBT, is premised on this theory: if my dysfunctional beliefs are exposed as illogical, I will come to understand that I have misconstrued a particular situation and, over time, will reorganize my perceptions.

What, then, if I already know that my worldview is dysfunctional but find myself unable to change it?

THE SCIENCE UNDERLYING CAPTURE

From Aristotle to James, those who have attempted to explain why we feel, think, and act as we do have operated within the limits of theory; their contributions were based on a philosophical perspective of the human mind. These thinkers acknowledged the difficulty of controlling thoughts, feelings, and actions, but were, for obvious reasons, unable to draw from scientific research to validate their conclusions. Today, however, advances in our understanding of the human brain have made it possible to explain such ideas in scientific terms.

But while we can view the brain with much greater accuracy, our emotional lives cannot be reduced to a single neurobiological function or theory. The human brain (and the role it plays in subjective experience) is far too intricate to reduce all subjective experience to biological principles. Our brains give rise to our abilities to learn, reason, feel, and remember, and those aspects of our lives cannot be found directly in our neurophysiology. No amount of research could track the countless stimulations, arousals, and physiological changes that affect us in a lifetime—not to mention the learning and memory that accompany these phenomena or the infinite possible responses to such variables. Our knowledge of the brain is still and always will be evolving.

Despite these constraints, however, a theory of mental illness is more relevant if it is grounded in neurobiological evidence. By tying subjective experience to physiology, we can see how the process of capture fits within our current scientific framework.

Capture arises from a vast and complex circuitry in the brain. The brain is composed of neurons (nerve cells responsible for relaying information through electrical and chemical signals) organized in discrete layers, networks, and regions. Every time we experience something new, a neural pattern (a kind of microcircuit that connects different areas of the brain) is created in response. Over time, those neural patterns become associated with anything that evokes that experience. The sound of the ocean, the feel of sand on our feet, the glare of sun overhead are all markers of childhood trips to the beach. When we remember this kind of experience, or something connected to this experience, or even when we do something that calls to mind a thought or feeling we associate with this experience— these neural patterns are reactivated.

The brain creates these neural patterns through an elaborate sequence of events. When we encounter a stimulus (an object, a thought, a sound, a smell), our neurons begin to "fire." When neurons fire, they transmit chemicals that stimulate neighboring neurons by binding to them, thereby opening channels that allow a positive electrical charge to flow through the cell. With this jolt of electricity, the neighboring neuron becomes unstable; when it can no longer hold back, it fires in turn, propagating its own electric signal. Thus, neural activation is a chain reaction; the firing of the first neuron initiates the response of others. And, generally, the more intense the initial stimulus, the more rapidly the neurons will fire.

There are many different kinds of neurons, and each subset, as well as each individual neuron, is highly selective in the kind of stimulus to which it responds. For example, our sensory neurons are coded to fire in response to stimuli with specific characteristics. Some sensory neurons detect pain and temperature, while others detect position, vibration, or light touch. Certain neurons in our visual system become activated only when a horizontal line passes through our visual field, while others respond only to vertical lines. There are neurons with a preference for thick lines, or for thin ones, or for lights that blink rapidly, or for those that blink slowly.

While individual sensory neurons contribute to how we perceive a stimulus, there are also "hub" neurons (highly connected neurons) located throughout the brain. These neurons receive input from multiple other subsets of neurons. In addition, there are groups of neurons in the brain that are programmed to respond in concert to the same stimulus. Together, they fire more frequently and strongly than they would if each were tuned to a separate stimulus.

The brain's response to stimuli, then, is fundamentally a collective enterprise, with different neurons activated by different kinds of sensory input and various networks of neurons activating one another. The more frequently neurons fire together, the more powerful their connection to one another becomes. This is nicely summarized in an axiom known as Hebb's rule: neurons that fire together wire together.

Likewise, when we are repeatedly exposed to a stimulus that triggers a particular neuron to fire, its response becomes more vigorous. That stimulus may be a drug, or a particular song, or a coworker's taunting glare. If two characteristics of a stimulus are present at the same time—a visual cue and an associated memory, for example—neurons tuned to one characteristic may gradually become associated with neurons tuned to another.

Once a connection has been established between two neurons, or two groups of neurons, one becomes primed to activate the other. Thus, when one group of neurons fires in response to a stimulus, the second group of neurons is more likely to fire as well. A similar association occurs in regions of the brain that process information from multiple sensory sources—for example, a neuron in the visual cortex that is tuned to red and a neuron in the auditory cortex that is tuned to a certain pitch can become associated in the prefrontal cortex, the area of the brain from which our thoughts and decision making are believed to arise. Couple these two sensory inputs with the memory of an ambulance and you've got a clear example of Hebb's rule.

While some neurons activate other neurons, roughly half our neurons do the opposite: they inhibit firing. Instead of transmitting a positive electrical charge to another cell, these neurons emit a negative electrical charge that counteracts any possible response from the other neuron. The chemicals capable of either strengthening or weakening the neural response, known as neuromodulators, include dopamine, serotonin, and norepinephrine.

Bundles of neurons create a network, or a neural pattern. These networks can fire together or in sequence and can communicate across different areas of the brain. Branching out even farther are networks linked to other networks, which are likewise stimulated or inhibited by neuromodulators. Each interconnected network serves a particular function, such as triggering movement or stimulating sensation within the body. These cortical networks are interconnected and never function independently.

Neurons in different regions of the brain oscillate at different frequencies. These movements are associated with different states of arousal, such as moving or sleeping. Oscillations can either allow or block the transfer of information between neurons. The complex communication that determines neural response to any stimulus is shaped by both the timing and the frequency of this oscillation within or across a network. There is increasing evidence that selective attention uses these oscillations for amplifying relevant and suppressing irrelevant and distracting information.

At the biological level, capture is the result of neural patterns that are created in response to various experiences. A first experience can result in the creation of a unique neural network that is associated with certain feelings and actions, and in turn, these neural networks can elicit emotions and physical responses. The neural response, and the gradual creation of neural patterns, is not a static or immutable process. Neurons can change their tuning based on experience; as the information coming into a neuron changes, so does its response. This offers opportunities for new learning to occur. Often referred to as neuroplasticity, this idea has largely replaced the once widely held belief that the adult brain cannot change.

Capture is predicated on our ability to selectively attend to specific stimuli. When something grabs our attention, such as when the lights come up on Bruce Springsteen in concert as he counts off the opening notes of "Born to Run," our neurons are responding more vigorously to that particular stimulus than to the push of the crowd around us or the heat in the arena. As Springsteen starts to sing, we focus on specific stimuli because the networks of neurons that allow us to make sense of the world do not respond equally to everything in our environment. Rather, they permit us to distinguish between what matters more or less at a given moment, what is out of place or what is threatening—and to channel our attention in the appropriate direction. This is called selective attention; it is the gateway to awareness.

Yet even when we are not consciously paying attention to stimuli, our neurons are still responding to them. Neurons that fire in the presence of visual stimuli, for example, keep firing as long as our eyes are open: even as our minds have wandered from the book in our hands, we continue to register the words on the page in front of us. So many objects fall within our visual field that we cannot process all of them at once: there are simply too many stimuli competing for our attention at all times. We use the mechanism of selective attention to amplify neuronal signals so that certain stimuli become more insistent—or we inhibit signals that we do not need to consider.

The triggers of neural response are not merely sensory or external stimuli—thoughts and feelings can also activate neural firing. Separate neural networks respond to external and internal stimuli, with controls in place that allow us to switch our attention from one to the other.

There are two separate yet highly connected ways in which attention is controlled, bottom-up and top-down processing, each with its own networks of neurons. The top-down processes that control attention take place in parts of the brain that deal with sensory input and in regions that have learned to respond to tasks. These top-down processes often play a role in decision making. Most parts of the brain are involved in some sort of top-down processing.

Bottom-up processing is best understood as involuntary and automatic, a tool for detecting and focusing our attention on unexpected stimuli. Just as the body's immune system continuously scans the body for microbial invaders, so, too, the bottom-up networks of the brain are always monitoring the environment for urgent signals. When such a stimulus is detected, these networks reorient our attention, allowing us to register it immediately. This is how we perceive a person running in front of our car even before we become consciously aware of him.

Memory also plays an important role in the process of selective attention. The brain processes memories in two different ways. Working memory, or short-term memory, is essentially a mechanism for keeping at hand information we've recently learned or experienced. Working memory houses information long after the initial sensory stimuli associated with an experience have disappeared: there is no object to see, smell, taste, touch, or hear, and yet we can draw on our earlier recollections. One part of working memory, lodged in the lower parts of the brain, holds the information while the higher parts of the brain manipulate and update the stored contents. The information that is currently being held in working memory determines what we are focusing on presently.

Working memory is vital for learning. For example, when you first learn to drive, you need to make an active effort to coordinate all the steps involved in safely operating a car. You need to actively think about what you are doing, a process that involves manipulating information, because the endeavor is new and challenging. After you become an experienced driver, however, the skill requires less active mental manipulation and places less demand on working memory. Indeed, once you have learned to drive, the process can become so automatic that it can be challenging to teach someone else each step in the process.

There is a close interaction between attention, which allows us to highlight relevant information, and working memory, which allows us to use that information. Both are selective processes influenced by how vigorously our neurons respond to an internal or external stimulus. When we pay attention to something, it is more likely to remain in working memory. Conversely, if something remains in working memory, there is a greater chance we will pay attention to it. The act of paying attention can also allow us to control the relevance of the contents of our working memory. Having shown a person a red object and a yellow object, we can take those items away and then ask her to think about only the red one. Immediately, the neurons tuned to red will respond more rapidly. Both objects remain in working memory, but the one to which she is giving her attention generates more activity within the circuits of the brain.

Paying attention is also the conduit to lasting recall. If an event, feeling, or thought remains in working memory long enough because it is consistently the focus of our attention, it will be processed for storage in the regions of the brain designed to house long-term memory. This pathway can also operate in the opposite direction: working memory has the capacity to bring long-term memories back into active use.

Whatever we are paying attention to at a given moment (whether it's an external stimulus, such as food, or an internal stimulus, such as self-doubt or regret) initially resides in working memory. Over time, unless a competing goal redirects our attention, our response eventually becomes so instinctive that our brains no longer mobilize to create a new or different reaction. Our response becomes automatic—and when our response is discordant with our conscious intentions, we begin to feel as if we're losing control. This loss of control is a key feature of capture.

There are certain stimuli that seize our attention, with obvious reason—it would be hard not to concentrate on a bear approaching us or, for that matter, Game 7 of the World Series, bottom of the ninth, bases loaded, full count. But we are not always conscious of the stimuli that command our focus and, ultimately, come to steer our thoughts, attitudes, feelings, and behavior. A good deal of new learning is implicit, meaning it is so subtle as to be imperceptible to the conscious mind. If you feel a surge of self-doubt every time you are about to speak publicly, it is very likely that your anxiety is based on a previous response to a similar experience. Subsequent related experiences, or thoughts and feelings that evoked those experiences, allowed this learned response to gain more traction along the way. This is how we come to develop patterns of behavior and emotional response without ever being aware of their taking hold.

At the core of implicit learning is the association that develops among networks of neurons as they respond to stimuli. The associations between two stimuli are strongest when both are highly salient—that is, when they have features that make them stand out from other stimuli in the environment. But even seemingly neutral stimuli can provoke a strong neural response once they become associated with a salient stimulus.

The neutral stimulus acts as a cue to the more salient stimulus, and in doing so becomes salient itself. When we encounter this neutral stimulus, we can experience strong thoughts, urges, and desires that seem to have come from nowhere. For some, merely seeing the neon sign of a bar may evoke the heady pleasures of alcohol. There is an involuntary component to this response. It is not surprising that the areas of the brain that register salience are the same areas, or are connected to the areas, that register physiological changes, such as increases in heart rate and skin temperature. This explains why our heart rate increases so soon after we encounter a salient stimulus.

How, then, does a stimulus become salient for us? At the simplest level, brightness, color, shape, motion, and novelty can all give an object salience. Sensory properties are a key determinant—a bright color in a black-and-white scene or a red balloon traveling through a blue sky are salient images. Something markedly different from surrounding objects is also salient. When researchers placed a printer on a stove, observers looked at the scene longer than they did when a cooking pot was placed in the same spot. Things that are unexpected invariably capture our attention.

There is also salience in powerful desires, immediate or distant goals, attitudes toward adversity or opportunity, and major life events. The salience of a stimulus is heavily influenced by how closely it relates to our own emotional state or personal experiences. The same stimulus can be profoundly salient to one person, but generate no response at all in another. Crossing a bridge is just part of the daily drive to most commuters, but to a few it can trigger a sweat-inducing terror. Sometimes the same stimulus will be highly salient for two people, but in opposite ways, depending on their past learning and memories. Boarding a flight back to one's hometown, or merely thinking about getting on the flight, for example, can easily have a positive valence for one person and a negative valence for another. (Valence is the emotional dimension of a stimulus that prompts us to view it as positive, neutral, or negative.) While we all process stimuli differently, there is also a lot of commonality in how we come to assess stimuli. For example, violent or dangerous stimuli and sexually arousing stimuli are salient for many people.

Ultimately, the stimuli with the greatest salience—often those that are associated with past emotional events—capture our attention most fully. When this occurs, we pay less and less attention to other stimuli, which gradually cease to influence our emotions, or even to remain vividly in our working memory. Capture thus changes how we feel and, in turn, influences our responses to our environment in the future.

Though the exact circuitry of our brains has yet to be clearly mapped, researchers have confirmed that there are interconnected neural pathways that play a primary role in processing emotional stimuli. An essential component of this circuitry is the amygdala, the almond-shaped part of the brain's temporal lobe. The amygdala plays a role in instigating many of our emotions (fear, arousal, and disgust, among them), but it also functions to create "relevance detection," an individual process by which we appraise as particularly meaningful a stimulus that takes place outside consciousness. Specific portions of the amygdala are also involved in refocusing attention; still other parts of it are engaged for associating memory with particularly salient stimuli. The amygdala thus serves as a crucial neural circuit hub that prioritizes our conscious awareness. Our emotional circuitry, in turn, links the amygdala with other networks that instigate learning, memory, habit, motivation, and decision making.

This transaction wires us to pay closer attention to stimuli associated with past emotional events—which is a pivotal moment in the neurobiological narrative of capture. These very salient stimuli—the voice of an ex-lover, the handwriting of a parent, the restaurant where a crucial conversation took place—capture our attention and activate an emotional response, triggering the neural circuitry of the brain that spurs us toward motivation and action. Every time a stimulus triggers particular neurons to fire, a neural pattern emerges. Over time, this pattern is strengthened. As the intensity of the neuronal firing increases across the networks, our attention is drawn ever more acutely to the voice, the writing, the restaurant. Consequently, our thoughts and behavior are reinforced.

There is an equally important feed-forward mechanism in capture. The feed-forward loop is one of the basic elements of engineering and physics: the output of a given process becomes the input for the next stage, which then determines the next step in the process. For David Foster Wallace, this feed-forward mechanism was the source of his infinite regress: feeling broken, dispassionately scanning his mind for weakness, identifying it but finding himself still powerless to fix himself. This sequence caught him in a perpetual loop of negative thoughts and feelings.

3

WHAT CAPTURES?

A CONTINUUM FROM THE
ORDINARY TO MENTAL ILLNESS

The phenomenon of capture encompasses a broad spectrum of psychological concepts, including attention, learning and memory, triggers, sensitization, reinforcement (when salient stimuli inspire repetitive thoughts and behaviors), obsession, motivation, the lasting effect of childhood experiences, attachment, and perhaps most notably, emotion and mood.

The feelings associated with capture can take many forms, and can vary in intensity, duration, and quality. Sometimes the mental commotion that is provoked lasts for only a few moments. But capture can also take hold with increasing intensity over a protracted period. It can come to resemble something more like single-mindedness. It changes how we feel and can limit our worldview.

The actions we take as a result of capture also differ wildly. The effect can be temporary, with only mild consequences or changes in behavior. Or it can be so long-standing as to wholly change the course of a life. Always this process begins outside awareness. As capture takes hold and narrows our attention, we may begin to feel as if our thoughts are beyond our control—a sensation that may induce fear or even panic. Alternatively, capture can allow for an experience of flow, the sense that all consciousness is channeled in a single direction in an uninterrupted manner for positive effect.

The following stories chart a range of human experiences, from romantic yearning to addiction to artistic passion to despair. Some show how a single thought or idea can seize us in a more commonplace way, even as it gives rise to a powerful shift in perception; others depict the ways in which inflexible thought patterns lead to destructive and painful actions. The first is the only narrative told from the point of view of a fictional character—but Edith Wharton's depiction of Darrow's sudden capture is one to which most of us can relate. As we will see, art is often born of capture. At the same time, it might be said that art is a productive expression of the attempt to release us from capture's grip.

REJECTION

Darrow could not wait to see his beloved Anna. And then a telegram arrived.

"Unexpected obstacle. Please don't come till thirtieth. Anna."

With no explanation, without the "shadow of an excuse or regret," as Edith Wharton writes in her 1912 novel, *The Reef*, Anna had postponed their highly anticipated reunion for two weeks. The telegram was tossed into Darrow's compartment as the train pulled away from the station. Though he had boarded the train looking forward to an exquisite exchange—he was planning to propose to Anna—Darrow now felt robbed. His world had been suddenly upended, and he could think of nothing else.

"All the way from Charing Cross to Dover the train had hammered the words of the telegram into George Darrow's ears," writes Wharton. At the end of his rail journey, as Darrow wonders whether to continue on to France, nature itself seems to turn against him: "And now, as he emerged from the compartment at the pier, and stood facing the windswept platform and the angry sea beyond, [Anna's words] leapt out at him as if from the crest of the waves, stung and blinded him with a fresh fury of derision."

Clearly, he'd been a fool to think Anna would ever marry him. How else could he interpret such a directive?

Three months earlier, when he'd run into her at a dinner party in London, he'd been overwhelmed by a sudden rush of memories. He could still feel the surprise of coming "upon her unexpected face, with the dark hair banded above grave eyes; eyes in which he had recognized every little curve and shadow as he would have recognized, after half a life-time, the details of a room he had played in as a child. . . . All that and more her smile had said; had said not merely 'I remember,' but 'I remember just what you remember.'"

The two had had a brush with love twelve years earlier. Their attraction was instant and undeniable, but they had been too young, neither of them quite able to recognize the rarity of their feelings. Anna had slipped away—in Darrow's view, without much sense or reason.

When, more than a decade later, they found each other once again unattached—she was widowed, he'd never married—they felt incredibly fortunate, as though they'd been granted a second chance. Darrow, roused by circumstances that seemed to him like nothing less than fate, concluded she was the woman with whom he would finally settle down.

Upon receiving the telegram, however, Darrow replayed this history in his mind, examining it with a newly critical eye, searching for his point of miscalculation. Perhaps Anna had withheld just enough to allow him to paint the strokes that he'd needed or wanted on her blank canvas. Perhaps she never felt the magnitude of emotion that, just moments earlier, he'd been certain that they had shared.

Everything was now subject to reinterpretation.

Gradually Darrow's entire world became muted, indistinct, colorless. The passing scenery blurred into wet streaks of color, and the broken lines of conversation that reached his ears ceased to be intelligible. He was restless, wildly annoyed by the sounds of the locomotive, by everything that was not Anna.

The sting of rejection reduced the world to a hostile place: "Now in the rattle of the wind about his ears, Darrow continued to hear the mocking echo of her message: 'Unexpected obstacle.'" As he walked out onto the pier, he was jostled and shoved by the crowd around him. A jeering chorus began to echo in his ears: "'She doesn't want you, doesn't want you, doesn't want you,' their umbrellas and their elbows seemed to say."

Somewhere, deep in the recesses of his mind, Darrow knew that the message might not be a rebuff, that hundreds of perfectly plausible explanations, all benign, might justify Anna's sudden postponement. Still he could not silence the echo of those three words: "Please don't come."

Wharton's character embodies a very real phenomenon: the painful rumination that can seep between the lines of even the briefest noncommittal message.

A BRUTISH FATHER

Children are especially susceptible to capture, as they struggle to make sense of an adult world that seems irrational or cruel. Memories from childhood, or any particularly formative experience, can take hold and gain significance, often to an individual's detriment.

In a 1919 letter to his aging father, Franz Kafka reflected on his early sense of awe, fear, and confusion. "For me as a child, though, everything you barked at me was as good as God's law," he explained. And yet, the man issuing those commandments had been far from divine: "In your armchair you ruled the world. Your opinion was right; every other was mad, wild, *meshugge*, abnormal." Even in adulthood, the writer found himself haunted by the memory of his father, unable to escape the emotional orbit of his childhood. "My writing," he conceded, "was about you: all I did there was to lament what I could not lament on your shoulder."

Raised by a large, domineering, brawny man, the young Franz was painfully aware of his diminutive stature. Every misstep became a disappointment, a failure to live up to his father's masculine ideal. "I had in fact gained a little independent distance from you," Kafka wrote, "even if in doing so I slightly resembled a worm, its tail pinned to the ground under somebody's foot, tearing loose from the front and wriggling away to the side." Kafka's letter to his father was an attempt to explain a lifetime of simmering resentment; it records a series of painful memories, which imbue the everyday slights and petty tyrannies of family life with a sense of overwhelming helplessness and despair.

Eat first, talk afterward: Kafka's father had a list of rules for the dinner table. There was to be no discussion during the meal, no cracking of chicken bones, though Kafka's father flagrantly disregarded his own edict, tearing through a carcass of meat like a caveman, cutting his fingernails at the table, cleaning his ears with a toothpick. Vinegar was to be sipped daintily, even as his father slurped with greedy abandon. The bread was to be cut straight, though his father sawed away at it, his knife dripping with gravy. No scraps were to fall on the floor, but by the end of the meal, an impressive pile had formed around the patriarch's chair. Above all, every last morsel was to be consumed. Why, then, did his father call his mother's cooking "swill," accusing "the beast" of ruining perfectly good meat?

In another episode, the young Franz remembers spending hours on a winter night begging for a glass of water, "partly to be annoying, partly to amuse myself." As the boy continued to howl, Kafka's father lifted him out of bed, carried him in his cotton nightshirt to the balcony of their small Prague apartment, and locked the door behind him. Shivering just outside the family home, Franz suddenly saw his existence as precarious, dispensable: "Years later it still tormented me that this giant man, my father, the ultimate authority, could enter my room at any time and, almost unprovoked, carry me from my bed out onto the pavlatche, and that meant I meant so little to him."

This belief only grew as Kafka aged. Memories of childhood shaped not only the dark, surreal landscapes of his fiction but also his emerging understanding of his worth as a writer. His father's criticism trailed him through adolescence and into adulthood, despite all his worldly success:

> I had only to be happy about something or other, be inspired by it, come home and mention it, and your response was an ironic sigh, a shake of the head, a finger rapping the table: "Is that what all the fuss is about?" or "I wish I had your worries!" or "What a waste of time!" or "That's nothing!" or "That won't put food on the table!"

What is telling about Kafka's letter is his sense of utter humiliation at the hands of a testy, demanding, sometimes hypercritical parent. Though he admits that his father "hardly ever really beat me," Kafka recounts with dread the threat of violence that never actually occurred: "But the shouting, the way your face got all red, the hasty undoing of the braces and laying them ready over the back of the chair, all that was almost worse for me."

What begins as a moment of domestic strife is soon transformed into the scene of a hanging:

> Imagine a man who is about to be hanged. Hang him and he is dead, it is all over. But force him to witness all the preparations for his hanging and inform him of his reprieve only once the noose is dangling in front of his face, and you can make him suffer for the rest of his life.

The clarity of death seems, if only momentarily, preferable to the endless threat of humiliation. In this way, much of Kafka's adult life was shadowed by the certainty of impending pain. This certainty began to color not only the present moment but the entire realm of future possibilities. Like the Condemned of his short story "In the Penal Colony," the writer lived in a state of permanent suspension, as if awaiting his own execution.

Even as a successful author, Kafka remained in thrall to his father. Having so consistently deflated the aspirations of his son, the elder Kafka seemed to embody all that was flawed beyond redemption in

Franz. Eventually, the young writer found that he could not even accept a rare compliment from his father: "it became a permanent habit, even when your opinion was for once the same as mine." As the years passed, Kafka rewrote the entire narrative of his life in terms of a struggle to free himself from tyranny.

One particular image from childhood came to embody the entire weight of the writer's suffering. Belittled by the mere physical presence of his father in the bathing hut—"There was I, skinny, frail, fragile; you strong, tall, thickset"—and ashamed at his awkwardness in the water, the young Franz felt a painful pang of recognition: "at such moments all my past failures would come back to haunt me."

DRINK

"I loved the way drink made me feel, and I loved its special power of deflection, its ability to shift my focus away from my own awareness of self," writes Caroline Knapp in her bestselling 1996 memoir, *Drinking: A Love Story.*

Few writers or scientists have better described the particular experience of the addict.

"I loved the sounds of drink: the slide of a cork as it eased out of a wine bottle, the distinct glug-glug of booze pouring into a glass, the clatter of ice cubes in a tumbler.

"I loved the rituals, the camaraderie of drinking with others.

"I loved . . . the warming, melting feelings of ease and courage it gave me. . . .

"A love story. Yes: this is a love story," she wrote. "It's about passion, sensual pleasure, deep pulls, lust, fears, yearning hungers. It's about needs so strong they're crippling."

For some of Knapp's readers, the description of ice cubes clattering in a glass evokes an entire world. The images that Knapp so lovingly describes revolve around a potent experience: the anticipation of alcohol. These anticipatory cues have the power to arouse interest, focus attention, spark desire. Knapp writes, "I still don't know, today, if that hunger originated within the family or if it was something I was simply born with."

In fact, capture is the result of learning and memory. Cues have no significance in the absence of associations with past experience. If you've never been a smoker, then the crinkling of cellophane, the throat scratch, the curl of smoke, the image of a camel or a cowboy will have no resonance; they certainly won't prompt you suddenly to desire a cigarette.

Knapp finally answers her own question about the origins of her love affair with alcohol. It was not simply "love at first sight." As she writes in *Drinking*, "the relationship developed gradually, over many years, time punctuated by separations and reunions."

Dear Dad,

I got drunk the night you died. I also got drunk the night of your funeral, and the next night, and every night after that for one year, 10 months, and 13 days.

This is stuff we never talked about, at least not directly: your drinking, my drinking, our drinking, the way they all got tangled up together, in such subtle and seductive ways. So I thought I'd bring it up. I suspect that you—a psychiatrist, an analyst, both in and out of the family—would have a special appreciation for what this means, what it means to broach a secret like this. On the surface, it all looks so unseemly: The daughter of an analyst, an alcoholic.

Doesn't sit well, does it? Doesn't feel right. But you knew a lot about how surfaces deceived, how turmoil could bubble and roil beneath even the loveliest facades, and how complicated this business is, living and trying to love people and coping with difficult feelings.

I don't remember my first drink, but I suspect I had it with you. I suspect we were sitting in the living room, where you and mom had cocktails every evening. . . . I loved drinking, for a long time. I loved it so much I could have died for it, literally. . . . I learned how to drink from you, from watching you. I saw you come in after work and mix up the pitcher of martinis, and I saw

how the tension began to drain out of you after you drank the first
one. . . . There was always such an edge of sadness to you.

Knapp identifies, as only an addict can, the central tenet of addiction: a firm, undeniable, unalterable conviction of need. This "I need
it" feeling is cued by some stimulus: our attention is diverted, gliding
as if propelled by gravity toward the glass. In this sense, it is not the
substance itself that captures us; rather, the feelings that the substance produces narrow our field of attention until it is occupied entirely by the object of our craving. As Knapp writes, "I sometimes
think of alcoholics as people who've elevated that search to an art
form or a religion, filling the emptiness with drink, chasing drink
after drink, sometimes killing themselves in the effort. They may
give up liquor, but the chase is harder to stop."

Knapp compares the alcoholic's near-religious search for peace to
her childhood fixation on the perfect "pair of patent-leather party
shoes" or "horseback-riding lessons"—complete, of course, with
"knee-high riding boots and exactly the right kind of black velvet
riding hat"—or the "tallest Christmas tree in the lot." In each of
these cases, underlying a particular desire was something other than
its ostensible object—the promise, however illusory, of contentment.
For Knapp, there was always "some spiritual carrot on a stick promising comfort and relief."

PHYSICAL PAIN

Throughout adolescence, Caroline Kettlewell was a cutter. In her
book *Skin Game*, she tries to make sense of the episodes of self-harm
that lasted through her twenties: "When I started . . . I had no idea,
really, why I was cutting. I just knew it was what I had to do. I wasn't
blind to the fact that it looked pretty crazy to be cutting yourself like
that." Alongside the normal-seeming teenager drifted a shadowy
presence of anxiety and dread: "Every day I got up and went to
school, ate my meals and did my homework, walked and talked

through my life, but every day I felt I was spinning a little bit further out of my own control, as if I had only the most tenuous connection with that Caroline."

Kettlewell describes the moments before she cut herself as a "terrible, itching, twitching restless unease." This unease became an "oppressive tension, like something crawling on my flesh, and I wanted to shake it from my skin the way a horse shakes flies. I sat on my bed, digging my fingernails into my face, wanting to tear the skin away. What do you do with a want like that?"

In the wake of these experiences, Caroline tried to sort out how an ordinary person might end up behaving in ways that were not only self-destructive, but also utterly bizarre. What was it that drove her to hurt herself? None of Caroline's "penny ante" teenage concerns provided an adequate explanation: "How many troubles *should* equal a legitimate reason for self-mutilation?" She drew up a list of possible explanations: cranks and eccentrics in the family lineage (drinkers, depressives, and suicides); extreme homesickness bordering on panic; fear of the night; nightmares. But none of these explanations seemed sufficient, or true to her experience. All she knew was that she somehow had to satisfy her need, and that when she cut, the anxious restlessness was gone. Cut, and you could get through the day. Cut, and you could do your homework.

I let the razor's edge kiss the pale skin near my left elbow, and then drew it slowly—so slowly that I could feel through the blade the faintest tug of resistance and the sudden giving way of the flesh along my arm. There was a very fine, an elegant pain, hardly a pain at all, like the swift and fleeting burn of a drop of hot candle wax. In the razor's wake, the skin melted away, parted to show briefly the milky white subcutaneous layers before a thin, beaded line of rich crimson blood seeped through the inch-long divide. Then the blood welled up and began to distort the pure, stark edges of my delicately wrought wound. The chaos in my head spun itself into a silk of silence. I had distilled myself to the immediacy of hand, blade, blood, flesh.

She would cut when she felt frustrated, or humiliated, or insecure, or guilty, or lonely. "I wanted to cut for the cut itself," she explains, "for the delicate severing of capillaries, the transgression of veins. I needed to cut the way your lungs scream for air when you swim the length of the pool underwater in one breath."

Caroline's search for the genesis of her cutting ultimately led to a dead end. The clues she had hoped to uncover, some initial emotional charge or circumstance pressed into memory, never emerged. She craved the emotional release, the unraveling of some invisible knot in the moment of "elegant pain," but the feelings seemed to carry no intelligible meaning, nothing beyond the fullness of sensation in the moment.

In this sense, Caroline found a way to draw her focus away from what was distracting or troubling her and toward a moment that she could orchestrate. This very orchestration becomes a type of capture in and of itself. It leads to a feedback loop that is difficult to break. One of the conundrums of understanding the genesis of these behaviors is determining whether they represent antecedents or consequences, causes or effects. Was Caroline captured by cutting and therefore drawn to it in moments of discomfort or anxiety, or was cutting the result of being captured by anxiety and needing release? The fact is that capture can beget capture.

CHILDHOOD TRAUMA

As a child, Nora was confused by the disparity between the superficial image her family presented and the reality she experienced, which rendered her unable to make fundamental distinctions—between love and aggression, safety and danger, rationality and chaos. "All families have protocols," she reasons, "systems that are designed to create illusions and obscure realities."

For a long time, Nora lived alone with her terror within family life. "In one of my earliest memories," she tells me, "I am sitting at

the dinner table—our household was a linen on the table, sixty minutes of adult conversation, and elbows off the table kind of dinner
place—and I threw up. I don't know if I was sick or nervous. I have
no idea. But I vomited everywhere. At this point in my memory,
everything goes silent, my mother's mouth is opening and closing,
she's making noises, but I'm not hearing. I bolt from the table—
although I see this happening in slow motion now—and she chases
me. Some yards later, she finds me and grabs me. She tries to force me
to eat my own vomit. I pee all over myself. I remember little else for
some time after that."

There are many of these types of stories spanning the years—
operatic scenes in which Nora's mother yells at her for days on end,
locks her in her room, wakes her up in the middle of the night to
teach her a lesson, demands an apology for something that never occurred, tells her daughter that she has made a horrific mess of her
life.

Years later, Nora links her near-constant sense of panic in adulthood to these traumatic experiences. "I remember staring at my
mother and thinking, 'If I make myself as tiny as possible, there's
going to be some bit of me you're not going to get.'"

In adolescence, Nora found that some behaviors offered temporary relief from her overwhelming anxiety. Over the course of a
decade, she traded one compulsive behavior for the next in a continual effort to create a world in which she was in control.

"Cigarettes helped me do everything," she says. "They got me
up in the morning, they got me out of the house, they got me
through social interactions. They were the best camouflage because there was all that action involved and all those things you
could do with your hands. I felt protected. I could take a drag on a
cigarette and use that as a minute to collect my thoughts before
responding to somebody."

Whereas smoking promised a refuge from others, amphetamines
later offered Nora an escape from herself. "Oh, I loved speed!" she
says, genuinely brightening at the memory. "I mean everything
people find objectionable about being up all the time and being really

speedy felt really normal to me. I felt less fear—and just faster. I felt like it was me but in refined form."

Speed was her one and only; Nora was never interested in other drugs. Anorexia, however, would soon provide a way for her to let go of reality altogether. It was a sublime act of restraint and precision, and she carried it out with devout austerity. "I loved the counting. You know, how many slices can you cut an apple into? I felt morally right. I thought everybody else was wrong," she says, describing this time in her twenties. "And I liked being in a little bit of pain all the time; I think it made me feel purified."

This self-made universe was so suited to Nora's emotional needs that she lost sight of all else: "Despite needing to buy children's clothing, weighing seventy pounds, not menstruating, I was absolutely positive that I was the healthiest person in the world. I thought I was in perfect control."

And yet, one by one, these actions ultimately left her more vulnerable than protected. So she made a clean break. She quit smoking. She gave up amphetamines. And when she found herself in the hospital trying to rip IVs out of her arm because she was worried about how many calories they were pumping into her body, she let go of anorexia as well.

None of these behaviors, Nora admits, offered lasting relief from the constant terror of being inside her own mind.

"I feel so fragmented—there is not enough of me to rally to my defenses, to stave off the outside. Any request or demand can frighten me," she explains. "You know, going to the drugstore. Maybe that's going to be okay, maybe that's not going to be okay. And the first time it isn't okay—if I fail once standing on line because I got frightened—then drugstores are off the list until I can force them back on."

"What scares you about the line?" I ask.

"I might be unable to cope with the crowd. I could make a spectacle of myself. I'm always afraid of not functioning, so I'm always monitoring." When she was younger, she would drive someplace (to the dry cleaner, for example) and then sit in her car, unable to go in.

"I would have to talk," she says. "That last step would be too much for me."

This "self-modulating," the incredible energy she puts into assessing herself and the circumstances before doing anything, has held Nora back in innumerable ways. "I feel to some extent that I have wasted much of life fighting just to stay present," she admits.

"I'm always struggling, at least on some level, to quiet what I call noise, by which I mean an endlessly shifting series of anxiety-based distractions, everything from basic internal self-loathing babble to phobias. I always feel, even when at my best, that I am juggling."

Nora's attempts to "stay present" are exercises in control. Exerting control of any kind, even over things that are inconsequential, calms her and makes her feel better. The more she can control what is around her, the safer she feels. But these efforts are also obstacles to freedom, as her need for control becomes its own prison. Nora is caught in a never-ending loop. I ask her if she thinks she would ever be able to ignore the urge to keep herself in such tight check. What would happen?

"I can't imagine not exerting maximum control," she counters immediately.

I mention that her anorexia was a similar exercise in self-regulation, one from which she finally managed to free herself, although she experienced a profound sense of loss when she decisively ended the behavior.

"My life started with every manifestation of my not having control. That was the whole point, that I had no power," she reflects. "So now I think that controlling myself *is* power. It feels like the only way I can stay alive.

"I guess that's why it doesn't matter to me how small the things I'm controlling are; I don't care if I'm controlling the way the Q-tips look in a jar because, you know, I've mastered the Q-tips."

Nora admits that she'd be happier if she could loosen her grip. She feels that the constant striving for self-mastery is exhausting. But if she didn't count the Q-tips, where would that energy go? In her view, she has created an elaborate, and successful, ritual for fending off the larger panic, the one that beckons from her childhood.

BLIND LOVE

"A day doesn't go by when I don't think about him," explained Jackie, a successful book editor in her mid-forties who lives in New York.

Jackie knew that I wanted to talk to her because of her need to talk about *him*. Although she is exceptionally successful in her professional life, her affair with a married convicted felon brought her to the verge of emotional ruin, leaving her vulnerable to capture by other negative behaviors, including excessive drinking and compulsive sex.

Jackie looked around. "I didn't realize how much anxiety being in a restaurant causes me, with those beautiful glasses of wine over there."

Almost ten years ago, Jackie's younger sister, Laura, a lawyer in Oklahoma, called her to talk about one of her clients, a former finance wunderkind and oil executive named Paul. Paul had recently been indicted on a charge for which he faced up to thirty years in prison, and was appealing a conviction on another set of white-collar charges. Laura had become fond of Paul and his wife, Gloria. She was indignant on behalf of her client, and said that everyone involved in the case was convinced that the charges were overblown. Paul was interested in writing a book about his experience of the justice system. Would Jackie be willing to meet with him to share her knowledge of the publishing industry?

In October 2005, Jackie and Paul arranged to meet. A strapping middle-aged man wearing an immaculately tailored suit arrived in her office. "There was this twinkle in his eye, and I just thought, 'This guy's handsome.' I had to catch my breath." Jackie couldn't help acknowledging her immediate attraction. "Convicted criminal, married, but the way he looked at me, the way he smiled at me—it felt like the universe had moved."

Jackie had never reacted so viscerally to an introduction before. Her face flushed, and the office noises around her quieted and fell away. Still, she remained calm and professional. She offered Paul advice on structuring a book proposal and finding an agent, and then they parted ways. They exchanged a few follow-up e-mails, which were a bit flirtatious in tone, but not out of the ordinary on her end, given her gregarious, warm personality. Then she heard from her sister that Paul had been convicted and sent to a medium-security prison.

Jackie is buzzing with excitement as she recounts the story—racing through the narrative, then jumping back to insert details she left out. When I comment on her evident excitement, she says that she loves talking about *him*: "In the moment, I feel alive."

Nothing happened after their first meeting, as Paul would soon begin a thirteen-month term in prison. After his release, he scheduled a meeting with Jackie, with a renewed interest in writing his story. Seeing his face again, Jackie felt a familiar excitement; this time, though, she instantly committed to the project. They decided to discuss the book that night over dinner at a restaurant in Greenwich Village. Jackie remembers more specific details about that evening than perhaps any other moment in her ensuing three-year relationship with Paul.

"I remember seeing him from this crowded bar. I remember I had the sea bass. I remember he ordered a bottle of pinot noir. I remember that we were three tables from the door."

After dinner, which lasted a couple of hours, Paul offered to drop Jackie off at her apartment. While the cab was heading uptown, he gently placed his hand on her thigh.

"He didn't grab my thigh. He just touched it with his hand, and it lingered there for—it felt like three hours, but it was probably all of ten minutes."

Jackie describes this moment as "the opening of a door." The combination of dread and desire—dread perhaps that her desire, which she knew was wrong, might be fulfilled—was almost unbearable.

After their dinner, Paul began to e-mail her more frequently; for Jackie, what began as a confused aura of attraction gradually evolved into a deep need. Soon she was unable to think of anything other than Paul. When one literary agent declined to work on the project,

Jackie felt an inexplicable sense of obligation toward Paul and the book he had yet to write. "I knew how important doing this book would be to him. I felt such *injustice* happened to this man and *I* needed to help him. I thought, 'We *have* to do this book, I *have* to. I *owe it* to him.'" His only hope of redemption, she believed, was telling his story, and she was the only person in his life with the publishing acumen to help. He started sending her portions of the text, noting that he hadn't told his wife about some of the things he was revealing to her. She would edit his work, then spend hours on the phone discussing the project with him.

"It gave meaning to my little world," she says. "I got so enveloped in his story: the story and the man were completely enmeshed."

Over the course of a few weeks, the phone calls became longer, and the discussion veered toward things other than the book. Eventually, Paul confessed his burgeoning feelings for her. Despite her attraction to Paul, Jackie responded angrily, reminding him of his family, and even of her own nascent relationship with a young lawyer. Jackie resisted in part, she says, because she had been previously involved in a scarring relationship with a married man. But Paul was relentless: "He basically said he was in love with me. I went to a department store and walked around in a fog." When a song that reminded her of Paul came over the loudspeaker, she began to cry. She called her sister for advice. Laura told her that, although she was sure Paul and his wife would eventually divorce, Jackie should not get involved with him. "You are out of your mind," she said, and for the moment, Jackie agreed.

Paul eased up at first, and then reinitiated the wooing. "That's when he started sending me little notes about how much he loved even my fingers, my hands," Jackie recalls. He soon told her he was saving their e-mails to show their future children. He insisted his relationship with his wife was over and that Jackie was the woman he really wanted. Eventually, Jackie gave in. She describes the decision as an enormous relief. "I just surrendered," she says. "I couldn't take it anymore."

The two began a physical relationship, and also continued to work together on the book, which served as a convenient cover. They communicated almost every waking minute: by phone, by instant message, and by e-mail. Jackie jumped whenever her BlackBerry beeped, hoping it was an e-mail or a text from Paul. He begged her not to leave him during what would certainly be a trying period, insisting that when the time was right, he would leave his family and start a new one with her. "This is going to be the hardest thing in the world," he'd say. "You *can't* leave me."

Jackie found Paul's reliance on her thrilling. She was infatuated with his height, his freckles, his confident swagger, and she loved that he seemed totally enamored of her. It was a new experience for Jackie—to be treated as truly beautiful rather than merely sexually enticing. She also found Paul's professional achievements alluring, partially because wealth had always felt so unattainable to her, the child of a hotel maid and an electrician. This good-looking, rich man was for her an unattainable ideal.

After about a year of carrying on their clandestine romance, Jackie and Paul received devastating news. In November, Paul lost his appeal and was told he would soon have to return to prison. Jackie swears that long before he told her the news, she knew instinctively that something was wrong: "I remember feeling like I was immersed in this dark energy." After he relayed the news to Jackie, the two went to pray at the local Catholic church. But their prayers would go unanswered. Right after the New Year, in 2009, Paul returned to prison.

"That's when my insane dedication and love for him grew exponentially," Jackie says. "I lived for those two phone calls a week." Sensing that her friends were unsympathetic, Jackie isolated herself, communicating only with her priest and confidant, Father John. She would write lengthy missives to Paul, and spend the evenings drinking wine. Whenever possible, she flew to the midwestern town where he was incarcerated, and while she became friendly with others visiting their loved ones in prison, she was able to convince herself that she was ultimately different from them. Paul, she rationalized, hadn't committed a violent crime or sold drugs, like some of his fellow prisoners; he was the target of an "injustice of the greatest proportion," perpetrated by a spiteful judge. Jackie, too, was a victim of a grave injustice: "This was preventing me from getting on with my life," she explains. "This was preventing me from having children." The quest for Paul's freedom became her life, a cause to which she would devote herself.

Jackie readily admits that she is at her best when faced with an obstacle, and these years were no exception. The moment Paul returned to prison, she reached out to everyone she knew in a position of political power, or anyone in media who might be willing to publicize the case. She even had a friend approach a congressman and ask to have a certain sympathetic U.S. attorney's confirmation sped up in the hope of having Paul's conviction thrown out.

That May, Father John invited Jackie to go along on a church trip to Israel. She eagerly accepted: "My mission was to go there and pray for Paul." She beseeched God at the Western Wall, on top of Mount Horeb, and at prominent Catholic churches in Jerusalem. She acquired numerous prayer beads and other totemic objects to pass along to Paul. A visit to the prison where Jesus was said to have been held the night before his crucifixion was especially emotional; the parallel, between one wrongfully imprisoned man and another, was almost too much to bear. Throughout, she found comfort in her singular sense of purpose: "I felt this heaviness, but I also felt like I was doing the right thing and that all my prayers and efforts were going to get me closer to victory."

Paul was paroled late that autumn. First, he was transferred to a halfway house; a few months later, he was allowed to go home, albeit with a monitoring device. Finally, he got word that the recently appointed U.S. attorney, the one with ties to Jackie's friend, had dropped the case. It remained unclear whether Jackie's machinations were behind this outcome.

Not long after receiving this good news, Paul began to renege on some of the promises he had made to Jackie early in their relationship. He put off divorce proceedings against his wife, who had found out about Jackie and Paul's affair, and although he was looking at real estate in New York, he told her he didn't think they should live together right away. "I could feel him pulling away," Jackie says. They got into an argument one evening after Jackie, who had been drinking, made a sharp comment about Paul's wife. She suggested they take a break; when he agreed, she completely fell apart.

"I could barely breathe," she remembers. "My heart wouldn't stop pounding. It was physically painful."

Jackie describes herself in the wake of their breakup as barely functional—"like a zombie." Sitting in her office, she would inevitably succumb to the urge to look at pictures of Paul on her computer. This would lead to reading old e-mails and, inevitably, sobbing. "I would have to close my door at work and just heave," she says. She cried everywhere: at work, on the subway, at home at night listening to songs she and Paul had once loved. "My doctor put me on medication because all I could do was weep and weep and weep."

Jackie began to drink heavily. She had always relied on alcohol to calm her when she was nervous, but now her drinking "skyrocketed." Through the holidays, she consumed copious amounts of alcohol, and then on New Year's Eve, after a night of heavy drinking, she tripped on the sidewalk in front of her apartment and fell directly on her face. She spent the next day in the emergency room. The incident didn't scare her, though; she did not care about her pain. If only she were different, she thought—prettier, more confident, more *something*. If only she had fought harder for him. "I felt that the loss was so great," she explained, "that only sobbing and feeling completely depleted would honor it."

She continued to drink heavily for the next four years, hoping to drown out the memories of Paul. Often when she was drunk, she sought out the company of men, and soon she found herself using sex as a way to numb the pain of rejection. She slept with other married men, colleagues, strangers, men she didn't remember going to bed with when she woke up the next morning, head pounding from a hangover. "I wanted to get revenge on the world by behaving irresponsibly."

Finally, she decided to get sober. And though it's been five years since they've seen each other and one year since she stopped drinking, thoughts of Paul still continue to enter her mind, unbidden. "I still sit on the subway and fantasize that something horrible will happen to me and that Paul will feel guilty for the rest of his life."

OBSCENE FASCINATION

Richard Berendzen never planned the calls. Yet they always seemed to happen in the afternoon, when he felt most anxious. Once his thoughts strayed to the receiver on his desk, he could think of nothing else.

The then-president of American University in Washington, DC, Berendzen at first made the calls only sporadically. By January 1990, however, the frequency had increased. Having chosen his victims from newspaper ads for child-care services, Berendzen would ask his assistants to hold all incoming calls and would go into his private study, where he would draw the blinds and turn off the lights. Whenever he felt stressed—his "head spinning and temples throbbing"— Berendzen's mind would wander from his work to the telephone, and he would soon find himself reaching for the classified sections of the newspaper.

"When I made a call, I mentally left my book-lined study. I abandoned the presidency," Berendzen notes. "I forsook my wife and daughters. I turned away from all that I had studied. . . . Ninety-nine percent of me was the man the world saw: university president, husband, father. But that other one percent was a bomb about to explode, destroying everything I stood for and had worked to achieve."

Berendzen asked one of the women who believed she was being interviewed for a nanny position whether she had an "open family" and explained that his wife and children were accustomed to walking around their house naked. "We share everything," he allegedly said, "and I do mean everything."

Berendzen described to the women in graphic detail having sex with his children. He talked about his extensive collection of child pornography and sometimes even insisted that he kept a four-year-old Filipino girl as a sex slave. Learning that one of his victims had two daughters of her own, he suggested, in vain, that she put them on the line so that he could instruct them to perform sexual acts on

their mother. He also boasted that he used a collection of sadomas-
ochistic instruments to discipline his family. "He [said he] had a
wheel in his basement that he would strap his wife to," said one
woman. Sometimes he told his victims that he was masturbating as
they spoke.

The case made headlines throughout the country. Berendzen's
psychiatrists contended that his calls were linked to sexual abuse he
had endured as a child. Paul McHugh, who was chairman of the Department of Psychiatry at Johns Hopkins University, told *Nightline*,
"This behavior . . . is a kind of a foreign body imprinted in him earlier in his life, and we are very confident, looking at the behavior
itself, that it wasn't a kind of erotic enjoyment . . . but rather a kind
of horrible fascination about what happened to him as a child and
whether those things were happening to others."

Whether or not Berendzen derived pleasure from the calls, it is
clear that he could think of nothing but his erotic fantasies: all other
thoughts dissolved. Sexual counselors often refer to this "trance
state" as one of the emotional dimensions of sexual experience. In
Constructing the Sexual Crucible, David Schnarch traces this experience along a continuum of intensity: "Depth of involvement in
sexual trance increases as day-to-day reality fades, replaced by increasing concentration on the sexual reality of the moment." Initially, the individual remains rooted in day-to-day reality, "scanning
the environment unrelated to the sexual encounter." As the sexual
trance deepens, "daily reality fades and interest is focused on the
sexual interaction," though attention remains "vulnerable to minor
external interruptions." At its strongest, sexual capture completely
obliterates the surrounding world: "the individual becomes totally
absorbed in the sexual reality and loses awareness of extraneous
events."

"I've heard of people," Berendzen said, "who become so transfixed by a fire that they stick their arm into it. Such actions lie outside logic; caution and reason have nothing to do with them. . . . As
the conversations continued, I fell further and further into the
flames."

GAMBLING

With five or ten *louis d'or* in his pocket, Fyodor Dostoyevsky headed to the casino. Hours later, the novelist returned home, disconsolate and penniless, and begged his wife to forgive him his "abominable passion."

Night after night, Anna Dostoyevsky recorded her husband's pleas in her diary: "He said pathetically that he reproached himself for his weakness . . . that he loved me, that I was his beautiful wife, and that he was not worthy of me.

"Then," she added, "he asked me to give him some more money."

This pattern repeated itself for years, as Fyodor and Anna's household funds dwindled to almost nothing. Nightly rounds of breast-beating and penitence led only to further recklessness: "But Fedya implored me to give him at least two *louis*, so that he could go to the tables and get some relief." Anna no longer blamed her husband's gambling on an "ordinary lack of willpower"; rather, it was "something elemental, which even a person of strong character would be powerless to resist. It was something . . . one had to view as an illness for which there was no cure."

Every so often Fyodor would come home beaming, impatient to regale Anna with tales of his success at the roulette table. Though Anna hoped that these rare jackpots would finance Fyodor's literary work, the cache would all too soon be raided. Once his winnings had trickled back to the casino, Dostoyevsky would make the rounds of the local pawnshops, carrying his wife's winter coat, a pair of boots, even his wedding ring.

Anna did her very best to steer her husband away from his self-destructive habits, but "devilish gaming" held Fyodor in its thrall. In a letter to his brother Michael, Dostoyevsky described the financial tumult of those years: "Within a quarter of an hour I won 600 francs. This whetted my appetite. Suddenly I started to lose, couldn't control myself and lost everything. . . . I risked 35 *napoléons* and lost them all. I had 6 *napoléons d'or* left to pay the landlady and for the journey. In Geneva I pawned my watch."

This cycle of exhilaration and desperation left its mark not only on Dostoyevsky's marriage but on his writing. Early in his career, to avoid a stint in debtors' prison, Dostoyevsky signed a highly unusual contract with a Russian publisher. If the writer did not deliver a novel by November 1 of that year, the publisher would acquire the right to publish all Dostoyevsky's works for nine years without compensating him. Panicked, Dostoyevsky dictated a semiautobiographical novel, *The Gambler*, to Anna in less than one month.

Set in the fictional German spa town of Roulettenburg, *The Gambler* chronicles the misadventures of Alexei Ivanovich, a young tutor who gambles obsessively, both in the casino and in love. With every turn of the roulette wheel, Alexei tries to free himself from the drudgery of circumstance—in particular, his failure to win the affections of the beautiful but aloof Polina Suslova. As Alexei contemplates his first bet of the evening, the entire world fades away, and the steady march of time stretches into a glistening band. "Gamblers," writes Dostoyevsky, "know how a man can sit for almost twenty-four hours at cards, without looking to right, or to left." The novel's many gamblers all crave this opiated lull: When Antonida Vasileva sees "a truly splendid carriage" whirl by, she perfunctorily raises her head and asks, " 'What is that? Whose is it?' "

"But," Alexei says with a sigh, "I believe she did not even hear my answer."

As his chances at romantic success dwindle, the tables exert an ever stronger pull on Alexei. He no longer thinks of money; rather, he gambles for the pleasure of risk itself: "And then—what a strange sensation!—I remember distinctly how all of a sudden a terrible craving for risk took possession of me, now quite apart from any promptings of vanity. It may be that, in passing through so many sensations, the soul does not become sated but is only stimulated by them and will ask for more and even stronger sensations until utterly exhausted."

In the climactic scene of the novel, this state of blissful suspension is interrupted only when Alexei hears the other players marveling at his winnings. He later admits that, for those precious few hours, he didn't think once of his unrequited love for Polina: "I was then experiencing an overwhelming pleasure . . . scooping up and raking in the bank notes which were piling up before me."

"They couldn't tell where their finger ended and the screen started," explained Natasha Schüll. An anthropologist who studies compulsive gambling, Schüll was describing her research subjects, whose game of choice was video poker.

All Schüll's subjects described a moment of intoxicating release, "a state in which they no longer experienced themselves as subjects in the world. They had no social identity, there was no time—all of those structures fell away. Time, money, space, the existence of other people—they all dropped away. There was only this movement, this sort of continual motion."

For these gamblers, the goal of play is to flood their attention to the exclusion of all else. Schüll's subjects compared the force of their cravings to a magnet, pulling and pulling inexorably. When they sat in front of a poker machine, it was as if time had stopped; they were soon in a trance, on autopilot. They would do whatever they could to stay in this altered mental state, totally absorbed, numb to reality, for as long as possible. Many could remain transfixed for hours. Though Schüll's subjects came from all walks of life, their cultural background and socioeconomic status had little bearing on their experience of the poker machine; for these gamblers, all that mattered was being in the zone, maintaining the rhythm.

This desire for escape—from the painful, the familiar, or the disappointments of everyday life—lies at the heart of compulsive gambling. For many, monetary losses are quickly forgotten, but none forget how they felt in that moment in the casino. Ultimately, winning or losing is beyond the point; what matters is the experience of complete immersion. As Schüll explained, the makers of video poker machines understand this psychology all too well. The games are designed to ensure a smooth ride: nothing gets in the way of the player's flow or breaks the action.

"Some gamblers talk about their bodies disappearing into the screen," Schüll recounted. "They say, 'I wasn't present, I was gone.'"

THE BODY

Capture results in hypochondria when people cannot stop focusing on their own physical vulnerabilities. This psychological torment stems from a hyperawareness of bodily sensations.

Tennessee Williams was at work on the short story "The Accent of a Coming Foot" when, having arrived at a climactic scene, he suddenly became aware of his own heartbeat. He was certain it was "palpitating . . . skipping beats." Without any sedatives on hand—"not even a glass of wine," as he later wrote in his diary—he jumped up from the typewriter and rushed out onto the streets of St. Louis.

"I walked faster and faster as though by this means I could outdistance the attack. I walked all the way from University City to Union Boulevard in St. Louis, expecting to drop dead at each step. It was an instinctual, an animalistic reaction, comparable to the crazed dash of a cat or dog struck by an automobile, racing round and round until it collapses, or to the awful wing-flopping run of a decapitated chicken. This was in the middle of March. The trees along the streets were just beginning to bud, and somehow, looking up at those bits of spring-time green as I dashed along, had a gradually calming effect—and I turned toward home again with the palpitations subsiding."

Williams came to call episodes like this his "cardiac neurosis," physical symptoms that seemed both to stem from panic and fear and, counterintuitively, to quiet his anxiety. On Sunday, March 29, 1936, Williams wrote in his notebook: "What a week is *behind* me! Wednesday received poetry prize. Not as gruelling as I expected. In fact, it couldn't have been made any easier for me. No stage. No speech. Just a room full of tired, elegant old ladies, a couple of priests and some very young poets. Lovely sunny place. Nevertheless palpitations for about five minutes. Afterwards tea and talk."

A couple of years later, while working on a play, he wrote, "Soon my nerves began to pop—heart neurosis developed this week—a crisis last night—after several previous—This morning felt very weak and sickly—although I had gotten a fair sleep—Once again I feel dangerously cornered, cut-off—wonder how I'm going to fight my way through—Some external stimulus must be applied to snap me out of *this*. But what?"

Despite his restless search for an immediate cause, it seems that some part of Williams understood what was taking place:

"What makes all this so stupid is that the fear is so much worse than the thing feared . . . it is the fear that makes all the concomitant distresses, the dreadful tension, the agora and claustro-phobia, the nervous indigestion, the hot gassy stomach."

Williams came to rely on alcohol and sedatives to cope with these disturbing physical sensations: "Just now—a painful jolt in my heart. I got up to get my Scotch. . . . I suppose on the whole I am reasonably calm. I have taken a seconal and am having a drink. . . . There is one thing to do, one thing and one only—put it out of my mind and pass the night as though it hadn't occurred."

Frequent palpitations, jolts, and bouts of anxiety eventually made it necessary for Williams "to carry a flask of whiskey with me wherever I went." Doctors offered little help.

During one summer, while in Memphis, Williams recognized his attraction to a young man. Over dinner at the Peabody Hotel with his new love interest, he once again felt an attack of palpitations. Panicking, he summoned a doctor:

"A lady, an extremely bad doctor. She gave me a sedative tablet of some kind but informed me, gloomily, that my symptoms were, indeed, of the serious nature. 'You must do everything carefully and slowly,' said this gloom pot. She told me that with the exercise of care and a slow pace I would live to be forty!"

Several decades later, his *New York Times* obituary would read, "Tennessee Williams, whose innovative drama and sense of lyricism were a major force in the postwar American theater, died yesterday at the age of 71. He was found dead about 10:45 A.M. in his suite in the Hotel Elysee on East 54th Street. Officials said that death was due to natural causes, and that he had been under treatment for heart disease." The obituary labeled him "a monumental hypochondriac" who "became obsessed with sickness, failure and death. Several times he thought he was losing his sight, and he had four operations for cataracts. Constantly he thought his heart would stop beating. In desperation, he drank and took pills immoderately."

As Williams had predicted, fear itself did prove to be more dangerous than the objects of his fear. Six months after his death, New York City's chief medical examiner reported that Williams was "apparently trying to ingest barbiturates when he choked to death on a plastic bottle cap." According to Dr. Elliott Gross, "the cause of death was asphyxia." It is believed he used the lid of the bottle to take the barbiturates.

A WORK OF ART

For the eight years that American artist Jay DeFeo spent painting her masterwork, *The Rose*, one idea remained inviolable: the painting had to have a "center."

DeFeo built up the painting layer by craggy layer, only to tear it down multiple times, sometimes even removing all the paint from the canvas.

According to her friend Bruce Conner, the painting gradually came to fill DeFeo's entire San Francisco apartment: at first the canvas was framed by a bay window, with "hard wedge-like lines radiating from the center." After the paint had thickened, DeFeo realized that it needed to extend well beyond the edges of the window; the painting grew, becoming rounded, then once again sharp-edged, pullulating across the room.

Rising above the canvas at certain points to a height of eleven inches, the work was as much sculpture as painting—or it was many paintings, one built on top of the next. DeFeo "sharpened knives on a drill press and hacked and carved the surface" to achieve this effect. Ultimately the approximately eleven-by-eight-foot canvas contained some 1,850 pounds of paint.

DeFeo herself emphasized the organic process by which the work came into being—one stage of development followed by the next—

rather than the final product: "It went through, I would suggest, a life span, a chronology of different stages." The first phase, "almost like an infancy period," gave way to "a very geometric stage," when the work became "crystalline," with "no curved forms whatsoever." Then it became "more organic in character," finally entering "a super-baroque period." "I really wasn't aware of how flamboyant it had become," DeFeo explained. "I walked into the studio one day and the whole thing seemed to have gotten completely out of hand. I felt that it really needed to be pulled back to something more classical in character."

It seemed as if the painting would never be complete. "The room itself was the work," Conner explained. The floor of DeFeo's apartment was covered with "chunks of almost flesh-like paint," which she had scraped off the canvas. Entering the mica-speckled room, Conner explained, was "like walking into a temple." And still the painting grew, absorbing endless coats of paint, the walls, the ceiling, and the floor of the apartment. As time went by, *The Rose* gradually became DeFeo's only identity "with any kind of exterior reality." As the boundaries separating artist, artwork, and studio began to blur, it became clear that only an external event, some sort of violent interruption, could bring the work to completion.

Martha Sherrill, an art critic for the *Washington Post*, wrote of DeFeo, "The process of [*The Rose's*] magnificent accretion seemed to consume her. It was as though the work exerted a strange magnetism, pulling everything toward it—people, paint, needles from the Christmas trees in DeFeo's studio, and the artist herself—and refusing to let go. People joked that the work would be finished only when the artist herself died, and a myth grew up around them—DeFeo and *The Rose*: a religion, a relationship, a compulsion, an addiction, a fabulous love affair that would not end."

About seven years after she began work on *The Rose*, DeFeo and her husband were evicted from their apartment. The painting had to be removed. Because of its sheer size, movers cut out part of the wall of her apartment to transport it to a museum, chunks of paint falling off it in the process. DeFeo continued to work on the canvas long after it left her apartment. Only when others began to take an interest in *The Rose*, to recognize it as a work of true historic significance—one that DeFeo knew would be cared for—did she experience a sense of release.

When she was twenty-three years old, DeFeo wrote in a letter to her mother from Florence, "I believe the only real moments of happiness and a feeling of aliveness & completeness occur when I swing a brush. I don't think I can do without it."

For DeFeo, the creative impulse stemmed from an inner restlessness. She acknowledged that the experience of painting, as opposed to the painting itself, was "kind of a cliffhanger" for her, a way of suspending time. Bringing this large, radiant, unwieldy work into being required "an inner core of faith that this thing would emerge into an ultimate form, of which I had no knowledge. I just kept reaching for it intuitively."

DEATH

Suicide is, for most of those who choose it, a last resort: the ultimate negation of suffering. For David Foster Wallace, suicide offered the prospect of escape from an unbearably painful world. But for the American poet Anne Sexton, the experience of pain was eclipsed by a fascination, indeed an obsession, with the idea of death. Where many see death as the only escape from capture, Sexton was captured by death itself.

"It's like a person who takes drugs, and they can't explain why they want to do it. You know, there really isn't a reason," Anne Sexton explained to her therapist, Dr. Martin Orne.

"There always is: drugs are addicting," he said.

"Suicide is addicting," she responded.

Sexton shared with her close friend, the poet Sylvia Plath, an enduring fascination with death. Reflecting on Plath's death, Sexton said that Plath had "had the suicide inside her. As I do, as many of us do. But, if we're lucky, we don't get away with it and something or someone forces us to live."

In a letter to her intimate friend Anne Wilder, Sexton described her sense of complete alienation from the world of the living:

"Now listen, life is lovely, but I CAN'T LIVE IT. . . . To be alive, yes, alive, but not be able to live it. Ay that's the rub. I am like a stone that lives . . . locked outside of all that's real . . . do you know of such things, can you hear???? I wish, or think *I* wish, that I were dying of something for then I could be brave, but to be not dying, and yet . . . and yet to [be] behind a wall, watching everyone fit in where I can't, to talk behind a gray foggy wall, to live but to not reach or to reach wrong . . . to do it all wrong . . . believe me (can you?). . . . I want to belong. I'm like a Jew who ends up in the wrong country."

More than twenty of Sexton's poems deal explicitly with wanting, or perhaps needing, to die. Many critics found fault with Sexton's poetry because it exemplified all too plainly the vogue in midcentury American poetry for confession. Indeed, it often reads as if the poet were talking to her psychiatrist. "Wanting to Die" takes the form of a letter to Dr. Orne: Sexton tries to explain her darkest ruminations, which hide just below the surface of the everyday. Responding to her therapist's incredulity at her hopelessness, his vain attempts to revive her will to live, she traces the arc of her suicidal desire:

> Since you ask, most days I cannot remember.
> I walk in my clothing, unmarked by that voyage.
> Then the almost unnameable lust returns.

Here, the rhythms of lust, the predictable rise and fall of sexual desire, become a metaphor for her underlying drive toward death.

For a mind so thoroughly consumed by its pain, the question *why* matters less than *how*. In "Wanting to Die," Sexton insists that suicides "have a special language. / Like carpenters, they want to know which tools. / They never ask why build."

Sexton's preoccupation with the mechanics of death revolves

around the moment when the burden of living dissolves into nothingness:

> In this way, heavy and thoughtful,
> Warmer than oil or water
> I have rested, drooling at the mouth-hole.
> [. . .]
> To thrust all that life under your tongue!—
> that, all by itself, becomes a passion.
> Death's a sad bone; bruised, you'd say,
> and yet she waits for me, year after year,
> to so delicately undo an old wound,
> to empty my breath from its bad prison.

Suicide becomes a perversely rational option when emotional
pain becomes too great to bear. But how had Sexton become locked
in that "bad prison," where the irrational seemed all too inviting?

The term "breakdown" was commonly used by doctors and lay-
people alike to describe a prolonged period of severe mental or
emotional pain. It implies that the human being is, in an important
sense, like a motorcycle or a washing machine: parts can wear down
or rust over; fuses can blow. But like mechanical objects, those who
had experienced a breakdown could, at least in theory, be repaired.

For Sexton, however, the promise of recovery proved hollow.
Though she experienced episodes of both mania and depression
during adolescence, her first complete "breakdown" appears to have
been sparked by the usual demands of caring for young children.
When her daughter was suffering from croup, Anne spent an entire
night convinced that the girl was about to die. Fear for the safety of
her children soon devolved into fear that she would hurt them. In the
years that followed, she suffered from increasingly destabilizing
mood swings and, in between bouts of literary productivity, spent
many months in psychiatric wards.

Her notes from this period record a near-daily descent into an
abyss of pain: "Nothing seems worthwhile—I walked from room to
room trying to think of something to do—for a while I will do some-

thing, make cookies or clean the bathroom—make beds—answer the telephone—but all along I have this almost terrible energy in me. . . . I sit in a chair and try to read a magazine and I twirl my hair until it is a mass [of] its snarls—then as I pass a mirror I see myself and comb it back. . . . Then I walk up and down the room—back and forth and I feel like a caged tiger."

What begins as listlessness and confusion gathers terrible strength until Sexton's mind has turned completely on itself, and the outside world of cookies and telephones and magazines has all but ceased to exist. Eventually her mind would steady itself, allowing her to disguise the rawness of her pain with humor: "Well, I'm not gonna kill myself in the doctor's office, all over his beautiful carpet."

Over time, suicide came to represent a radical form of self-alteration, the only way Sexton could reclaim control over her errant mind: "It's as though I wanted to kill someone else, but that someone else is me." When her therapist asked what death would feel like, the poet described a cozy, pastoral scene: "Spring. Warm. Leaves."

But for Sexton, not all paths to death were equal. Indeed, Dr. Orne was surprised to discover that she had a paralyzing fear of flying, and in particular, of jet engines. Why would someone so intent on killing herself be so terribly afraid of dying in a plane crash? In a husky, hurried voice, Sexton pondered her morbid fascination with "those great, powerful motors that can take this impossible weight off the earth and put it up there."

Then her voice rose in pitch: "This heavy thing, you know, can only be lifted by a great power, and if that power should fail . . ."

Orne was nonplussed: "You've told me over and over that you want to die."

"Not like that." Only by controlling her own death—the timing, the manner—could Sexton assert some degree of control over her life.

"Why are you afraid of that kind of death? You know, [with] sleeping pills, you'd be just the same way dead."

But Sexton's attention had drifted away from the consultation room, rising into the air: "And we're going up . . . and I hear the motor stop. And we're in silence."

A THREAT

"I never would have admitted that I was afraid of my students when I started teaching," Charlotte told me. We were talking about her first job as a teacher, which began just one month after she graduated from college. Charlotte decided to teach English shortly after suffering through American literature as a high school junior. Her instructor was young; she gave out worksheets and study questions, but never led the class in discussions about literature—"which made me wonder if she had ever read the books she assigned," Charlotte adds. As a result, Charlotte realized that she wanted to help teenagers think about their lives as they read and analyzed fictional characters. Even now, as a fifteen-year veteran teacher, she relishes this aspect of her job.

Still, Charlotte worried that she lacked the wisdom, or the gravitas, of a respected teacher: "I think many of my young colleagues shared in my first school worry: that students would mock our lack of experience or authority. At twenty-two, I was only a few years older than the seniors that I passed in the hallway."

Five months before she began her first job, two high school students had opened fire at Columbine High School in Colorado. "I was a college senior in a teaching practicum class when I heard that news. It rattled all of us," she explains. "I remember crying days later, when Amy Grant sang at a televised memorial service. But I wasn't afraid then that a similar act of violence might happen in my future. Perhaps this was because the demands of student teaching kept me otherwise occupied, or because, at that time, the Columbine shooting was an anomaly to us Generation-Xers. I had witnessed fistfights as a high school student, but nothing graver."

Even as she mourned the Columbine victims, Charlotte continued along her chosen professional path. For those first few years, she pinned her hair up, traded contacts for glasses, and sifted through Macy's clearance racks for suits. She avoided making personal connections with students, and if they asked her age, she added a few years and said she was twenty-seven. It seemed a proper and sophisticated age.

By the time she actually turned twenty-seven, Charlotte indeed felt more confident in her position. "I let my hair down and started participating in casual [attire] days," Charlotte recalls. "I gave the students honest answers when they asked about my background." She still felt anxious at the start of every school year, but several years of experience had taught her that to deal with her insecurity about her young age, humor worked better than hairpins and glasses. She found that wit deflected teenage attitudes better than detention could.

Had she not begun to realize her professional identity by then, she says now, she might have felt even more displaced when there was another shooting.

"On April 16, 2007, I left school just before four o'clock," Charlotte recounts. "As I pulled out of the main entrance, I heard NPR's hourly update. A gunman had opened fire inside lecture rooms at Virginia Tech.

"I hadn't gone to Virginia Tech. I didn't know anybody who had gone to Virginia Tech," Charlotte explains. "But my heart stopped. I felt cold, and helpless, and scared. My hands shook. I drove straight home, listening for more information. I wanted to know how many people had died. I wanted to know why the shooter had targeted a college campus, and why nobody could stop him."

When she got home, she went for a run along a three-mile loop that led by a river. Halfway through, she had to drop to a walk. Her throat felt like it was closing. She stopped and stood still.

"I panted and worried that my throat would seize up. I held it with one hand and started walking again. I wanted to go home. I wondered if I had developed allergies. I thought maybe I was having a reaction to an insect bite."

When her husband came home from work, she told him what had happened on her run. He asked if she had heard about the Virginia Tech shootings; she didn't want to talk about it.

"We watched the evening news. I went to bed early but couldn't fall asleep—I worried that I would struggle to breathe in my sleep and suffocate."

The next day, she went to work. Some of the students wanted to talk about what had happened at Virginia Tech. She encouraged them to write a free-response entry in their notebooks, and to express their immediate reactions to the news. She doesn't remember what she said to those students who may have shared their entries, but she does recall avoiding conversations in the English department office about the shootings.

That afternoon, she again experienced shortness of breath while running. But this time, it occurred while she thought about Virginia Tech. She wondered what she would do if someone pulled out a gun along her running route. She scanned the path, looking for a place to hide. She imagined dodging bullets. These thoughts didn't subside until she reached home. Later that evening, her heart began to race. She sat down, pushed her hands against the hardwood floor, and tried to catch her breath.

"When my husband came home, I told him that I needed to go to the hospital. I ran outside. He followed me and started to call 911. I told him I felt better. Being outside helped me to calm down, slow down. I told him to hang up. That is when we both realized that I had been having panic attacks."

The nightmares started that night. She had a recurring dream in which she was standing in front of a classroom full of teenagers and a gunman would burst into the room. "Holding a rifle, he would instruct me to choose which students he would shoot," she recalls. "If I refused, he would kill everybody, including me."

The dream varied in length, but it always included the same ultimatum. "Sometimes I saw students die," Charlotte says. "The kids in the classes consisted of students I recognized and those I didn't. I still remember one girl from the dreams. She was skinny with long, brown hair. She wore a colorful top, and tears ran down her face as she looked at me."

For months after the murders at Virginia Tech, Charlotte was plagued by nightmares of a school shooting. "The scenario haunted me on my drive to school, during my classes, and on my afternoon walks—I physically could not bring myself to run again for a while," she explains. "I didn't think about which students I would name, but about how I would thwart a shooter in a similar situation. I thought about locking my classroom door during every class. Then I realized that it was possible the shooter could come from my class roster, in which case locking the students inside would cause them more harm. I had inherited my classroom from a teacher who had retired; either she or someone else had stored a golf club atop a built-in cabinet along one wall. I thought about moving the golf club closer to my desk so that I could use it as a weapon. But I didn't know how I could hide it."

Though anxiety takes many forms, salient stimuli play an active role in all of them. The stimulus is almost always some kind of perceived threat; anxiety is, in other words, a distortion of selective attention, the survival mechanism that alerts us to danger. Like many sufferers of anxiety, Charlotte lived in a near-constant state of hypervigilance: she remained on high alert, expecting signs of menace at any moment. Scanning her environment for a perceived threat, or any hint of one, only heightened her perception of danger and increased her feelings of vulnerability.

Charlotte grew up in a nurturing, well-educated lower-middle-class family. They often moved for her father's work. Charlotte didn't mind changing schools, but she was very aware of how other children perceived her.

"I remember observing, again and again, the happiness of classmates. I was good at pretending I was happy; I didn't want to look abnormal and I didn't want to worry my parents, who each worked more than one job most of my childhood," she recalls. "I wasn't unhappy, but I always felt a little bit sad. I liked rainy days best. They gave me an excuse to withdraw. Other kids would complain about the weather hampering recess, but I preferred indoor activities like board games and movies to playing tag or tetherball.

"I didn't like bedtime. I didn't fear sleep as much as I dreaded the tasks I set for myself before I fell asleep. I would turn the light switch on and off in my bedroom a certain number of times, and then I would take a certain number of steps from my bedroom door to my bed. I would then kneel beside my bed and count to five hundred as quickly as I could for each member of my family—my parents, two brothers, and two grandparents. I worried that if I didn't do this, something bad would happen to them."

She got so tired of this ritual that one night she cried and prayed for help in breaking it. The severity of her tasks, she says, did soften after that prayer.

She wondered sometimes why she counted so much: the number of holes in the tiles on the ceiling at her doctor's office, the number of cracks on her backyard basketball court, the lines in the panels on the wall of her back porch.

"I didn't like counting, and when I caught myself doing it, I tried to stop," she says. "I tried to stop other things as well—like in the eighth grade, after I learned to type, I would take lines from conversations and type them in my imagination, looking at the words as they appeared on a screen in my head. Sometimes I could stop myself from these repetitive behaviors."

Charlotte was soon diagnosed with obsessive-compulsive disorder. In OCD, routine features of daily life take on great consequence: sensory phenomena lead sufferers to become profoundly disturbed by things that don't look, feel, or sound just right—coat hangers turned in different directions, large and small cans of food intermingling on a shelf, books not arranged by height.

Correcting what feels out of order (say, placing shoes so that all the toes point in the same direction) or giving in to the physical sensation briefly relieves discomfort, but this relief does not last. Soon enough, the long-established cues again trigger the urge to act, and that urge again engages the motivational and movement circuitry of the brain, driving the repetitive behavior and making it ever harder to resist.

Although her need to count finally subsided, Charlotte was plagued by recurrent nightmares throughout childhood. "The first nightmare I remember occurred before I turned eight. A bunch of cats were clawing at my family's screen door, trying to get inside to jump on me," Charlotte recalls. "The content of my nightmares became more mature as I aged: my parents died again and again, a wizard chased me through a parking lot, a group of mean college boys chased me through a park. Even now, I still pray that God will keep bad dreams from my mind as it rests.

"I thought about talking to a school counselor, or the social worker, about my fears," Charlotte recalls. "But when I was most haunted by my visions, I didn't want to tell anybody. After all, I wasn't incapacitated. I could get up and go to work every day. I could do my job. I just couldn't control my thoughts. I wanted help, but I didn't know anybody who would understand the depth of my illusions. It was like having a nagging pain but not going to see the doctor."

Though she found some comfort in religion, Charlotte found greater solace, albeit transitory, in nursing her fears: "My mind kept running through violent scenarios that I was becoming increasingly dependent upon accepting as reality. I was conscious of observing two selves: the one engaged in the reality of life, and the observant other, who was constantly preparing for something inevitable."

TWO ADDICTS

I. JOHN BELUSHI

People who knew John Belushi well recognized that drugs were robbing him of himself. "I never stopped loving him," said Michael Klenfner, who signed Belushi and Dan Aykroyd's Blues Brothers at Atlantic Records. "When he was clean and sober he was the most wonderful man I've ever known. When he was fucked up, he was as big a horror as you ever could imagine." His friends simply couldn't comprehend his self-destructive behavior. Belushi himself struggled to understand why he couldn't give up cocaine and the other drugs that were slowly killing him.

Bob Woodward, who chronicled the events leading up to Belushi's death in *Wired: The Short Life and Fast Times of John Belushi*, recounts a time when the actor was filming *The Blues Broth-*

ers. On the last day of shooting in Chicago, Belushi refused to come out of his trailer to shoot the next scene. The director, John Landis, who had previously directed Belushi in *Animal House*, decided to confront the actor about his aberrant behavior. Landis entered the trailer and found Belushi disheveled and disoriented, looking like a real-life version of Bluto, the "my advice to you is to start drinking heavily" character he played in *Animal House*. On the table lay a pile of cocaine alongside a bottle of Courvoisier, its contents splattered all around. A puddle of urine was trailing along the floor.

"You're killing yourself," Landis said. "This is economically unfeasible. Do not do this to my movie."

Belushi, barely able to respond, simply bobbed his head up and down.

"Don't do this to me! Don't do this to yourself!" Landis shouted. He threatened to bring in one of the photographers on set to take pictures so Belushi could later see just how pathetic he looked. But Belushi—who had once hired a bodyguard to help keep him away from cocaine—would likely not have been shocked.

Landis took the cocaine off the table and flushed it down the toilet. Belushi lunged toward his director. Landis threw a punch. Belushi ended up on the floor and broke down in tears.

"I'm so ashamed, so, so ashamed," he cried. "Please understand."

Landis asked Belushi why he was using again.

"I need it, I need it," he replied. "You couldn't possibly understand."

Earlier that year Belushi had written a letter to his wife, Judy:

I'm afraid of myself because of what I'm capable of doing to you. If I was the kind of person I want to be, at least I say I want to be, then why do I hurt you? Is it because no one can *be* that person? Do I kid myself? When I fail, who am I?

I want to take care of you, but I feel I'm not capable. The most difficult thing to deal with is disappointing you or finding myself unable to help you. When I'm sick, you help. When you are [sick], I freak. Is it because I feel helpless? What can I do? Hide in . . . drugs?

Please, please, please don't think I'm going back to my old pattern. I may have slipped the patch but not a pattern. What I really want is forgiveness that may not be deserved.

I'm going to beat this thing, God damn it! I know I can now. I may cause you pain, but I love you. You are my soul, my heart, my eyes to this life. This may happen again. Whatever it is, I can only say I'll try with all my heart, which may not be enough. And if that is the case I respect your decision to live your life without the pain and confusion I bring to it.

Belushi offered a similarly desperate plea to the actress Carrie Fisher, who had starred alongside him in *The Blues Brothers* and was engaged briefly to another costar in that film, Dan Aykroyd. Fisher tells a story about going with Belushi to a club in West Hollywood: "I lost John for a second, then he came back to me with this panicked look and said, 'Oh my God, I just did some coke.' I said, 'John, we can leave right now. If we leave now everything will be fine,'" Fisher recalled. "And he just stared at me in the state of total fear and said, 'I can't.'"

Belushi's cravings, and his compulsive behaviors, were automatic: addiction had commandeered his attention so insidiously and so thoroughly that he could think of nothing else.

The power of addiction lies in its grip on the reward circuitry of the brain. Whenever we encounter a salient stimulus, our neural response conditions us to behave in the same way over and over again. With every hit of cocaine, the reward became encoded more indelibly in Belushi's memory. While cocaine (as one of many substances) is inarguably and empirically a powerful stimulant for many, for the addict, it gets connected to a broader and deeper network of neural connections. These associations get inextricably entwined with the user's understanding of who he is. His very sense of self cannot be separated from the feelings triggered by the drug: he is mired in a negative feedback loop, which is reinforced when he succumbs to his cravings.

Shortly before his death, Belushi, who was living in New York, decided that he needed to go to Los Angeles. Judy knew if he went to Los Angeles, he would binge on cocaine. Both she and Aykroyd tried to persuade him not to go.

"Your body can't take it anymore," his wife said.

"Oh, is that so? What are you saying? That I am out of control?" Belushi countered.

"Yes!"

"It's a classic pattern," director Harold Ramis, who cowrote *Animal House*, said of Belushi's drug habit. "It's what makes an addict an addict, the expectation that if you just get more, then you'll finally cross over and find some permanent satisfaction."

As with most people, Belushi's early experiences with drugs had been experimental—a way, he thought, to fuel energy and stimulate creativity. Of course, these bursts of productivity were short-lived.

Judy understood that a number of cues could trigger her husband's need to get high. "He could be triggered because there was a scene on *Saturday Night Live* that he really wanted to be in, but that he wasn't going to be in," she explained. "Or he could have a great success that was so exciting that it triggered him."

Belushi could offer only this by way of explanation: "I've got to have it."

2. NED

When it comes to drug addiction, Ned is ostensibly one of the lucky ones. Unlike Caroline Knapp or John Belushi, he has survived to tell his tale.

At the age of three, Ned was taken from a Russian orphanage, the only home he'd ever known, and adopted by a successful journalist from the Washington, DC, area. He was afforded all the luxuries of a suburban American childhood: his own bedroom, summer camp, swimming lessons, and a good education. But none of this could stave off the periodic emotional outbursts and obsessive thoughts that had tormented him for as long as he could remember.

Ned started experimenting, as many teenagers do, with pot, but soon he began using pills; by the time he was twenty-two, he was injecting himself with up to seven bags of heroin a day. In order to better understand how his drug addictions crept into and eventually took over his life, I called Ned one evening at the North Carolina recovery house where he lives, and asked him to tell me his story.

Though he doesn't have any clear memories of the orphanage, Ned believes his time there profoundly shaped his childhood. "I was very attached to things. I was very emotional." Even after he had settled in with his adoptive family, he remained overly anxious, mostly when he felt he wasn't getting his way or was going to be left alone. A lot of this nervousness centered on his adoptive mother, about whom he constantly worried. If she wasn't around, he soon became convinced that something bad had happened to her. If the family had a babysitter for the evening, he would worry that his mother was never coming back. If she was late getting home from work, he'd think she'd been in a car crash.

Though he was a decent student in elementary school, Ned was often in trouble for being disruptive in class. But rather than dissuading him, the attention he got from teachers for his volatile behavior made him popular among his peers. Before long, he had developed a reputation as irascible and rebellious, which cast him firmly in the "bad kid" role. His behavior followed suit: He tried his first cigarette in fifth grade. Three years later, his older sister offered him marijuana, which he smoked with her in the backyard of their house while their mother slept. He said he remembers the moment "like it was yesterday."

By this time, Ned was excelling as a competitive swimmer, often spending seven days a week practicing on a team comprising mostly older, larger athletes. He wanted to perform well and also fit in with the older kids, and the stress of the situation often led to friction with his mother. But smoking weed made him forget the pressures of swimming.

"Smoking [pot] made all my anxieties just evaporate. I stopped worrying about being the new kid on the swim team, and about my mom embarrassing me in front of my friends. I went back to my room, where I had these glow-in-the-dark stars pasted on my ceiling, and as I stared at them, they started moving. I was just in my own world."

For a while after this first experience, Ned continued to get high at night with his sister; she would procure the weed, smoke a bit, and then give the rest to her brother, who would devour a whole gram himself. Soon he started to find himself in trouble. In eighth grade, after he bragged to friends that he was smoking pot, school officials searched his locker. Later that year, he and a friend were picked up by the police for vandalizing cars at 3:00 a.m. in downtown Washington, DC. At that point, his mother decided to send him to a new school, hoping he'd use the opportunity to reinvent himself.

It didn't work out that way. Adrift in a new environment, Ned immediately found acceptance among the school's crowd of drug users, and his drug use escalated rapidly: "It went from smoking at lunch to smoking six, seven times a day. I would barely show up to school. The school would call my house when I was absent, and I would try and get home before my mom did just so I could delete the message, or else I'd come up with an elaborate excuse for why I'd missed class." To some extent, smoking was integral to his identity as a budding criminal; it was a prop, like the gun he later carried while hanging out with gang members. At the same time, though, marijuana made Ned feel physically better. When high, he felt more affable, more articulate, and better able to face even the most routine social situations, such as having dinner with his mother. "I felt like I was the way I should be," he says.

As he passed through tenth and eleventh grade, he began experimenting with other drugs, notably PCP. "I spent the rest of the day in the bathroom puking my brains out . . . then two days later I'm smoking it again." Soon he started using opioids such as Percocet and Vicodin. Ned began spending more and more time hanging out, and buying drugs, in dangerous neighborhoods of DC. He started robbing people at gunpoint and dealing drugs to fund his habit. But marijuana was his first and most consuming need. When he bought larger quantities, hoping to sell it, he often worried he would succumb to temptation and smoke the whole stash himself.

Though he relished the lifestyle of the petty criminal, he still envisioned attending college and swimming competitively in his future, and his lack of control over his smoking frightened him. "I had a calendar where I tried to count the days I could go without it, but eventually I couldn't even last a day. It opened my eyes at that moment. I thought to myself, 'Jesus, Ned, this is bad, you can't stop. . . .' I knew my life was getting out of control and I was losing it. I wanted to die."

In May 2011, when Ned was a junior in high school, he was arrested for dealing drugs at school. He went into a forty-five-day residential treatment program, where his fellow patients were mostly wards of the state or multiple offenders. When he was released, he stayed clean for about a month, but relapsed shortly before he was to appear in court. "I lied my ass off. I said I wanted to clean my act up and I was happy being sober." Some parts of his statement were true: shortly after his court hearing, he admitted to his mother that he was using and got serious about recovery. He enrolled in a local community college and started to see a therapist, who would become his long-term confidant. He decided to get honest about his sexuality, too, and told his mother he was gay. To his relief, she was supportive. He began a long-distance relationship and worked as a swimming coach. Ned seemed to be piecing his life back together. But soon he would encounter heroin, which would capture him in a way that not even marijuana had.

One night, worried about his rocky relationship and missing his old neighborhood buddies, Ned accepted a friend's invitation to attend a rave. He loved the fast music, the lights, the throbbing energy. When his friend offered him ecstasy, he took it without a second thought. "I wanted to get out of myself," Ned said, and the ecstasy certainly helped him do that. "It was fucking great. It makes the music a hundred times better—everything sounds good, everything feels good, you're happy. And everybody becomes your friend." Soon he made a group of friends in the rave scene, one of whom was into OxyContin. Ned tried it. "I was numb in every joint." The high was so pleasurable that it made the inevitable vomiting that followed worth it. But when OxyContin became too expensive, he followed an increasingly common path and began snorting heroin.

At first, it was once a week or so—mostly because he recognized heroin as a dangerous turning point. "This is a drug that I told myself I would never do." Perhaps it was the fact that he only snorted it, or that he was still intermittently smoking weed, that allowed him to use infrequently at first; once he began to inject it, however, he needed it constantly. Just three months after first trying OxyContin, Ned was spending every day in a friend's bedroom shooting up, smoking cigarettes, and watching TV—and when the high wore off, shooting up again.

"Nothing compares to that warm rush," he says. "I lived for those first five seconds after being injected with heroin."

Because he shot up before he went to sleep, he often didn't start to feel the effects of withdrawal (aches, nausea, sweating) until around noon the next day. His appetite for it grew more voracious. Once, after a day spent shooting up at his friend's house, he went to "hit himself" again only to have his friend express concern.

"That's too much," he said, "you're really fucked up." Ned just brushed him off and went right back to injecting himself. Immediately after, he blacked out. When he came to, his friends were staring at him in horror.

After his first overdose, he "cut back for about three days," frightened by his brush with death. But once withdrawal set in, he couldn't hold off any longer. He began pawning things to pay for dope, shooting up, and spending the classes he sporadically attended staring at his veins. The heroin paradoxically gave him some respite from the guilt he felt about lying to his mother and his therapist, who believed he was clean. Doing heroin had negative consequences, but it wasn't as painful as living with the knowledge that he was a fraud because he was only pretending to be sober.

Then one evening he and a friend shot up in their car, parked in downtown DC. Again his friend warned him that he was about to do too much, and again Ned shrugged off his concerns. After they were done, Ned started the car and had begun to drive down Pennsylvania Avenue when things went black. When he came to, he was surrounded by policemen and paramedics demanding to know what he had taken. "They took me and threw me on the stretcher, and I'm looking out and Pennsylvania Avenue was backed up for miles, and I turn around, and the back end of my car is completely blocking traffic and the front end is pointing toward the curb." Ned would later learn he had passed out while driving and his friend, afraid of being caught with heroin on him, had called 911 from the bushes. The paramedics had administered Narcan, a drug that reverses the effects of an overdose, and brought him to the hospital.

"One of the doctors came in and told me, 'You're lucky to be alive. If we hadn't gotten to you in the next five minutes, you'd either be dead or brain dead.' I just started crying."

Not long after, Ned agreed to go to rehab. Though he's had some slipups in the year since his accident, he is trying to stay committed to recovery, working hard at his job detailing cars, and strengthening his relationship with his mother. But he tentatively admits that drugs will always be attractive to him.

"Yeah, I think they're always going to appeal to me," he says. "Once heroin gets a hold on you, it never lets go."

It is not surprising that Ned feels that way. We often feel nostalgia for things that have held us in their orbit and made us feel better, even when their effects were detrimental. For the addict, the salient substance plays a paradoxical role: it provides easily accessible relief from his cravings even as it reinforces them. The result is a vicious cycle. This physical need can be seen as similar to Kettlewell's cutting, or Berendzen's phone calls. In addiction, however, this sensation of need comes at a high cost to the body, one that is ultimately not sustainable.

CONTROL

In May of her sophomore year in college, Frances, a successful writer now living in New York, had an exam in her Indian Civilization class.

"I couldn't focus too well—I just wanted to sleep, and I didn't really care so much about the grade, which made me feel ashamed.

I looked at my two friends frantically memorizing the names of Indian states and hated myself for being so undisciplined," Frances told me.

Frances also had a Modern British Literature test the following day. "I was still sleepy from getting in bed in the wee hours. I got an Adderall off a friend and took it just before the test." Hours later, Frances was bouncing with energy.

"I had no appetite, but I picked a few dried cranberries out of a plastic container, pleased at how indifferent I was to the food. Nostalgia pulsed through me; I reminisced about the days when my body felt always so light, so needless. I wondered if perhaps I shouldn't consider dieting a little.

"I wouldn't let it get out of control like when I was in high school, of course. If I told myself the truth, I had *wanted* to get out of control. I had been rebelling—passively, fantastically—and had been hoping that others would notice my miniature wrists or scarred arms and see them as a statement of my despair."

But Frances believed she was beyond all that.

"I told myself it was just about being comfortable in my body, and healthy. Yes, healthy."

When the semester ended, Frances went home for two weeks. Both her parents worked full-time, and she was blissfully alone all day.

"In a way, I felt like I was free of food anxiety at this time. I still believed it was in my power to be a little underweight and remain there. Most evenings I ate pita and hummus for dinner, but soon the hummus began to worry me. As I ate it straight from the tub, I couldn't help wondering if I wasn't somehow indulging more than I expected. 'You're eating way more than a serving size,' I told myself. 'To be sure, you ought to use the tablespoon. Otherwise, you should probably just forgo the hummus altogether.' I started measuring the hummus every evening, and writing down how much I ate. The rules still felt too vague, and so I allotted myself a certain number of hummus-specific calories. Soon it seemed necessary to record my entire caloric intake. I used a blue spiral notebook that I carried with me everywhere."

In July, Frances moved back to a massive old house near campus, so she could be within a close commute to Manhattan, where she would be doing her summer internship. Her focus on her weight was spiraling out of control, even as she doggedly told herself she had ultimate control. Dinner became a routine of punishing calorie counting. She ate with painstaking slowness, carefully choreographing her every bite so as not to appear ravenous.

"Often I had frozen yogurt for dinner. I was terrified of pizza, because I loved it so much and was positive I wouldn't be able to control myself if I got a taste. Even walking by the local pizza place, the smell of the melted cheese made me woozy with want."

On weekends, Frances invented errands downtown and reasons for needing to walk there. If she rode the subway, she had to stand even if there were seats (she knew standing burned more calories than sitting), and she did makeshift exercises by pulling her body toward the pole with her arm. She spent hours grocery shopping on Saturday afternoons, but only rarely bought anything.

At the end of the summer, Frances went home for a few weeks before school began. She often caught her mother looking at her sideways. "Once or twice, she told me I looked *thin*, and there was an edge to her voice. I aggressively avoided the topic. I told her I was running errands, but instead I spent hours in the local enormous mall trying on clothing I thought might help me achieve the look I was going for: ethereal yet organized, delicate but guarded. I invested in thin, sheer sweaters, baggy vintage-style T-shirts, and slim-fitting pants. Nothing was very revealing—I figured when I lost enough weight, I'd be comfortable baring my body—but most tops were low-necked, so as to reveal my collarbone, which I liked to examine daily."

Frances moved back to school a week early to get away from her suspicious parents. "This didn't help me much, as my friends at school were also wary of my increasingly spindly frame. When they hugged me hello, I could feel their hands stroke my back to feel how prominent my shoulder blades were. I wore two pairs of jeans to make my legs look plumper, and sometimes tights underneath the two pairs. All summer, I had allowed myself a few bits of sugar candy (like bubblegum or Mentos) per day, but when my weight loss had stalled, it seemed necessary to eliminate these last vices, so I trained myself to envision my teeth rotting when I even thought of processed sugar."

At the simplest level, eating disorders are the result of selective and undue attention to food-related stimuli. Anorexics are captured by the promise of control. Indeed, many see only two poles: total control over eating or total loss of control. The "perfectionism" that is often ascribed to patients with anorexia may signal their way of rigidly coping with any lurking anxiety around the loss of control. Control over eating becomes a sign of achievement and safety. Take away that control and they feel unstable, often wildly so; their eating can become chaotic, swinging to episodes of binges and bulimia. Anorexia is so hard to treat because such patients are extremely determined to control their food intake and are extremely effective at resisting change.

For people with eating disorders, food cues become highly salient. Densely caloric, or "bad," foods are perceived as particularly threatening stimuli. An attentional bias develops to such cues, which are perceived as a threat to the person's control over his or her shape or weight. The person avoids these stimuli in order to decrease fear or negative affect. Restricting one's food intake becomes a way of feeling better.

But that pain inevitably returns, leading some people to binge in response to food cues. Like food restrictions, the binging diminishes bad feelings, albeit temporarily, as the person focuses completely on the act of eating, to the exclusion of other stimuli. He or she soon comes to rely on binging as a method for avoiding negative emotions. The problem is exacerbated, however, because binging can lead to feelings of shame or disgust; purging and restrictive behaviors become a means of regaining control.

When she began her junior year of college, Frances found herself more focused on her food intake than on her schoolwork. "My blue notebook was now entirely devoted to calorie counts, with the occasional reminder to myself that I was greedy and obese. I almost never went out. The few times I went to restaurants with friends, I defiantly ordered plain grilled fish or egg white omelets and practically dared them to confront me. I exercised every day, though not as much as I wished I would. I was good about taking the stairs up to my dorm room at least twice a day—no small feat, because I lived on the thirteenth floor."

Her behavior got worse as the fall went on; she was by then seeing a therapist and wanted desperately to eat better, but consistently failed. "Even the process of figuring out dinner—what to eat, when and how—was torturous. I usually had already planned it out, but I managed to agonize over it anyway. 'What you should do,' I thought, 'is not eat at all, but you are far too weak for that.' Usually I'd have an energy bar, or some cottage cheese, or whatever I'd allotted myself. Let's say it was an energy bar one night. I would give myself certain windows of time, and if the window passed, I was out of luck. Dinner I had to eat by nine, so I would start the decision-making process around eight or so. Every evening, I would pace back and forth, debating the advantages and disadvantages of eating the bar. I would tear the wrapper all the way open and leave the whole thing lying there. I would start pacing again. I returned to the bar and used the tips of my fingers to rip off a small piece. I ate it as I paced. Repeat. Repeat. Repeat. Twenty-five minutes later, the bar was finally gone. Then I would sit down and plan out what I would have the next day."

SADNESS

"Describe what it feels like," I ask.

There is a long, deliberate silence.

"It's a debilitating unease and anxiousness to the point of almost nausea, and then it's a seesaw back and forth as you're trying to understand it," he says finally, carefully crafting his description of the physical sensation caused by his depression. "It's anxiety, but you're not anxious about anything. Say you can't find your child and there's a pit in your stomach. Your body is telling you something and you automatically know what you need to do. Well, what if that pit in your stomach is your entire body weighing you down for no reason? Nothing has happened—or nothing that you can perceive has happened. And there is nothing you can actively do in the moment to make it stop."

"Which comes first?" I ask. "A thought process or the feeling?"

"I try to figure that out—but it happens so fast," he replies. "It's impossible to distinguish."

As we talk, Wes narrates his experience and, at the same time, tries to understand how depression captured him.

"You can't simply act in order to change. You're telling yourself if something doesn't change in day-to-day life, in the status quo, I'm not going to be able to snap out of this. But you're not helping that process along, because there's an inability to focus on anything else."

I press him to elaborate on this, to tell me what goes through his mind when he feels this way. What are the intrusive thoughts that keep him from concentrating on anything else? But he won't answer this question because he doesn't view his struggle along these lines; he is captured, he explains, by the *way* he thinks, rather than by specific repetitive fears or doubts.

"It's not one thing or another or many things; it just is. It's not a singular thought I'm trying to stop; it's the process. It's not that I don't like myself; it's that the quality of life, and how it makes me feel, is not where I would like it to be."

Wes explains that he has always had a negative baseline feeling— "it's just how you wake up in the morning"—and he has made efforts to change his emotional life in an attempt to raise this baseline. "The way I describe my current state for my doctor and my therapist is that I am at a zero. I have improved to a zero," he continues. "It's easier to build on zero than it is to build on a negative. But zero is good only because it's not negative—not because there's anything inherently good about zero."

Often the original trigger for emotional vulnerability fades from memory, but this seems not to be the case with Wes. He does not believe that there was one thing—an early trauma or self-sabotaging perception or difficult experience—that instigated his depression. "It started at consciousness. It started at thought. It started at the formation of the self," he says decisively. "The existential obsessive thought process started at my earliest memory."

While he is able to articulate some of the feelings this thought process evokes, he maintains that these emotions are not the source of his unhappiness. "*Loneliness* is the wrong word, because of the amount of activity that I enjoy on my own, but it's some type of missing validation of that activity," he offers when asked to describe something that makes him feel a sense of discomfort. "If whatever is enjoyed is in a solitary way and nobody knows that it happened, does it matter?"

We do not know if depression results primarily from emphasizing the negative or ignoring the positive, but both processes are clearly involved. Technology, however, has allowed us to study emotional responses with much greater precision than we could in the past. We now have tools that directly track eye movement and thereby reveal where we are focusing our attention. Likewise, medical imaging allows us to watch the brain as it reacts to stimuli. These images offer evidence that there is increased activity in certain regions of a depressed person's brain, including the amygdala, hippocampus, and prefrontal cortex, and decreased activity elsewhere.

In depressed people, negative stimuli are not only more arousing than positive stimuli, but also more likely to become locked in memory. Even a stimulus that might objectively be considered neutral, such as the expression on a colleague's face, is viewed in a negative light. The encounter suddenly becomes contorted into something worthy of grim analysis.

Another distinguishing feature of depression is the tendency to overgeneralize autobiographical memory—that is, to highlight and revisit negative experiences from the past and see them as representative of an inevitable pattern. Regardless of subsequent professional success, for example, the depressed person who was once chastised by his supervisor internalizes the idea that "people at work don't like me." The article rejected by an editor, the end of a relationship, even a careless word from a store clerk—all are perceived not as isolated incidents but as an indictment of personal worth.

The tendency to ruminate is also frequently part of depression. Rumination is a repetitive, involuntary, and almost compulsory return to specific thoughts. In the case of depression, they are negative, self-focused thoughts such as "I'm a terrible person," "I fail at everything," and "I don't deserve to be happy." There is a two-way interaction between depression and rumination. Rumination predicts more and longer-lasting episodes of depression and relapse; it is also a by-product of the capture associated with depression.

Characteristic features of depression include a sense of worthlessness and hopelessness, which can lead to self-destructive behaviors, such as withdrawing from social interaction, overeating, or self-medicating with alcohol or other substances. These feelings also engender physiological responses, such as insomnia. They reinforce and exaggerate the negative biases that underlie depression: the disorder becomes a feedback loop that essentially feeds on itself.

No one, of course, is immune. Stressful events can become salient to anyone, and previous stressors increase the likelihood of subsequent stress in response to a life event. But some people react with much more intensity than others, and their sensitivity makes them more likely to become depressed. Only by tracking experiences over a lifetime and detecting how one experience shapes subsequent responses might we uncover what makes this negative salience unmanageable for some people and not for others.

"Fear of regression." Wes explains to me with a long pause, "That's the worst. Even when there's a temporary reprieve from that feeling—from being less than zero, being back in that negative place is constantly on your mind." When I ask for an example, an even longer silence follows. "A stable relationship is impossible. Because of that fear, any bump in the road feels catastrophic. There is no foundation. The positive has nothing to build upon."

Wes describes feeling regret over what he has surrendered to depression. "I think often about lost time," he says. "Wasted years. Well, you know, really large amounts of time. All of a sudden, years have passed and you still feel the same way."

The deep-down feeling of nauseated sadness, he says, is just that: sadness. This loop—feeling dissatisfied with the state of his life and then feeling capsized for having lost time to feeling this way—gives rise to a sadness that captures again.

GRANDEUR

In 1949, Robert Lowell suffered, in his own words, "an attack of pathological enthusiasm." The poet was observed running wildly through the streets of Bloomington, Indiana, "crying out against devils and homosexuals."

"I believed I could stop cars and paralyze their forces by merely standing in the middle of the highway with my arms outspread. Each car carried a long rod above its tail-light, and the rods were adorned with diabolic Indian or Voodoo signs. Bloomington stood for Joyce's hero and Christian regeneration. Indiana stood for the evil, unexorcised, aboriginal Indians. I suspected I was a reincarnation of the Holy Ghost, and had become homicidally hallucinated."

Every sight and sound became further evidence of Lowell's divine vocation, his quest to save a damned America from its own putrefying sin. While Lowell lurched from one epiphany to the next, a theater manager reported him to the police for having stolen a roll of tickets. Lowell soon came to blows with an officer. Finally, a general alert was sounded: a disturbed poet was "on the rampage in downtown Bloomington."

Soon Lowell rang a doorbell and, to his surprise, found himself face-to-face with an off-duty policeman. "You must be Robert Lowell," the officer said. Lowell "nearly fainted." An irrefutable sign of "divine intervention—this stranger knew his name."

Though he is widely hailed as one of America's greatest twentieth-century poets, Lowell battled bipolar disorder for most of his adult life and regularly descended into florid mania.

Born into one of Boston's oldest and most prominent families (their Boston Brahmin lineage could be traced back to the *Mayflower*), Lowell was just seven when he first imagined himself as Napoleon. "Bristling and manic," he compulsively memorized the names of "two hundred French generals . . . from Augereau to Vandamme." A 1936 letter to Ezra Pound suggests that for much of his youth, Lowell was prone to bouts of mania: "All my life I have been eccentric according to normal standards," he confessed, with a certain pride. "I had violent passions for various pursuits usually taking the form of collecting: tools; names of birds; marbles; catching butterflies; snakes; turtles etc.; buying books on Napoleon." But these collections were soon abandoned, as Lowell flitted from one obsession to the next: "I was more interested in collecting large numbers than in developing them. I caught over thirty turtles and put them in a well where they died of insufficient feeding. I won more agates and marbles than anyone in school, and gradually amassed hundreds of soldiers; finally leaving them to clutter up unreachable shelves." Despite his retrospective insight into his own erratic behavior, the Harvard undergraduate ended his letter to Pound with a passionate plea. Having enclosed a few of his poems, Lowell begged Pound to anoint him his protégé: "Again I ask you to have me. You shan't be sorry, I will bring the steel and fire, I am not theatric, and my life is sober not sensational."

In fact, Lowell's life was rarely sober. His childhood passion for collection transformed, over the course of his undergraduate years, into a series of all-consuming convictions about his life mission. Lowell saw himself variously, and sometimes simultaneously, as an inheritor of the English poetic tradition; a spiritual prophet; a political crusader; and an ardent lover, ready to leave behind the world in pursuit of the ideal union. He was at once "Christ, Satan, Ahab, Moby Dick, America, and God." Suggestions to the contrary were met with unrepentant rage. When Lowell's father, Commander Robert Traill Spence Lowell III, caught word that his son's girlfriend had been in his room unchaperoned, he sent an angry letter to the young woman's father. Lowell's revenge was swift and spectacular. After driving his girlfriend to his house, he left her in the car, went into the house, and punched his father.

Lowell's remarks about American poetry reflect his manic exuberance and depressive plummets: "We have some impatience with the sort of prosaic, everyday things of life, that sort of whimsical patience that other countries may have. That's really painful to endure:

to be minor and so forth. We leap for the sublime." Lowell leaped again and again, after one sublime and then another. In 1936, when he first kissed his girlfriend Anne, he immediately proclaimed his undying love. That kiss became a lodestone around which his entire life would be reorganized. To Lowell, it meant that the young couple had "become thoroughly and firmly engaged, almost married." He solemnized their union by giving Anne his grandfather's watch.

Though he soon broke off the engagement, Lowell continued to scan his environment for hidden clues about his destiny, often finding meaning in the vagaries of the everyday. During Lowell's fellowship at Yaddo, a retreat for artists in upstate New York, two FBI agents visited the retreat in order to investigate the alleged Communist sympathies of the writer Agnes Smedley. Lowell became convinced that Yaddo's director, Elizabeth Ames, was "deeply and mysteriously involved in Miss Smedley's political activities." He soon launched a campaign to have Ames removed. At a board meeting convened to investigate Lowell's charges, Lowell served as chief counsel and grilled the witnesses. At one point, he announced to the board, "If action is not taken by the Board that we consider adequate, I intend to confer with certain people in New York . . . and immediately to call a large meeting of the more important former Yaddo guests; at this meeting we will again press our case at great length."

In the end, the board found that Lowell's entire case rested on nothing but gossip and hearsay. He was censured for making unsupported accusations.

After his campaign backfired, Lowell became consumed by religion. His friend Robert Fitzgerald remembers his gradual descent into mania: "That morning . . . [Lowell] filled his bathtub with cold water and went in first on his hands and knees, then arching his back, and prayed thus to Thérèse of Lisieux in gasps. All his motions . . . were 'lapidary,' and he felt a steel coming into him that made him walk very erect." A few weeks later, Lowell was found running through the streets in Bloomington. Soon thereafter, he was admitted to a private psychiatric hospital in northeastern Massachusetts. In a letter to George Santayana he described himself as "having rather tremendous experiences." To William Carlos Williams, he wrote, "I think the doctors are learning about as much as I am." In yet another letter, he proclaimed, "I'm going through another Yaddo, but with flying colors."

Episodes of mania punctuated the rest of Lowell's life. In 1958 he wrote in a letter to Pound, "Do you think a man who has been off his rocker as often as I have been could run for elective office and win? I have in mind the State senatorship."

Rewarding stimuli, such as accomplishments and praise, have a heightened salience for those who, like Lowell, are at risk of manic behavior. People with bipolar disorder anticipate or respond to rewards with unusually positive emotions. Everyone enjoys praise, of course, but if you tell a manic person that he has performed well on a task, his pleasure will generally last much longer than it would for others.

As a result, the manic person is motivated to pursue achievements, and the consequent praise, with focused intensity. He might imbue with special meaning an A on a geometry test, for example, or a compliment from a friend or colleague, and conclude that he possesses some rare gift. As it did so often for Lowell, this insight becomes a self-fulfilling prophecy, as the person begins to scan his environment for additional signs of his genius. Gradually the capacity to synthesize such clues, to weave a story, however incredible, out of disparate elements, allows someone to disconnect completely from reality, which gives rise to delusions and even hallucinations.

For most people who suffer from bipolar disorder, mania is coupled with sharp swings to the other end of the mood spectrum: depression. They are captured by both negative and positive stimuli, although mania is the defining feature of the disorder.

ABANDONMENT

In 1950, while in her late teens, the poet Sylvia Plath wrote in her journal, "Frustrated? Yes. Why? Because it is impossible for me to be God—or the universal woman-and-man—or anything much . . . I want to express my being as fully as I can because I somewhere picked up the idea that I could justify my being alive that way." Even at this young age, Plath was able to starkly describe the challenges of being an ambitious woman during this era—the double standards, the frustrations of having to choose between a career and married life, not to mention the seeming impossibility of choosing both. And Plath had enough talent and drive that the grandiosity of her aspirations hardly seemed delusional.

But she was also tormented, primarily by an early loss that had come to reside in her as a feeling of profound abandonment: when Plath was eight, her father died. This loss colored her relationship with her mother, whom she partly blamed for her father's death, and her marriage to the British poet Ted Hughes. She would later describe Hughes as filling the "huge, sad hole I felt having no father."

Her steep fear of abandonment also gave rise to an anxiety about betrayal. "Images of [Hughes's] faithlessness with women," Plath wrote in 1958, "echo my fear of my father's relation with my mother and Lady Death." This mistrust continually beckoned her back to the emptiness she felt after her father's death, creating a demanding contradiction: even as Plath was ruthlessly blazing new paths with her writing, she was perilously constructing her identity as a wife.

Plath met Ted Hughes when she was a student on a Fulbright scholarship at Cambridge University in 1956. They were married within months. Hughes had already begun to build a reputation as a poet—in addition to being regarded as the most charismatically seductive man on campus; he would later become the nation's poet laureate. Even in the earliest years of his marriage to Plath, he enjoyed a measure of fame and power on the London literary scene. Plath longed to share idyllic hours writing alongside her husband—when she wasn't busy typing his manuscripts, acting as his agent, or raising their children. She felt he was a genius, and whatever she sacrificed to him of her own time and psychic energy was, she believed, for a worthy cause.

At the end of August 1961, the couple left their cramped flat in London, having found another young couple, David and Assia Wevill, to take over the lease. The Hugheses moved to an ancient

market town in North Devon, four hours away from the social distractions and urban headaches of London. Despite having doubts
about leaving city life, Plath threw herself into creating a perfect setting for her perfect life: decorating, working in the garden, getting to
know the community, and giving birth to their second child.

Best of all, though, for the first time in her life she had her own
private space to write and a huge desk fit for the task. She began writing her most sophisticated poems to date, preparing a collection of
them for publication, selling short stories to magazines, and editing
a draft of her first novel, *The Bell Jar*, a fictionalized account of her
stay at a psychiatric hospital following a suicide attempt during her
college years. Her mental health had been relatively stable since that
time, though she was unusually sensitive to both setbacks and joys,
small and large. "It is as if my life were magically run by two electric
currents: joyous positive and despairing negative," she once wrote in
her journal, "which ever is running at the moment dominates my
life, floods it." This same raw-edged vulnerability defined her poetry.
She lived with exceptionally porous boundaries between her sharply
observed external world and the inner world of emotion and imagination.

The following May, the Wevills, the young couple who had taken
over the Hugheses' London flat, came to stay for a weekend visit.
The sexual charge between Hughes and Assia was immediately palpable to Plath.

This awareness surfaced in a poem Plath wrote the day after their
arrival:

> *I felt a still busyness, an intent.*
> *I felt hands round a tea mug, dull, blunt,*
> *Ringing the white china.*
> *How they awaited him, those little deaths!*
> *They waited like sweethearts. They excited him.*
>
> *And we, too, had a relationship—*
> *Tight wires between us,*
> *Pegs too deep to uproot, and a mind like a ring*

Sliding shut on some quick thing,
The constriction killing me also.

Yet when her mother (in whom Plath routinely, if somewhat problematically, confided) came to visit soon after, Plath insisted on projecting an illusion of bliss: "I have everything in life I've ever wanted—a wonderful husband, two adorable children, a lovely home, and my writing."

Six weeks after the Wevills' initial visit, however, Plath's suspicions were still churning. One afternoon, upon returning from a shopping trip, she heard the phone ringing and raced in to answer it before her husband could intercept, though he stumbled on the stairs in his effort to get there first. She recognized the disguised female voice that asked for Hughes and then listened, stunned, to the briefest of conversations between the guilty parties.

Denial was no longer possible. When the call was over, Plath yanked the phone wire out of the wall. This was the very rupture that had always terrified her. And she experienced this betrayal as nothing less than the destruction of her identity. She was enraged. She had sacrificed so much of herself for Hughes. Not only had her husband betrayed her with another woman, but he had also lied to her about their life together. To her friend Elizabeth Sigmund, she agonized, "When you give someone your whole heart and he doesn't want it, you cannot take it back. It's gone forever."

In her journals, letters, and poetry from this time, Plath dwells often on what she experienced as public humiliation, whether prying questions from village gossips or snide commentary from the London literary scene. She foresaw that Hughes's fame would torment her for the rest of her life: she would never be able to avoid hearing of his latest book, his most recent award, his newest flame. "I think now my creating babies & a novel frightened him—for he wants barren women like his sister & this woman, who can write nothing, only adore his stuff," she wrote, consoling herself.

Hughes was now her fiercest competition, and she was determined to compete at his level: "Ted may be a genius," she wrote, "but I'm an intelligence. He's not going to stop that." Plath began relying on sleeping pills to counter insomnia, but in the dark hours of the early morning, when the sleeping pills wore off, she wrote. She described this also as the time of day when depression was most threatening to her. It was her midwife's suggestion that she use these hours, while her children were still asleep, to write, instead of tossing and turning fruitlessly. And she began writing, often a poem a day, her best work.

By October 1962, Plath had produced an astounding twenty-five poems, and she described to friends how they were bursting from her in a way that she could not have ever anticipated. In numerous letters, she explained how she had always believed she needed a steady and settled domestic life in order to write, but now that all that had been thrown into chaos, she was writing like never before, "as if domesticity had choked me."

Living life at this heightened pace, however, was not sustainable. In the last weeks of her life, this rush of creative energy faltered and a perfect storm of setbacks darkened her horizons on all sides. Wanting to free herself further from Hughes, she attempted to start a romantic relationship with a close friend, an influential editor, and was devastated when he shunned her advances. *The Bell Jar*, published in England in January 1963, was rejected by her U.S. publisher, and her first collection of poems received lukewarm reviews in the States. Her strikingly original poems were turned down by one editor after another, many of whom described them as "alarming." Meanwhile, London was suffering the coldest weather in recent memory, with pipes bursting and illness raging. Plath and her children were repeatedly sick with the flu, and frequent visits from Hughes, separated from Plath, to see the children created profound disruption.

In the final four days of her life, Plath's mood shifted starkly. She had argued with her au pair, relations with Hughes had become increasingly strained, and he was threatening her with legal action for slander. Plath was recovering from an exhausting respiratory infection, and the children had once again been ill with flu. At the start of February, her doctor diagnosed her as "pathologically depressed" and placed her on antidepressant medication. "Barren" Assia was also by then pregnant, though she would have an abortion a couple of weeks later. It is not known if Plath learned of this pregnancy before her death.

One of Plath's last poems portrays her imminent suicide:

The woman is perfected.
Her dead

Body wears the smile of accomplishment,
The illusion of a Greek necessity

Flows in the scrolls of her toga,
Her bare

Feet seem to be saying:
We have come so far, it is over.

GOING MAD

What finally captured Virginia Woolf was her fear of the inevitable oscillation between stratospheric highs and paralyzing lows. She was fifty-nine when she committed suicide in 1941, but there had been at least two attempts earlier in her life, and according to her nephew and biographer, Quentin Bell, the fear of madness had been with Woolf for decades, at least since her first significant breakdown, at age thirteen, following her mother's death. "From now on," he writes, "she knew that she had been mad and might be mad again." Some years and bouts of illness later, a nerve specialist diagnosed Woolf with neurasthenia, a catchall term implying weakness of the nerves. She suffered from symptoms characteristic of anxiety and depression: worry; insomnia; irritability; excitability; headaches; tingling sensations; lightheadedness; tormenting feelings of failure, shame, and guilt; and hopelessness about the future. Disgust with her body, aversion to food, paranoia, and even occasional hostility were recurring symptoms of her illnesses as well. Her husband, Leonard, believed she suffered from manic depression, and Quentin Bell agreed.

Despite many wretched stretches, some prolonged, Woolf was neither depressed nor manic most of the time. Her letters and diaries, her fiction and essays, her prodigious reading and reviewing, and of course the recollections of colleagues, acquaintances, and friends attest to the energy and optimism, and the extraordinary work ethic, that characterize her prolonged periods of health, as well as her brilliance (or perhaps even genius), her considerable capacity for joy, and her sociability, humor, and charm. Her friend the author Lytton Strachey once suggested that, current loves aside, Virginia Woolf was the person anyone would most want to see coming up one's drive for a visit. Even when well, however, Woolf was always more sensitive to certain stimuli, such as noises, and she was often plagued by the same headaches, insomnia, excitability, and difficulty with concentration she experienced when ill. Leonard believed these symptoms were unmistakable signs of potential danger. The lines between her inventiveness and flights of fantasy, between her being gregarious and uncontrollably garrulous, were not always clear.

That Virginia might become either overstimulated or overtired was a constant source of concern for Leonard, who was initiated into just how treacherous her illnesses could be even before their marriage and, more devastatingly still, in the years that immediately followed. He knew how quickly and utterly her thinking could become disordered—irrational, dangerous, morbid. For a breakdown to be avoided, Leonard said, Virginia needed to retire immediately to a "cocoon of quiescence," a state of near hibernation. She balked at his insistence on a highly regimented life and his parceling out of fun and freedoms. Fear of sleeplessness was a cause of anxiety for Virginia as well. She was terrified, rightly, of finding herself ungovernable: "It is odd why sleeplessness, even of a modified kind, has this power to frighten me. It's connected I think with those awful other times when I couldn't control myself," she wrote. Innocuous criticism or laughter at her expense would reverberate and ricochet within Woolf's mind, but she understood that her emotional response was often blown out of proportion. "I bring home minute pinpricks which magnify in the middle of the night into gaping wounds," she wrote.

Woolf's instinct to express herself in writing and her deep pleasure in doing so were curtailed only when she was fully incapacitated, and her diaries illuminate, among myriad literary, metaphysical, political, and personal concerns, her changing moods. First the depression: "Sank into a chair, could scarcely rise; everything insipid; tasteless, colourless. Enormous desire for rest . . . only wish to be alone in the open air. . . . Thought of my own power of writing with veneration, as of something incredible, belonging to someone else; never again to be enjoyed by me. Mind a blank . . . No pleasure in life whatsoever; but felt perhaps more attuned to existence. Character and idiosyncrasy as Virginia Woolf completely sunk out. . . . Difficulty in thinking what to say. Read automatically, like a cow chewing cud. Slept in a chair."

Slowly over time her mood would pick up: "Sense of physical tiredness but slight activity in the brain. Beginning to take notice . . . much clearer & lighter. Thought I could write, but resisted, or found it impossible. A desire to read poetry set in on Friday. This brings back a sense of my own individuality. Read some Dante & Bridges, without troubling to understand, but got pleasure from them. Now I begin to wish to write notes, but not yet novel."

She then recorded a marked upswing in mood and perception: "The suggestive power of every sight & word is enormously increased." Weeks later she would begin to anticipate and sense the onset of the pain again: "Oh it's beginning it's coming—the horror—physically like a painful wave swelling about the heart—tossing me up. I'm unhappy unhappy! Down—God, I wish I were dead . . . I've only a few years to live I hope. I can't face this horror anymore—(this is the wave spreading out over me). This goes on; several times, with varieties of horror. . . . Why have I so little control? It is not credible, nor lovable. It is the cause of much waste & pain in my life."

Woolf summed it up herself: "I wish you could live in my brain for a week. It is washed with the most violent waves of emotion.

What about? I don't know. It begins on waking; and I never know which—shall I be happy? Shall I be miserable?"

Woolf's diaries, and to some extent her letters, give a sense of the complexity of her feelings about her temperament and vulnerabilities. She understood, for example, in a not insignificant way, the literary value of her madness: "As an experience, madness is terrific I can assure you, and not to be sniffed at; and in its lava I still find most of the things I write about." The episodes were "how shall I express it?—partly mystical. Something happens in my mind. It refuses to go on registering impressions. It shuts itself up. It becomes chrysalis."

Woolf despised the prolonged rest-cure treatments that were mandated for her illnesses. She felt that month after month of excessive feedings and enforced inactivity of the body and mind (particularly being denied books and the option of writing when she felt herself again able to do so) were dehumanizing. "Never has a time been more miserable. . . . I don't expect any doctor to listen to reason," she wrote during one such period. The treatments made her only more secretive about and fearful of her symptoms, for their onset predicted a course of treatment that was worse, in her mind, than the symptoms themselves. "I really don't think I can stand much more of this," she wrote during the eighth month of one such treatment. "What I mean is that I shall soon have to jump out of a window."

Given Woolf's decades-long, if intermittent, history of breakdowns and her recovery from them, and the ensuing years of extraordinary creativity and productivity, we have every reason to imagine that she might have recovered from her final illness as well. Why was this breakdown different? Certainly the times were enormously difficult, bleak and ominous, the threat of impending doom probably both amplifying and seeming to reflect Woolf's state of mind. The issue in 1941 was not only Woolf's fear of madness, but also her, and the world's, well-founded fear of German occupation. She and Leonard, along with many others, had discussed specific plans for suicide in the case of invasion, assuming that their deaths were imminent regardless. She wrote to a friend in early March 1941,

"Do you feel, as I do . . . that this is the worst stage of the war? I do. I was saying to Leonard, we have no future."

Yet it was, of course, the chaos in Woolf's mind, rather than in the world at large, that was ultimately intolerable to her, that robbed her of any faith in the possibility of recovery. The physician who was attending to her wrote on March 12 that Woolf had admitted to "feeling desperate—depressed to the lowest depths, had just finished a story. Always felt like this—but especially useless just now." On March 21, Woolf told her physician that her biographical works were "failures," saying, "I've lost the art," and admitting that (presumably by then already agitated and unable to sleep or think) she had taken to scrubbing floors when she couldn't write. Woolf saw her doctor, at Leonard's insistence, for the final time, on March 27. Although she eventually acknowledged her fears that she would lose the ability to write, "The interview was difficult. Virginia at once declared that there was nothing the matter with her. It was quite unnecessary that she should have a consultation; she certainly would not answer any questions." She committed suicide the next day.

Woolf left three suicide notes, two to Leonard and one to her sister, Vanessa Bell, which offer our best chance of understanding Woolf's mind: "I am certain now that I am going mad again. It is just as it was the first time," she wrote to Vanessa. "I am always hearing voices, and I know I shan't get over it now. . . . I can hardly think clearly any more . . . I have fought against it, but I can't any longer."

ACCUMULATION OF BURDENS

It is the most harrowing mistake a physician can make. He was wrong—"wrong about the risk, and the loss is irreparable."

Dr. Howard P. Rome, the chair of psychiatry at the Mayo Clinic, had discussed the topic of suicide at length with his patient. They had talked about the patient's father's suicide, the patient's suicidal "musings," and even his prior attempts.

In July 1961, a few days after Dr. Rome had discharged Ernest Hemingway from the hospital, the writer killed himself with a shotgun. Four months later, Dr. Rome tried to answer in writing the question that he had repeatedly asked himself and that had now been squarely put to him by Hemingway's wife: what more could have been done?

In the letter, Dr. Rome said of Hemingway, "He was obsessed with the idea that he could never again meet his obligations and therefore would be unable to work." Dr. Rome observed the classic "features of an agitated depression: loss of self-esteem, ideas of worthlessness, a searing sense of guilt at not having done better" by his family, by his friends, and "by the myriad of people who relied upon him." These symptoms had ultimately led Dr. Rome to prescribe electroshock therapy.

Hemingway had also demonstrated an intense melancholic preoccupation. He didn't feel that he could trust his lawyers or financial advisers: "For weeks we talked about the meaning of his unfounded insistence that if he weren't legally declared a resident of Idaho he would be broke. When we had all of the factual information, it seems to me that he came to see that what he was really saying had less to do with money as money and more to do with him as a person with productive assets.

"Our conversations repeatedly got back to the future. What were the pros and cons of a permanent residence in Idaho as against someplace in Europe or even Africa?"

At the same time, Hemingway became preoccupied with his health. He would often double-check the nurse's recording of his blood pressure and kept very detailed records of his weight and what he was wearing when he stepped on the scale.

Spontaneously, Hemingway brought up the subject of suicide in his conversations with Dr. Rome.

He said that I could trust him; in fact, he pointed out that I had no alternative but to trust him. As demonstration of this, he pointed out the many ways potentially available to him, saying that if he really wanted to destroy himself there were mirrors of glass, belts, ways in which he could secrete medications and the rest.

You asked what more could have been done. . . . I can't see that you could have done anything more. He often said that he knew he was a difficult person to live with and that you had somehow or other acquired the knack. He was especially proud of the fact that you had been able to share him with what he frequently referred to as that thing in his head—tapping his forefinger against his temple.

I think I can appreciate what this has meant to you; the whole ghastly, horrible realization of its finality now. And all of the endless echoes of why why why why. And the totally unsatisfying answers. This kind of a violent end for a man who we knew to possess the essence of gentleness is an unacceptable paradox. . . . In my judgment he had recovered sufficiently from his depression to warrant the recommendation I made that he leave the hospital. You accepted this in good faith. I was wrong about the risk.

I contacted a close confidant of Hemingway in his last years. "When I met Hemingway, I was very struck; his personality was lighthearted, he was interested in everything, he was always drawing people out and he wasn't morose at that period," says Valerie Hemingway, who was nineteen years old when she first encountered the writer in Spain in 1959. Soon she had taken on the roles of his daughter, muse, and employee. (After Hemingway's death, she would marry his son Gregory.)

"It was perfect in Spain because you don't sit down to the table until eleven p.m. and getting up from dinner at two a.m. was perfectly normal or you might talk on until three," she continues. "In other circumstances, it became noticeable that one of the reasons [Hemingway] kept people around him and kept talking was that he did not want to go to bed and try to sleep."

Despite his successful writing career and literary fame, Hemingway had faced debilitating setbacks: he was on his fourth marriage; he'd suffered serious physical injuries in multiple car and plane accidents; many of his close friends had died. He knew depression intimately. Still, he'd generally been able to press on by traveling, keeping good company, and above all writing.

Soon after he left Spain, however, things changed dramatically. "When I went to stay with him in Cuba, in January of 1960," Valerie Hemingway recalls, "I saw a completely different Hemingway." It was in Cuba, Hemingway's adopted homeland, that he began his "accumulation of burdens," as she put it, causing him to fall a little harder each time and straining him that much more as he struggled to get back up again.

Upon his return to Finca Vigia, his beloved farm just outside Havana, Hemingway found himself in an uncomfortable, and unlikely, position. Philip Bonsal, then the American ambassador to Cuba and a longtime friend of the Hemingway family, paid a visit one evening in April. "Hemingway was told by Phil that he needed to get out of Cuba," Valerie remembers, noting that she was sitting at the table with the two men as they had their conversation. Hemingway resisted, declaring that he was living in Cuba not as a political statement but because it was a cherished home and sanctuary for his writing. Bonsal went one step further, declaring "that Cuba was now an enemy country and . . . that the word 'traitor' had come up" among Washington officials with regard to Hemingway. He ended the conversation by repeating his original message: Hemingway needed to leave right away.

Hemingway focused on the word "traitor"—"that played into his fears a lot," says Valerie. "He understood the enormity of that accusation. . . . Once a word is bandied about, it's picked up and suddenly it becomes a defining [label]. That was very, very disturbing to him."

The threat, too, of having to let go of Finca, where he had a staff he considered close friends, some of whom had worked with him for decades, quickly toppled Hemingway's sense of stability. "When you're very precise and things change, that bothers you," Valerie Hemingway remarks. "I noticed [this in Hemingway] so clearly. . . . The possibility of losing Finca was one of the real burdens, because that was the place he could go and feel he was among friends. He always liked to have a group of reliable people around him who would ward off the outside world."

Later that spring, Hemingway suffered another blow: his vision began to fail. "He literally read three books every week," Valerie recalls. "But when his eyesight was failing, it became hard for him to read. Then, his relaxation was going out twice a week at least, fishing on the *Pilar*, and of course this required [good vision] . . . to see the fish in the water, the ripples, and so on. Everything that he did—that he loved to do—really required the use of his sight."

Convinced that his life had become precariously uncertain, Hemingway returned his attention to an old anxiety about his finances. The seed of this fear had been planted in 1928, when Hemingway saw his father's mood plummet when a bad real estate investment swallowed up a large portion of the family savings; his father committed suicide not long after. Hemingway had to rummage through a slew of unbalanced checkbooks, uncollected bills, and unpaid back taxes to sort out the financial ruin his father had left behind. Now his worries about money deepened when his own seemingly secure economic circumstances were suddenly turned upside down: in the late 1950s, his accountant reported that he'd been underpaying Hemingway's taxes and the writer would need to pay an unanticipated sum of $50,000 to the U.S. government. As it turned out, the income from Hemingway's writing and royalties that year was less than he had expected, and for the first time he found himself strapped for cash. "Ernest felt very poor. And he felt untrusting of those around him," Valerie Hemingway explains. "One year you could feel rich, and then the next year, it's all taken away from you. He became attuned to the unpredictability of what could happen."

All these issues exacerbated Hemingway's insomnia, and long restless nights only made it impossible for him to write the following day. This inability to write, in the end, was at the core of Hemingway's despair: "He had this obsession that writing was more important than anything else," Valerie says. "He felt that if he couldn't write better than he had written before, then there was no point in living. This particular spring, when he got up to write—that really agitated him. His whole personality changed: he became grumpy and depressed."

Hemingway's fears soon began to multiply: at various points, he worried that people were trying to kill him, that his wife would leave him, that his lawyers hadn't reported four thousand dollars in gambling winnings, that there would be dramatic recriminations for his negative portrayal of a famous Spanish bullfighter in a two-part *Life* magazine story.

When Valerie went to stay with Hemingway for a second time in Spain, in the summer of 1960, she found a gaunt and abandoned figure, a ghost of the man she'd met the year before. "I felt that the person wasn't there," she says. "I simply couldn't get through to him in any way."

In fact, some of Hemingway's paranoia was rooted in reality. The FBI had been tracking his movements since the 1940s, when J. Edgar Hoover decided to keep an eye on his activities in Cuba. But focusing on his vulnerabilities only amplified Hemingway's paranoia. The inventiveness that had once allowed him to gain an edge on these fears, his writerly imagination, had turned on him.

A UNIFIED THEORY

As he drove home one night with a take-out order of steaming curries nestled in the passenger seat, Ralph Hoffman noticed that every single stoplight he encountered was green: across Chapel Street, down to Elm, all the way home.

Hoffman was barely aware of the string of coincidences until, all of a sudden, he found himself zooming through the fourth and then the fifth green light. This was no longer in keeping with his ordinary expectations of traffic patterns in New Haven. "It started to feel uncanny. What is going on here? What is this force controlling the green lights? And so, for about a half a second, I had a paranoid fantasy—a conviction that some outside force was manipulating the traffic lights."

Hoffman, a psychiatrist and medical director of Yale Psychiatric Institute, uses this anecdote to describe the experience of psychotic

patients, who regularly find hidden meaning in everyday coincidences. "If I had a propensity toward psychosis at that time," says
Hoffman, "I would have somehow grabbed on to this remarkable set
of coincidences."

When I explained my theory of capture to Hoffman, he recognized in my description not only the experiences of his patients,
many of whom suffer from delusions or hallucinations, but also his
own mind-set that evening. Capture, he suggested, becomes an inevitable chain reaction in our lives: we attach meaning to certain
stimuli, our attention becomes progressively biased, and we scan our
environments for similar stimuli. "This," explains Hoffman, "is an
autocatalytic process." Once our attention is primed by a particular
stimulus, we begin to recognize previously invisible stimuli, and
sources of meaning, in our environments, which only intensify our
hypervigilance. Or, in Hoffman's words, patients "seek out the very
things that drive them crazy."

This cycle of narrowing attention is so vicious that psychiatrists
often wonder how some patients ever manage to recover. What
allows us to break the cycle?

Psychosis, Hoffman suggests, can be understood through the
lens of capture. Research suggests that many of us (perhaps 70 percent of the population) hear voices just before we drift into sleep.
(The same research has shown that sustained sleep deprivation
makes us particularly prone to such experiences.) What differenti-
ates these "normal" moments from the all-consuming delusions and
hallucinations experienced by the schizophrenic? For people prone
to psychosis, Hoffman explains, a voice heard before one drifts off
to sleep possesses an irresistible authority; it is salient. After all, if a
psychotic patient is not captured by his hallucinations, he is unlikely to continue hallucinating. "That," says Hoffman, "is the
single most important indicator that a patient is on the road to recovery. If he is able to ignore [the voices], then they will probably
soon melt away."

Bipolar disorder is, in Hoffman's words, "an addiction to a certain
affective state, one that is highly compelling and motivating." He
continues: "Bipolar patients are captured by this sense of their own

specialness, and then recaptured by any kind of evidence whatsoever that supports this self-conception." The manic phase of bipolar disorder is characterized by a collection of hypersalient cues that reinforce the individual's extraordinary sense of self-worth. At the same time, bipolar patients regularly ignore potentially relevant evidence that would challenge their grandiose understanding of themselves. This attentional bias may lead patients to believe that they possess some rare gift: a radically innovative insight into the origin of mankind, or a particularly intimate connection to God. As manic convictions gain solidity—Robert Lowell's sudden belief that he had been entrusted with a divine mission, for instance—any suggestion that might puncture this certainty is ignored or dismissed out of hand.

Not surprisingly, the major pharmacological treatments for both psychosis and bipolar disorder, the so-called second-generation antipsychotics, target the neurological circuitry of salience. By muting our investment in particularly salient stimuli, these medications allow patients to regain some measure of control over their attention. Unfortunately, patients often find that these drugs mute all their responses, rendering their emotional landscape flat and their days listless.

In depression, the process is largely analogous, albeit in reverse. We become hypersensitive to a very different set of cues in our environment: "I've disappointed my father. I'm a terrible failure. No one could ever love me; I will always be lonely." Alternatively, we may project our interpretation of cues outward onto the heartless, unjust world: "Why do bad things always happen to me?" To assuage this overwhelming sense of self-loathing, depressed people regularly turn to drugs and alcohol or food, which only compounds their sense of worthlessness.

While depression is a function of being captured by negative stimuli, what is felt or experienced as depression is exhaustion, emptiness, and defeat—capture's sequelae. The learned helplessness of depression is a by-product of our attempt to protect ourselves from painful stimuli. We respond by altering our behavior, while our thoughts reflect a constant attention to negative stimuli: "I don't want to get out of bed. I am in a void infinitely deep and black." All personal initiative disappears, all desire, all sense of being a person; feelings of unreality come to permeate our entire outlook. Depression sustains itself as these feelings perpetuate the capture.

As we discussed these conditions, Dr. Hoffman returned again and again to a key paradox. We are all vulnerable to capture; we all need to find meaning in the random flux of stimuli that bombard us at every moment. Yet, for most of us, this susceptibility is offset by our ability to redirect the wanderings of our attention. Hoffman insists that this countervailing force pulls many sufferers out of the rabbit hole. After all, he has seen it at work in his own patients.

But the battle is uphill. Indeed, as Dr. Hoffman acknowledges, the brain's drive to discover salience likely holds as much sway as the drive for pleasure or reward. "Salience," he said, "is sought through experience, either via our senses or via internally generated states such as beliefs, images, or memories. Mental illness ensues when the brain gets stuck on these high-salience experiences to the exclusion of everyday mental processes."

We have seen capture at work in healthy individuals without any discernible form of mental illness, and in those paralyzed by psychic pain. The process of capture is not merely synonymous with psychopathology; nor is it the only force behind mental illness. Rather, it is a common mechanism through which we can better understand an array of mental suffering, from the everyday to the extreme.

This underlying similarity across seemingly unrelated experiences should challenge us to rethink our current understanding of mental distress. For the last thirty-five years, psychiatry has based its diagnostic categories on patients' symptoms. Over time, the name attached to a given cluster of symptoms has become accepted as the cause of the patient's suffering. Consequently, these labels, such as "bipolar disorder" and "depression," have been accepted as explanations rather than descriptions. Unfortunately, such confusion has deflected attention from the crucial link between psychological pain and brain function.

Yet there is mounting evidence to support a common mechanism behind seemingly disparate mental conditions. For example, various forms of psychiatric treatments, including medication and psycho-therapy, have proven effective in a range of disorders. Both anxiety and depressive disorder respond equally well to the three classes of psychotropic drugs; indeed, antidepressants such as Zoloft and Prozac have been approved for major depressive disorder, obsessive-compulsive disorder, panic disorder, posttraumatic stress disorder, premenstrual dysphoric disorder, and social anxiety disorder. In all these conditions, too, salient stimuli activate key neural circuitry. Research has increasingly shown that antidepressant drugs work by reducing emotional reactivity to such provoking stimuli—in essence, these medications help to dampen the arousal caused by capture.

For too long we have neglected this evidence of a neural mecha-nism underlying diverse forms of psychic pain—and even of artistic and spiritual transcendence. My hope is that by examining the impact of capture in individual lives we might begin to understand how and why certain stimuli commandeer our attention. Only then might we begin to develop more effective, and safer, strategies for treating the most debilitating forms of capture.

4

WHEN CAPTURE TURNS ON THE SELF

DAVID FOSTER WALLACE

To understand the full impact of capture, we must view it as the result of a lifetime of experiences. David Foster Wallace's life offers an example of what can happen when capture is directed toward the self: when extreme sensitivity becomes striving perfectionism, which in turn evolves into relentless self-criticism and becomes coupled with an uncanny ability to analyze the flaws in one's own analysis. Suffering, in other words, becomes indistinguishable from our frantic attempts to evade it. Wallace was caught by this very loop, which resulted in a despair that ultimately he could not conceive of ever escaping.

By the age of sixteen, as a junior in high school, Wallace had begun an intense struggle to keep his emotional balance. He had been notably shy as a child, but as a teenager, he found his distress had sharpened. His impulse was, as his younger sister, Amy Wallace-Havens, described it, "to see the world and human behavior as pretty ridiculous." It wasn't mean-spirited, she said; it came across more as if he were from a different planet and he'd lost his tribe.

Nevertheless, Wallace sought community and found it, primarily within a local tennis team, starting lessons at twelve and soon becoming good enough to qualify for the sectional championships. In those same years, he also discovered that smoking pot helped to still his whirring mind, and he indulged frequently.

Much as these two activities, playing tennis and smoking marijuana, helped Wallace feel better, neither could fully arrest the steadily growing fear that was overtaking him. He told his lifelong friend Mark Costello of a deep feeling of powerlessness during high school; he described not being able to leave his bedroom for a certain period of time and the terror of not understanding what was happening to him. He suffered from agoraphobia and panic attacks—turning "ashen gray," as his father said, "vomiting, trembling"—and bouts of fury, once even going so far as to drag his sister through dog feces after they'd had an argument at school that landed them both in detention.

"He considered [suicide] in high school," recalled a college friend. "We talked about that because we [agreed] that it was tempting, that there just seemed like no way out of the personal hell, and that suicide seemed to be the only way to achieve relief, to stop the pain." Wallace was not the first among his family to think this way: his maternal great-uncle and his aunt took their own lives.

Throughout this period, however, Wallace remained unusually dependent on his mother. She had been the one who woke at the crack of dawn on Saturdays during tennis season to drive him to his tournaments. She visited her kids' bedrooms every night to talk to them about what was going on in each of their lives. When it came time to go to college, Wallace experienced such "misery about leaving home," Amy said, in large part because he was fearful about having to live without his parents'—perhaps most especially his mother's—support.

And yet when Wallace arrived for his freshman year at Amherst, he hoped for, and experienced for a time, a kind of renewal. On some mornings, according to freshman roommates, he would open the window of his dorm room and yell out, "I love this place!"

He formed friendships that offered the consolation of like minds. He met Costello early on—they lived down the hall from each other—and Corey Washington, a highly talented fifteen-year-old who lived at home with his parents because the college thought he was too young to reside on campus. "They were emotionally close for reasons of self-protection," Amherst professor Andrew Parker said of the three friends.

Dave Colmar, who befriended Wallace in his later years at Amherst, recalled, "We hit it off, talking about things that most people would consider ridiculous. I speak six languages, so at the time, I was working on a thesis that involved two foreign languages. I'm really into the minutiae of language and grammar, and Dave was as well. He wasn't arguing the split infinitive or his mom's views on 'less' or 'fewer' just for the sake of argument. It had real meaning for him. We shared the language of language. But what we were really sharing was a kind of camaraderie with things like loneliness, isolation—it was kind of a darker side. We smoked dope together a lot."

Charlie McLagan, a Chicagoan who met Wallace as a freshman, also appealed to this darker side. "We did a lot of drugs together," said McLagan. "I was doing a lot of LSD and mushrooms and ecstasy, and he and I tripped a handful of times together. He enjoyed it. We had many long overnight hours of listening to music and talking and enjoying the high. He'd show up at eight and leave at five in the morning." During his first year at Amherst, Wallace, always one for routine, smoked pot every day at about the same time, 4:00 p.m., with a group of guys who lived down the hall from him.

Forming a tight-knit circle of friends, in many ways, satisfied Wallace's intense desire for connection, an urge that would largely propel his writing later on. This group also helped to buffer Wallace from feeling like a misfit in college, as he'd so strongly felt in high school: "Some students nicknamed him 'mush face.' He was somewhat overweight. I wouldn't call him attractive," recalled McLagan. "I don't think he ever knew that he had that nickname, but I'm sure that he could perceive some disdain or outcast-ness from those people."

Perhaps most important, however, Wallace recognized, and became recognized for, his formidable intellect at Amherst. "He soon caught the attention of professors," said one of his classmates.

"They challenged him to excel as soon as they saw his first work—and he had great admiration for Amherst professors. He pushed himself to please them." Wallace never received anything less than an A-minus during his time there, and even managed to get straight A-pluses one semester.

Despite these triumphs, however, Wallace never felt a lasting sense of contentment. Initially, there were the social challenges of adapting to a new environment, particularly one so at odds with the life he had known at home. "Dave came from downstate Illinois, public school, no wealth; around us were guys from private schools like Andover and Deerfield Academy or old money from the Deep South or New York City or Boston or Washington, DC, diplomat kids," recalled one classmate. "They were used to being on their own; they were really comfortable with the girls, and socially very adept at that kind of situation. For Dave, that was not the case. Dave was obviously a keen observer of such differences—and although they were superficial, he was vastly outnumbered. I know this made him uncomfortable."

Wallace himself acknowledged this many years later, in a 1999 interview with *Amherst* magazine. "I wasn't in a fraternity and didn't go to parties and didn't have much to do with the life of the College. I had a few very close friends and that was it," he confided. "I studied all the time. I mean literally all the time. I was one of those people they had to flicker the lights of Frost Library to get out of there on Friday nights, who'd be out there right after brunch on Sunday waiting on the steps for them to open the doors. There were happy reasons for all this studying, and sad reasons. It was at Amherst, with its high expectations and brilliant profs and banzai workload, that I loved to read and write and think. In many ways, I came alive there. But I was always terrified. Amherst terrified me—the beauty of it, the tradition, the elitism, the expense."

On their own, these are hurdles faced by many college freshmen. What devastated Wallace was the dawning revelation that, in spite of the promising glimmer of a new beginning, he could not outrun his early vulnerabilities, and, as he ventured forth as boldly as he was compelled to do, new ones awaited him.

"When we were kids, neither one of us had a cavity until, I think, when David was in the eighth grade. He got a cavity and he just freaked out," Wallace's sister recalled. "And I remember hearing him carrying on and being stunned and disgusted. . . . I think he just wanted to be as perfect as possible. I think he liked to set himself apart from other people or from what normal is supposed to be. And that actually translated into not finding it acceptable to have a cavity. David loved being told how fantastic he was, no matter what it was. Everything from the dentist—'Here's this astonishing boy who doesn't have any cavities'—to having the best grades at school."

This was an early dilemma for Wallace: creating expectations that were so exacting that he was guaranteed to fail eventually. Wallace's grandiose intentions may have been meant to prove to himself his superiority, but instead they provoked a doomed inner narrative, one that would play on repeat throughout his life: he had not measured up; his life was a hoax; he was a failure. This extreme self-criticism, echoing beneath his every pursuit, constantly undermined him.

This sentiment extended to his personal life. James Wallace explained, "If somebody said, 'I love you' to David, his reaction to that was 'I've done it again. I've deceived this person into not recognizing my bad sides.'"

Yet, swimming upstream through his own torrent of disapproval, Wallace always hoped for more: more achievement, more recognition, more love.

From a very young age, Wallace was bent on impressing his parents, two highly intellectual people with their own rigorous standards. His father was a dedicated academic who, with a quiet passion, worked on the same set of difficult philosophical problems for decades. Sally was a professor of English at Parkland College, a community college in Illinois, and a commanding presence in the household. She loved wordplay and the sounds of certain phrases. She was also fiercely passionate about grammar, and in 1980 she wrote a textbook called *Practically Painless English*, designed to teach grammar and composition in a lighthearted manner. She shared these interests with her young children; Wallace responded enthusiastically. The two would make private jokes to each other, playing with the meaning and spelling of words. "David was fond of coming home from high school and yelling up at me, 'Any thin natal coverings?' also known as 'cauls,'" his mother offered by way of one example. "He did that, I think, probably until he went off to college."

"He really did seem so badly to want to make my parents proud of him," Amy said of her brother. "He thought that my parents didn't appreciate how intelligent he was."

The revitalizing hope that things would be different for him in a new environment began to wear off by Wallace's sophomore year at Amherst. His mind, always powerful, always processing, had been somewhat calmed in his first year of college, distracted as he was by the novelty and progress of his life there. Soon after he returned to school in the fall of 1981, however, his anxieties began to ramp up once again.

"When you said something to him, he would be sensitive both to the overt meaning of your utterance, but also to the valences of the utterance: what might have been the strategic intent, what deeper emotional valence that way of formulating the utterance might carry," explained William DeVries, one of Wallace's Amherst professors. "I don't think he seemed perceptually more aware of his physical surroundings, but I think he was more sensitive to the normative and emotional aspects of human activities and situations. . . . I don't think he thought this stuff out. I think he saw these deeper dimensions. Just as we perceive meaningful words, not patterns or pixels, he saw with equal immediacy the social and emotional meaning of people's actions."

Or, as one of Wallace's former roommates describes it, "He had the most unquiet mind I have seen in a person. His mind was working five times faster than anyone else's. And the problem is it wasn't five times more effective than other people's. It was just faster. His mind was in absolute overdrive."

Habits that had once offered Wallace relief or gratification began to harden into something more problematic. His marijuana use, for example, which had once allowed "his hypervigilance to dissipate and made him comfortable in his own skin," as his friend Dave Colmar described it, now made Wallace feel out of control. Initially, he believed he'd discovered a reasonable solution for easing his nerves; he even attributed some of his better writing—he'd started experimenting with fiction in his sophomore year—to smoking pot. "But it's all, of course, a lie because the control gradually goes away," Wallace would later say in an interview, "and it stops being that 'I want to do it' and becomes that 'I feel I need to do it.' That shift from 'I want something' to 'I feel I need it' is a big one."

In addition, his desire to impress his parents coupled with the high praise he received from professors, once a satisfying affirmation of his towering intelligence, now created a daunting sense of determination and a fear of failure: "Every waking moment, I'm obsessed with doing well, trying to do well," he told an old high school friend while at Amherst.

And, as ever, he failed to meet his own impossible expectations. "I didn't have the sense that David was just there for As," recalled another classmate. "It's not like he was trying to look good on paper to get to the next step. He was different. He was struggling and striving against his own standards, which were much harsher than those of Amherst College. He wanted to solve a problem, a puzzle, that he'd created, that was unsolvable in some ways."

By November of his sophomore year, Wallace began to have trouble functioning. His anxiety increased and then narrowed, oppressively attaching to seemingly benign matters. He began, for example, to worry excessively about losing his favorite pen. Wallace would get out of bed each night, just as he was falling asleep, to check that the pen was in his bag. It was around that time that he began taking, as Costello described it, "the dreaded three shots" of liquor before going to bed. "His ability to reassure himself was broken," Costello explained—and as a result, he began desperately to cast about, trying to find ways to numb the fear and regain control.

"The fear and the tightening circles, the anxiety, the inability to enjoy things or even to see things or notice them—that was how Dave's symptoms presented," Costello explained. "I've known people who are depressed, and for them, it is just more like the air goes out of the balloon and they can't get out of bed. But for Dave, it took the form of racing energy in this sort of negative and circular way. Dave was built to rev at a very, very high level, and his sort of emotional drivers were very closely harnessed by a really powerful intellect, sort of an epical intellect."

"You could smell [the fear] on him," another friend remarked. "He terrorized himself. There was the sense that there was this monster looming over him, this sense of inadequacy."

Wallace recognized the person he was becoming—or had been all along. "The realization that you can lose your brain in a super powerful way, I think that was a primarily terrifying experience for him," said Costello. "He had this secret tendency, this secret sort of wrong turn. He had strategies to deal with it and still function and go to Amherst and do very well, but he was always worried that the secret beast would come out." Wallace would wrestle with both these notions—that he was defective in a fundamental way and that he should hide this weakness from others—for the rest of his life.

One night, not long after he began ritually keeping track of his lucky pen, Wallace lost his favorite hoodie, a ratty, smelly sweatshirt he slept in every night. He spent the evening trying to find it, enlisting Costello to help him in his frantic search. Costello noticed that Wallace's humor was absent—during prior panics, he'd managed at least to be grimly funny, but now he was solely and spookily focused on the sweatshirt. He and Costello retraced his steps through the libraries he'd visited—he'd studied in every one on campus that night—but to no avail. This marked a turning point for Wallace: "I can't do it," he told his friend. "I can't keep going."

Three days later, he moved back home, and into his old bedroom, for the rest of that semester and the following summer. He read and wrote a bit; he did a short stint as a bus driver in the spring. Hoping he'd once again tamed his "secret beast," he came to feel steady enough to go back to school.

Wallace returned to college in the fall of 1982, technically still a sophomore. He enrolled in Ancient and Medieval Philosophy with William Kennick, a professor with whom Wallace would study for the rest of his time at Amherst. Through Kennick, he began to seriously examine the theories of Ludwig Wittgenstein. Wallace was already somewhat familiar with the philosopher's arguments; a disciple of Wittgenstein had, in fact, taught Wallace's father. Now, however, the major theme of Wittgenstein's work—broadly, whether language can provide an isolated or shared experience—caught Wallace's attention and commanded it for some time afterward.

Though the year may have started out unevenly, it ended well for Wallace: He had become something of a sensation in the philosophy department (this was the semester he received straight A-pluses). He'd even cofounded a humor magazine with Costello called *Sabrina*, garnering many admirers among his fellow students and, better yet, eliciting the flattering attention of some of the young women on campus. He went home in a good mood—hopeful once more that he'd hit his stride. But that summer, Wallace's parents, after having spent the past year living apart, told their children that they were getting a divorce. Two weeks later, however, Wallace's mother moved back into the house and the marriage resumed; neither child asked for an explanation, and none was given. Wallace sensed his dread creeping back up on him again; he began to suffer frequent anxiety attacks.

The problem was not only the profound shift in family dynamics, but also a burgeoning identity crisis. "For most of my college career I was a hard-core syntax wienie, a philosophy major with a specialization in math and logic. I was, to put it modestly, quite good at the stuff, mostly because I spent all my free time doing it. Wienieish or not, I was actually chasing a special sort of buzz, a special moment that comes sometimes," he would explain in a 1993 interview with Larry McCaffery. "One teacher called these moments 'mathematical experiences.' What I didn't know then was that a mathematical experience was aesthetic in nature, an epiphany in Joyce's original sense. These moments appeared in proof-completions, or maybe algorithms. Or like a gorgeously simple solution to a problem you suddenly see after half a notebook with gnarly attempted solutions. It was really an experience of what I think Yeats called 'the click of a well-made box.' Something like that. The word I always think of is 'click.' Anyway, I was just awfully good at technical philosophy, and it was the first thing I'd ever really been good at, and so everybody, including me, anticipated I'd make it a career. But it sort of emptied out for me somewhere around age twenty. I just got tired of it, and panicked because I was suddenly not getting any joy from the one thing I was clearly supposed to do because I was good at it and people liked me for being good at it."

Wallace didn't last long when he returned to school in the fall of 1983. He told Costello he thought he had a functioning strategy for making it through—presumably fortified by the academic success of his prior year—but soon he realized he'd been deluding himself anew. Not long after the semester began, he moved back home, taking a second leave from Amherst.

Wallace wrote a letter to Corey Washington earnestly detailing the circumstances leading up to his departure (until this point, he'd confided only in Costello). He explained to Corey that he had suffered from "a number of unpleasant neuroses" during his last two years of high school. Since then he was having "depressive episodes" that were increasingly severe and more frequent. "I figured, ultimately dumbly, a semester off on my own to walk in the woods and stroke my beardless chin would clear all this up but it just became clear eventually that it didn't and wouldn't." He went on to explain he'd begun seeing a psychiatrist, without telling his parents, who had prescribed him the antidepressant Tofranil, a medication that, as it turned out, only worsened his condition. By the end of the summer, Wallace wrote in the letter, his depression had become "the worst thing I'd ever experienced," and yet still he'd left for his junior year at Amherst "all happy-seeming (an insane desire to hide the festering pus-swollen canker at the center of my brain is a recalcitrant symptom)."

Wallace was also, however, constantly and fervently seeking his vitality, some version of the emotion he'd felt when he'd cried out the window, "I love this place!"—particularly in his intellectual life. So even as he battled his depression that summer, he was simultaneously cultivating a new passion: writing.

"Probably most kinds of art have this sort of magical thing," Wallace would explain to the talk show host Charlie Rose in 1997. "For a moment, there's a kind of reconciliation and communion between you and me that isn't possible in any other way." In 1983, however, he described this transcendent feeling to friend Mark

Costello in simpler terms: "I can't feel my ass in the chair." Writing, for Wallace, held out hope for both extraordinary connection and a reprieve from the pain he felt in his everyday existence.

"At some point in my reading and writing that fall I discovered the click in literature, too," Wallace continued in the McCaffery interview. "It was real lucky that just when I stopped being able to get the click from math logic I started to be able to get it from fiction."

While at home from Amherst for the second time, Wallace wrote a short story about a mentally ill student who withdraws from college, called "The Planet Trillaphon as It Stands in Relation to the Bad Thing." In it, an anonymous narrator suffers from the "Bad Thing," a concept Wallace would explore repeatedly in his fiction, giving depression a generic term as a way of expressing frustration with the bland uniformity assigned to the mood simply by naming it.

"The way to fight against or get away from the Bad Thing is clearly just to think differently, to reason and argue with yourself," Wallace wrote. "Just to change the way you're perceiving and sensing and processing stuff. But you need your mind to do this, your brain cells with your atoms and your mental powers and all that, your *self*, and that's exactly what the Bad Thing has made too sick to work right."

Recognizing that he is trapped in a vicious cycle, Wallace's narrator asks, " 'How the heck is the Bad Thing able to do this?' You think about it really hard, since it's in your best interests to do so—and then all of a sudden it sort of dawns on you . . . that the Bad Thing is able to do this to you because *you're* the Bad Thing yourself! The Bad Thing is you."

Wallace published this story in the student-run *Amherst Review* after returning to school in January 1984. That year, writing became the primary, more productive, way in which Wallace absorbed and expressed his restless energies.

"The things he would say in class and the times he would be up writing, really banging out the manuscript, and then other times when he would help me with my writing—with all three of those things, I had a clear sense that he was in his element," Colmar explained. "He was grounded. But at the same time, there was a certain timidity about it, a kind of bashfulness about it—'bashful' is probably the best word. I don't use that word very often. But definitely a nonboasting quality."

In his last year at Amherst, with his decision to be a writer firmly in place, Wallace began one of two theses, a novel called *The Broom of the System*, the story of a young woman named Lenore Stonecipher Beadsman whose great-grandmother, once a student of Wittgenstein, advises, "All that really exists of my life is what can be said about it."

The Broom of the System was Wallace's attempt at a language game, his tribute to Wittgenstein. The picture theory of language set forth in Wittgenstein's *Tractatus*—that words are like little images, one-to-one representations of experience—ultimately traps us within our own private worlds, in which we can never be certain whether language is building a bridge or a prison. In short, Wallace determined, "you're looking at solipsism."

Wallace's great regard for Wittgenstein stemmed largely from the fact that the philosopher decided he couldn't live with this ending— "he realized that no conclusion could be more horrible than solipsism"—and so he did an about-face: in *Philosophical Investigations* Wittgenstein argued that language is possible only if it is a function of relationships between people, if it is "dependent on human community." As Wallace pointed out, this approach still didn't solve the problem that we can't know if we're referring to the same images when we speak, though "we're at least all in here together."

Wallace also found the idea that only the self exists—nothing else can be known or understood beyond it—to be a hideous concept, both philosophically and personally. "Stemming from Dave's life was this terror of solipsism, this sort of devouring god," Costello said. "It's more solipsism as protagonist. This thing that goes out and systematically attacks, solipsism as virus. Dave was terribly concerned with mapping what it's like to be in the hands of this—and of course the act of mapping is an act of striking back in self-defense. That is one of the things that's most moving about his work to me: the very act of describing, and describing it for a reader's edification, as a reaching out. . . . I think at certain times his writing kind of glows with a sense of confidence that we'll get there, that victory is ours finally. Other times, not. But there is no question that this was what he was trying to wrestle with."

Wallace graduated from Amherst *summa cum laude* for his theses in philosophy and literature; *The Broom of the System* received an A-plus.

Soon after graduation, David fell apart when his college girlfriend broke up with him. He could not get her out of his thoughts, and those around him noticed he began to withdraw. "You got the impression he was just sort of slipping away from you," Corey Washington described. His parents took him to the psychiatric unit of a local hospital, where he stayed for several weeks. It was here that Wallace was first prescribed Nardil for his depression, the medication that stabilized him for most of his adult life.

By August, Wallace was well enough to go to the University of Arizona to begin an MFA in creative writing. "The poetry program here is uniformly just awesome. Most of the students are just, they— their poems are stomach-punchingly good. The fiction runs the gamut from fairly inferior . . . to really, really good," he remarked in an audio recording he made for his old Amherst friends soon after arriving. "I mean there are people here who've—who are publishing a lot and who have won national awards and stuff. And the professors are on the whole very good. So I love it here."

Perhaps more indicative of his mood, however, were the tears that erupted when he mentioned Corey Washington's name (one of the people he was addressing on the tape)—"Yeah, boys, you can tell even the thought of Cor and I get choked up," Wallace confided. "It's always hard for me to talk about emotion."

In fact, Wallace had begun to experience serious difficulties with the program. "Of all the MFA programs he could have gone to," his sister, Amy, said, "the one at Arizona was probably the worst fit for his style of writing." As a result, Wallace sped through his studies there and finished the program early.

Wallace had a tendency to romanticize places soon after leaving them, even if he'd had a gloomy experience while there. He built such a strong and superstitious mythology around having been able to write productively in certain environments that he would gloss over the agonizing emotional struggle that had occurred alongside the work. So, in 1987, Wallace went back to teach for a semester at Amherst. ("Please, please get me out of here," he wrote in a letter to his literary agent Bonnie Nadell soon after his arrival. "I had forgotten how much I hated Amherst College. . . . And these profs who thought I was either a madman or a troublemaker or both when I was a student now tousle my hair and talk tenure tracks and Guggenheims.")

He then returned to Arizona the following year to teach an undergraduate course. While there, feeling increasing desperation and fearing yet another emotional unraveling, he joined a recovery group and began the twelve-step process of Alcoholics Anonymous. As part of the program, he wrote letters to his former professors William Kennick and Dale Peterson apologizing for such past infractions as using secondary sources while writing his papers. "I tried to argue (without much success) that the very idea that the PSYCHE was alive, much less immortal rested on the specious premise that it renders a person alive, one no more true than that a virus that renders a person ill is itself ill," he wrote in his letter to Kennick before sheepishly admitting that his take on *Phaedo* was his father's, not his, and had come out of an argument or discussion with him. "I didn't have the courage to come to your office and tell you this."

Wallace also decided to quit taking Nardil, aware that the ardent among his fellow AA members would consider this yet another narcotic and a crutch. Within a month, however, Wallace had called his mother and asked her to come get him. Despite his efforts, he was in the midst of his fourth acute crisis. Back at home yet again, and back on the Nardil, he talked to his sister about his fear of life in general— "just the most basic stuff: How do you make yourself get out of bed in the morning, find a purpose, fulfill that purpose, do what you're supposed to do."

His father tried to coax him out of his ongoing despair. "I took him out to Allerton Park, which is about ten miles away from here— beautiful place," he recalled. "We walked around, and I tried to talk to him. It was like there was this wall between us. I just hoped I could use this nice open-air experience and maybe he'd talk a bit and that would help. But there was just no getting through to him. Whatever he was locked into—whatever anxiety and panic, whatever was going on at Amherst, needing to come home, needing to withdraw from whatever was making him feel bad there and then needing the same thing coming back from Tucson—I was really worried about suicide at that point."

The night before Amy was set to drive to Charlottesville, Virginia, where she was living at the time, Wallace pleaded with his sister to stay with him at their family home a little longer. She explained that she had to get back to start a new job.

"When I walked into my house in Charlottesville, the phone was ringing. [It was] my mom, who told me that [David] was in intensive care and had tried to kill himself." He had overdosed on Dalmane, a sedative he'd been given for his insomnia. "This is going to sound weird," Amy said, "but that was more purely painful than when I actually got the news that David had killed himself."

Although *The Broom of the System* was published as a novel by Viking Press in 1987 to critical acclaim—"Clearly Mr. Wallace possesses a wealth of talents," Michiko Kakutani wrote in her review in the *New York Times*, "a finely-tuned ear for contemporary idioms; an old-fashioned story-telling gift; a seemingly endless capacity for invention and an energetic refusal to compromise"—Wallace continued to struggle for recognition and a steady paycheck.

With these considerations in mind, and a more general desire to create a secure life, he decided to model his career on that of William Gass, a writer he admired who'd found success working as an academic philosopher while also writing well-regarded postmodern fiction. This, Wallace reasoned, would offer the structure of academic life and might also force him to write more regularly, with the only time to do so relegated to the margins of his life.

In the spring of 1989, Wallace moved into an apartment with Mark Costello in Somerville, Massachusetts—his old friend was working as a legal associate in Boston then—and prepared to begin studying at Harvard, where he'd been accepted in the doctoral program in philosophy.

Although he was reaching for stability, he was also counterintuitively reentering a world he'd already determined to be unsatisfying. "I never heard him speak of philosophy with a real urgency," Costello said of this decision. "He liked philosophy, did it easily and steadily and well. But it would be a job, not a passion."

"David found being a graduate student in philosophy uncongenial," James Wallace offered. "He had expected, perhaps naïvely, that the faculty and students would talk with one another more or less as intellectual equals. David never did cotton much to subordination."

It was not a surprise to his family or friends, then—or ultimately to Wallace—when his experience at Harvard did not go smoothly. "I think there were some pompous guys there. Some injured vanity, too. They were utterly unimpressed with Dave's fictional achievements," Costello recalled. "But did he really expect otherwise? Dave was a shrewd guy. He was only there about six weeks, and not in the best shape for said weeks. The fact that he had dumb reasons for going wasn't Harvard's fault."

By then, Wallace had given up the twelve-step program he'd started in Arizona and was smoking pot and drinking excessively— more, in fact, than Costello had ever seen before.

Wallace would later explain that this doubling down on alcohol and marijuana was deliberate—however illogical his rationale was for doing so. "For most of the late '80s, my method for 'quitting' drugs was to switch for a period from just drugs to just alcohol," he wrote. "Then I'd switch back to drugs in order to 'quit' drinking."

Costello recalled a similar vicious cycle from this period: "I can remember in Somerville, he got increasingly more depressed. . . . One of the things he was most focusing on was that his condition was never going to change. It's never going to get better. It can't get better. He thought if he wasn't sober he would lose his mind. But he was stuck with this out-of-control compulsion. He would get caught and caught and caught."

Wallace told Costello that he planned to kill himself in the marble entrance to Harvard's philosophy department. He wanted, as he put it, to "hang himself in the halls of philosophy."

Instead, he walked over to University Health Services and confessed he was having suicidal ideation. They insisted he go immediately to McLean Hospital, a psychiatric facility in Belmont, Massachusetts. He called Costello from the student health offices, "sounding hollow, detached, and explained the situation. He asked me to bring him certain things. He had composed a list: cigarettes, bathrobe, and notebook."

David was hospitalized at McLean for a month. He was diagnosed with depression and alcoholism. These illnesses were deeply interwoven for Wallace and at the time seemed as intrinsic to his being as his heart or lungs. "The Bad Thing is you"—the message of his first short story—still resonated, and he described addiction as feeling like a cage he had entered.

In November 1989, Wallace moved into the Granada House, a halfway house, and began attending Alcoholics Anonymous meetings. This was quite literally a last resort: the doctors at McLean had made it clear, and Wallace had finally accepted, that he had to quit using drugs and alcohol or he was going to die. "Nobody ever Comes In because things were going really well and they just wanted to round out their P.M. social calendar," Wallace would later write of AA in *Infinite Jest*. "Everybody, but everybody comes in dead-eyed and puke-white and with their face hanging down around their knees."

It would take some time before Wallace believed he had the ability to get better. The addict must reckon with a paradox: "The enslaving Substance has become so deeply important to you," Wallace wrote in *Infinite Jest*, "that you will all but lose your mind when it is taken away from you. Or that sometime after your Substance of choice has just been taken away from you . . . as you hunker down for required A.M. and P.M. prayers, you will find yourself beginning to pray to be allowed literally to lose your mind, to be able to wrap your mind in an old newspaper or something and leave it in an alley to shift for itself, without you."

His early resistance to treatment brought with it the rigorous intellectual manipulations that Wallace had practiced at school and in his writing. "I made it hard for anyone to help me. I could go to a psychiatrist one day in tears and desperation and then two days later be fencing with her over the fine points of Jungian theory," he wrote in a short essay called "An Ex-Resident's Story," which he posted anonymously on the Granada House website. "I could argue with drug counselors over the difference between a crass pragmatic lie and an 'aesthetic' lie told for its beauty alone; I could flummox 12-step sponsors over certain obvious paradoxes inherent in the concept of denial."

In the end, however, it was the simplicity of the program, which allowed him to let down his well-constructed guard, that won Wallace over. When he came to know his fellow members, people who were like "a weird combination of Gandhi and Mr. Rogers with tattoos," as he would write of the fictional characters that comprised the rehab group of *Infinite Jest*, "with enlarged livers and no teeth who used to beat wives and diddle daughters and now rhapsodize about their bowel movements"—but who also earned Wallace's begrudging respect for winning a battle he'd been losing for years—he gave in.

"David had to find his admiration for a working person—a human being, not an intellectual," confided a close friend who saw Wallace through recovery. "There was something very placating for him in that. The rules changed, he no longer needed to compete with the best of the best of the best. Suddenly, the measuring stick was about being a human being and doing the right thing." He was able to connect without the fear, judgment, or competition he had felt in many of his prior peer relationships. For David, AA offered a more comfortable sense of connection.

Costello referred to this capitulation as Wallace's "having to turn to something other than smart to live."

After publishing *Infinite Jest*, Wallace would describe his feelings, and the mood of his novel, with tempered wisdom (while simultaneously managing to skirt the depth of his own experience): "The sadness that the book is about, and that I was going through, was a real American type of sadness. I was white, upper-middle-class, obscenely well-educated, had had way more career success than I could have legitimately hoped for and was sort of adrift. A lot of my friends were the same way. Some of them were deeply into drugs, others were unbelievable workaholics. Some were going to singles bars every night. You could see it played out in 20 different ways, but it's the same thing. Some of my friends got into AA. I didn't start out wanting to write a lot of AA stuff, but I knew I wanted to do drug addicts and I knew I wanted to have a halfway house. I went to a couple of meetings with these guys and thought that it was tremendously powerful. . . . I get the feeling that a lot of us, privileged Americans, as we enter our early thirties, have to find a way to put away childish things and confront stuff about spirituality and values. Probably the AA model isn't the only way to do it, but it seems to me to be one of the more vigorous."

Wallace prayed twice a day, in the morning and again in the evening, for the rest of his life. No matter where he traveled, for nearly two decades, he would find and attend the local AA meeting. In one of his journal entries from after he'd begun recovery, he made a list called "What Balance Would Look Like":

> 2–3 hours a day of writing
> Up at 8–9
> Only a couple late nights a week
> Daily exercise
> Minimum time spent teaching
> 2 nights/week spent w/ other friends
> 5 AA/week
> Church

By age twenty-eight, Wallace was sober and starting to write "arguably his best stuff using this spiritual program," one friend speculated, "surrounding himself with nonintellectuals for safety."

Without the distraction of drinking or smoking pot, however, his compulsive energy began to surface in new ways. "He would feel really shitty and then not write and then beat himself up for not writing," said a friend. "Or he would feel really shitty and then go have sex with someone he shouldn't have sex with and then feel really shitty. . . . He was promiscuous, he used sex for power or pleasure. . . . He wouldn't use 'cause he knew that would kill him. But he'd overeat or he'd throw up . . ."

Still, Wallace held fast to the tenets of recovery; it was the only beacon he could see ahead of him. At one of his AA meetings, he became reacquainted with Mary Karr—they'd met once before at a party—then a little-known poet living in Belmont, Massachusetts, with her husband and young son. (Karr would publish her first memoir, *The Liars' Club*, in 1995.) They began a friendship—Karr considered it a literary kinship; Wallace saw it as a potential romance. In 1990 he started to teach as an adjunct professor at Emerson College, a job that Karr helped him get. As the year progressed, the two grew closer, and Wallace grew increasingly more focused on pursuing her. He began to tell friends that he and Karr were together, even going so far as to get a tattoo with her name on his biceps, but she denied any involvement. At the end of that year, Karr accepted a job at Syracuse University and moved there with her family.

Wallace was devastated. "I just think it's an addict's mind," one friend observed. "I think that rather than sitting with the pain of the fact that he may fail, the pain that he has to get up and write every day or the pain of dealing with what he's writing about, he's fixating on and having escapist obsessions. . . . He definitely had an obsession with her. Major obsession."

"He had her on a pedestal—it was like a fantasy," said another friend. "She was married, and there was this whole drama that was very immature. You would think they were in high school the way that they carried on. He was getting sober at the time and having an affair with a married sober person who had a child, and it was very tumultuous. He somehow decided that he was going to write this book to win her."

With Karr on his mind, Wallace, after a yearlong period of not being able to produce fiction at all, began in earnest to write the novel that would become *Infinite Jest*. "The key to '92 is that MMK was most important," Wallace, himself, scribbled into the margin of one of his books. "IJ was just a means to her end (as it were)."

By the spring of 1992, he had a partial manuscript of *Infinite Jest* to send to his agent. Also that spring, he heard that Karr's marriage had come to an end. In May he moved to Syracuse. By August his persistence finally paid off: Wallace and Karr became a couple.

Even with Karr by his side, however, Wallace continued to struggle. Why did he have to work so hard to feel content? What held him back from rounding the corner on maturity? Why did he continually veer off course? Propelled by the teachings of AA, he reached back and back, seeking an essential truth, a ready answer, until he arrived at what he believed was the origin of his perpetual unrest.

"He was dissecting the way that his psyche had been affected. In the twelve steps, if you are dishonest with yourself and others, then you create a lot of shame and you're going to use," said an ex-girlfriend. "If you have secrets, you're going to use. That's not to say that you're supposed to write a novel about it. But you have to tell someone, your sponsor. So he was approaching it from that angle."

Wallace felt that recovery was saving his life and "he knew that the only chance for him to be the man that he wanted to be, which was someone who was more accountable, would be to put it out there, be honest."

Wallace had come to believe that his mother was an unhappy perfectionist, one whose lofty ideals had led to a keen sense of disappointment she was unable to express. He soon began to believe that she had raised him as an extension of herself. In the margins of the popular self-help book *The Drama of the Gifted Child*, by Alice Miller, in a section called "The Lost World of Feelings," he wrote, "Becoming what narcissistically-deprived Mom *wants* you to be—performer."

In his fictional account of a child's reaction to a similarly stringent mother (the story "Suicide as a Sort of Present," published in his collection *Brief Interviews with Hideous Men*), Wallace writes of a small child who loved his mother "more than all other things in the world put together," so much, in fact, that "if it had had the capacity to speak of itself truly somehow, the child would have said that it felt itself to be a very wicked loathsome child who got . . . to have the very best, most loving and patient and beautiful mother in the whole world."

Yet, the mother in Wallace's story was filled with "self-loathing and despair." She took "all that was imperfect" in her child "deep into herself and bore it all and thus absolved him, redeemed and renewed him, even as she added to her own inner fund of loathing.

"She could not, of course, express any of this. And so the son—desperate, as are all children, to repay the perfect love we may expect only of mothers—expressed it all for her."

Wallace's relationship with Karr was uneasy from the start. She had a son who took priority. Wallace wanted more of her time; she found him self-absorbed. He asked Karr to marry him; she turned him down. Finally, when he got an offer in the spring of 1993 to teach at Illinois State University in Bloomington-Normal, he reluctantly accepted, realizing that it was time to move on. He relocated to Bloomington in July and started teaching that fall.

Meanwhile, he continued his work on *Infinite Jest*—he had sold it to the publishing house Little, Brown based on a partial manuscript, and the writing was going well. After a little more than a year in Bloomington, he finished the book.

Wallace drew heavily from his own experience for the novel, examining with ferocity and humor everything from family relations to depression to film theory to tennis to corporate greed to substance abuse and recovery programs—and producing an epic text that both comments on and mimics the culture's compulsive need for entertainment.

"There are a lot of little Daves in that book, different versions of himself: The Dave AA warrior I think is much more like [Don] Gately," Mark Costello suggested. "But the pre-AA Dave, the sick Dave, the broken Dave, I think, is much more like Hal [Incandenza]."

The publication of *Infinite Jest* in 1996 brought Wallace, then thirty-three years old, the exceptional recognition he had long yearned for, with critics describing it as "genius," "a masterpiece," and "a virtuoso display of styles and themes." Not every critic unrestrainedly fawned over the book in this way, but nearly everyone agreed it was the work of a masterful and generous writer. Wallace "is one of the big talents of his generation, a writer who can seemingly do anything . . . a pushing-the-envelope postmodernist who's also able to create flesh-and-blood characters and genuinely moving scenes," Kakutani weighed in at the *New York Times*. She went on, however, with less enthusiasm: "As the reader plows through *Infinite*

Jest, it becomes clear that the subplots involving Gately, Hal and the Canadian terrorists also provide a flimsy armature on which Mr. Wallace can drape his ever-proliferating observations and musings. Indeed, the whole novel often seems like an excuse for Mr. Wallace to simply show off his remarkable skills as a writer and empty the contents of his restless mind."

The attention, nonetheless, was keen. On the one hand, it made Wallace's gargantuan ambition and effort seem worthwhile. But it also felt as if the adoring audience of his dreams had suddenly materialized and they were now sitting at his doorstep hungrily awaiting his next work of genius.

Fame became another preoccupation for Wallace—and another conduit for his inexhaustible fear and self-contempt. Despite so much public acclaim and critical affirmation, "there was always a strong sense of inadequacy," Mark Costello recalled about his friend. He mimicked Wallace's inner voice: " 'I'm a performing monkey. I'm just giving people what they want. There's this David Foster Wallace persona and I'm just sort of an organ grinder. I want to write something that goes beyond that to some level.' "

This sense of writing as performance corrupted Wallace's creative process by not only paralyzing him but also robbing him of its former value, its ability to transport him from earthly suffering. Now it made him more anxious than ever.

"I play endless games with myself. This discipline thing is really the last significant area of my life where I still indulge in a self-pity and hatred that I feel like are childish and wasteful and wrong," he wrote in 1997 to the novelist Don DeLillo, who had become somewhat of a mentor to Wallace. "I stopped thinking neurosis was cool a long time ago."

His success also compounded his innate sense of duplicity—the "fraudulence paradox"—because now there were thousands of people who revered and admired him without ever having met him. (Indeed, many of his admirers credit Wallace with saving them from their own isolating unhappiness. "During the few times I have experienced depression," wrote one such fan on his blog, "I felt less alone knowing that the fictional Kate Gompert knew how I felt.")

Wallace choked under the pressure of the public's expectations. "The thing is I get scared it won't come. I'm back to thinking IJ was a fluke," he wrote after the novel's publication. He admitted to feeling nothing—dead inside. There was nothing worthwhile to pursue. "I'm now starting to want to run. I don't know what to do. Go to Fedex and movie after tennis. Play tennis well. Sit and relax. This seems formless and poor. The fear and frustration and self-doubt."

Yet, by 2000, Wallace had published two collections—*A Supposedly Fun Thing I'll Never Do Again* and *Brief Interviews with Hideous Men*—and had been awarded the MacArthur Foundation Fellowship. His crisis of fame had clearly subsided. He had even started to work on a new novel. That same year, Wallace, who had been taking the year off to write, received an inquiry from Pomona College in Claremont, California, asking if he'd be interested in a newly created position in their creative writing department.

"When he first came here in the fall of 2002, his house was about four blocks down the street [from mine]. I met him, I think, the second day he was at an AA meeting, and we hit it off," an AA buddy recalled of Wallace's early days in Claremont. "At that time, my wife was dying. It was a terrible, lingering death. If I was going to keep on working then, we had to get some help, people to stay with her. And David volunteered to come up here, and I'd known him maybe two, three months, max. He volunteered to walk up the street, with his laptop, every Tuesday morning and stay with my wife so that I could go to work."

Wallace quickly established himself in the AA community of Claremont, making both supporters and friends of the men and women he met in meetings. Not long after moving there, he also met a visual artist named Karen Green, who, with his permission, created an illustrated panel of "The Depressed Person," a short story that Wallace had published in *Harper's* magazine in 1998. Green, who lived in a neighboring area, delivered her piece to Wallace in person, and the two struck up a friendship that, after the dissolution of Green's marriage months later, soon became a romance.

At the end of 2004, Wallace and Green were married. This initiated an unexpected phase of ordinary personal satisfaction in Wallace's life. The couple bought a house together. They traveled—to Hawaii; to Stinson Beach, California, for Christmas with Wallace's parents; to Capri with fellow writer and friend Jonathan Franzen. They adopted a dog.

In 2005, Wallace delivered the commencement speech at Kenyon College in which he spoke of the "terrible master." The speech, after making the rounds on the Internet, became an essay-length book published by Little, Brown called *This Is Water: Some Thoughts, Delivered on a Significant Occasion, About Living a Compassionate Life*. In it, Wallace speaks as a gentle human, a wise elder counseling his audience (and perhaps himself): " 'Learning how to think' really means learning how to exercise some control over how and what you think. It means being conscious and aware enough to choose what you pay attention to and to choose how you construct meaning from experience. Because if you cannot or will not exercise this kind of choice in adult life, you will be totally hosed."

Though Wallace had started to agonize over his writing again, often barely eking out pages, in February 2007 he published an excerpt of the novel he had been working on—what would eventually become *The Pale King*—in the *New Yorker*, called "Good People." The piece follows a religious young man as he wrestles with his feelings for his pregnant girlfriend, who has told him that she will keep the baby but, if he would like, will raise it on her own. It is a story as much about life as it is about faith. "What if he was just afraid, if the truth was no more than this, and if what to pray for was not even love but simple courage," the narrator asks himself in the end, "to meet both her eyes as she says it and trust his heart?"

In those early Claremont years, "there were moments when he would be down but these moments were short," a fellow AA member recalled. "But then he started to isolate and detach—and then those isolated periods became more extended. At AA meetings, he would pass. He never passed when he was feeling good. But when he would simply pass when it was his turn to share, that was a sign something was wrong. And those moments became more extended and more frequent the closer we got to springtime of '07."

In the summer of 2007, Wallace was eating dinner at a restaurant with his parents, who were visiting him in Claremont. Wallace began to have heart palpitations and sweat heavily at the table. When he visited his physician afterward, the doctor recommended that he shift from Nardil to a newer antidepressant, one with fewer restrictions and side effects.

With trepidation, Wallace went off his medication. He told friends he thought the Nardil had become less effective anyway, but it was also all he'd ever known by way of a stabilizing medication. Still, he felt well enough in the first few weeks that he decided not to start taking the other antidepressant. By late fall, however, he had to be hospitalized for severe depression. Doctors prescribed new antidepressants, but none seemed to work—or Wallace would quit taking them sooner than recommended, abruptly declaring this drug was making him anxious or that one was not helping at all.

Thus began a complex undoing of his fragile emotional equilibrium. Wallace, who had by this time been struggling for more than half a decade to move his novel forward, who had to fight continually the belief that he wasn't capable of being a writer anymore, who loved his wife but who was also a self-described high-romance, low-intimacy companion, who led more naturally with self-hatred than self-assurance, began to revert to all these notions more powerfully as his mind became increasingly focused on his failings.

"His image was 'I'm a piece of shit,'" said a friend and fellow AA member in Claremont. "'I'm married to this wonderful woman and yet I can't—I can't be the husband that I should be.' I don't think he knew how to be a husband, and he felt like a failure. 'I've had some great success in writing, but I can't write.' The thing that he loved, the teaching, he had to stop doing. 'I can't even do what I love the most, because I don't know what's going to happen.' That bothered him. He felt worthless. Not making any contributions. The number of people in AA that David helped who had no clue who this guy was, and David couldn't see that."

There were quick flashes of hope, when Wallace seemed to steady. "He kept showing up at meetings, even when he was down. There were very few times when he would miss meetings," recalled one AA friend. "He was in intense therapy, with both a psychologist and psychiatrist."

"There were times when it looked like he was going to shift directions and then he'd become optimistic," said another. "Very brief—but there were moments. . . . He seemed to enjoy the socialization with people. We would take walks, we would take rides, we would sit around and talk about anything, about nothing."

But in the summer of 2008, Wallace checked into a motel and overdosed on pills. He called his wife as he came to consciousness at one point and apologized, saying he wanted to live.

"He promised Karen after he disappeared—he promised her that he would never do that again," recalled a friend. "That probably weighed on him terribly, because I don't think there was an extended time period from that moment on when he wasn't thinking about taking his life."

Wallace's life shrank as he moved backward through the compulsions, both productive and destructive, that had once propelled him forward. He abandoned his fixation on his mother and unhappy childhood. He'd quit drugs and drinking, but now being sober held less promise. He let go of his driving ambition, his tug-of-war with fame, his pleasure in unexpected fellowship, and his against-all-odds pursuit of vitality and, perhaps the greatest blow, he surrendered his work.

"He couldn't write anymore," Costello surmised. "If you can't write—marriage, being someone's son, someone's brother . . . that isn't enough Velcro. There wasn't enough reason to stick around without the writing."

Finally, he was left only with his sense of failure, but even this was sacrificed in its way: his mood was quiet and still now, no longer rippling with his once-striving self-criticism.

"I honestly think that David found it very difficult and very frustrating simply to be alive," Wallace's sister said. "I think that from the beginning, he became aware, from early on, that you have to spend a lot of time doing things that you don't necessarily want to do: going to school, cleaning your room, basic things like that. And then you're supposed to grow up, get a job, and that's just what people do. David was very frustrated with the easy superficial existence, but I think a part of him would have given anything to be able to be happy with that sort of thing. I remember David saying to my parents when he would have been about fourteen or fifteen, 'Why do people have children? It's really very selfish. I didn't ask to be born, and now I am who I am.' "

Among the many papers in Wallace's archive—manuscript pages, journals, books with underlined passages and personal notes, scraps with stream-of-consciousness jottings—there is a letter on the back of which a conversation is recorded in handwriting alternating between a childish scribble and a grown man's handwriting. It is Wallace's dialogue with a younger self.

"I hurt. You never listen to me," he wrote in the childlike scrawl.

"I'm scared of you," he wrote to that child. "You manifest an addictive impulse, despair, depression."

Though Wallace's story is one of tragic extremes, the battle he faced, and ultimately lost, is also familiar to many. What began as self-consciousness grew, over the course of his life, into acute self-awareness, which eventually transformed into self-hatred. This led to a desire for release, redemption, attention—which he found, albeit temporarily, in his writing, in drugs and alcohol, and in his relationships—but which only led to further self-indictment. Wallace's focus rarely shifted from his tormenting thoughts. He felt as though he existed within a "dark world, inside, ashamed, locked in." What some might view as narcissistic behavior, the torment of an artist seeking praise and perfection, is more accurately understood, in Wallace's case, as an overwhelming, debilitating sense of anxiety and unhappiness.

There were many times when Wallace was funny, happy, and loving. Yet he was never able to shift his attention away from what made him feel bad for a sustained period of time. He could not, as he once wrote, "perceive any other person or thing as independent of the universal pain . . . everything is part of the problem."

PART II

5

WHEN CAPTURE
LEADS TO VIOLENCE

Though I began my research intending to study the role of capture in mental distress and everyday experience, I soon found myself drawn to realms of experience far from what I had contemplated at the start. The chapters that follow explore behaviors and states of mind so extreme—so seemingly distant from our own lives—that they confound almost any attempt at explanation, let alone empathy.

David Foster Wallace's final act was directed at himself. What happens when feelings of anger and hopelessness are instead directed outward, when capture leads not to hurting or killing yourself, but to hurting or killing someone else?

I recognize that the material ahead is disturbing. While it may transport us to largely unfamiliar and uncomfortable territory, it is important that we recognize the role of capture at the outer limits of human experience. The outward signs of capture are as varied as the stimuli that can capture us; violence lies at the far end of this spectrum. In acts of wanton cruelty or ideological violence, for instance, we can see the outline of a now familiar process: an object gradually commandeers our attention, coloring our view of the entire world.

Every successive exposure to a salient stimulus strengthens capture's hold.

It would be folly to reduce all aggression to a single underlying mechanism. Nonetheless, capture often plays a role in otherwise incomprehensible behavior.

STRIKING OUT

Ted Kaczynski was a brilliant mathematician, a child prodigy. His parents pushed him to succeed, and at first he made them proud. He earned his undergraduate degree at Harvard University and went on to graduate with a PhD in mathematics from the University of Michigan, where he wrote an exceptional and widely praised thesis. He secured an assistant professorship at the University of California at Berkeley. Yet Kaczynski never felt at home in the modern world. He longed for a kind of tranquility and privacy that the city could not provide. He began to struggle in his lectures and avoided contact with others. In reaction to his growing sense of agitation and awkwardness, he left teaching and abandoned his urban life to live as a survivalist in the Montana wilderness.

For twenty years, beginning in the early 1970s, Kaczynski lived in a cabin he built himself; it was from this isolated home that, eventually, he would mail bombs to various organizations and people. After his first few years of self-imposed isolation, he moved back to Illinois and worked in a Chicago factory where his brother, David, was a supervisor. He was hired in 1978 and fired that same year, after a friendship with a female coworker went bad. Kaczynski retaliated by writing vulgar limericks about her, which he posted around the factory. As a result, his brother had to deliver the news that he'd been fired. In defeat, he withdrew to his haven in Montana, this time for good.

When he returned to his cabin, Kaczynski struggled with the urge to strike back against the affront of modern life. It was in this same year, 1978, that he built his first bomb.

On July 24, 1978, Kaczynski wrote in his diary, "Yesterday was quite good—heard only eight jets. Today was good in the early morning, but later in the morning there was aircraft noise almost without intermission, for, I would estimate, about an hour. Then there was a very loud sonic boom. This was the last straw and reduced me to tears of impotent rage. But I have a plan for revenge."

Hypersensitive to particular sounds, he fixated on the occasional, faint Doppler shift of a passing jet engine and was unable to escape the way the noise gnawed at him. What others might have considered a mere reminder of what they'd left behind, Kaczynski could not tolerate. His feelings of persecution and grandeur only intensified. Back in Illinois, the noise of jet engines had been a small aspect of the repulsive environment of modernity, the grating, growling, whining, clattering contraption of a city. The noise of a single jet was nearly drowned out by the traffic, the beeping garbage trucks, the rumble and racket of elevated trains. In the woods, though, a jet punctured his ear. It broke into the home of silence and solitude he'd created by hand.

"By silence I don't mean all sound has to be excluded, only man-made sound. Most natural sounds are soothing. The few exceptions, like thunder and raven cries, are magnificent and I enjoy them, but aircraft noise is an insult, a slap in the face. It is a symptom of the evil of modern society . . . yet where today can one get silence? Nowhere—not even up here in these mountains."

The noise of the jets overhead proved there was no escape from the modern world. Technology touched everything, even the wilderness. He could escape from his own reactive mind only in brief intervals. Later that year, he managed to suppress this anger over the affliction of sound and immerse himself in the forest around him on a camping trip deeper into the wilderness. Yet even there the noise followed him. After a single sonic boom, capping an hour of continuous noise as jets and planes passed overhead, his rage erupted again.

"Things are spoiled for me now. . . . I will work on my revenge plan. I was so happy here. I had looked forward to staying out in the woods much longer than this. Isn't there any place left where one can just go off by oneself and have peace and quiet?"

It's a compelling plea when detached from the violence it ultimately inspired. Yet instead of writing poems or essays in protest, Kaczynski produced bombs to send to victims who, for him, were collaborators in technology's erosion of the natural world.

According to Kaczynski, his disillusionment began during his senior year at Harvard. Yet it wasn't until he moved to Montana and read Jacques Ellul's *The Technological Society* that his purpose crystallized in his mind. He felt as if Ellul had put into words everything he had been thinking and feeling. "The honest truth," Kaczynski said, "is that I am not really politically oriented. I would have really rather just be living out in the woods. If nobody had started cutting roads through there and cutting the trees down and come buzzing around in helicopters and snowmobiles I would still just be living there and the rest of the world could just take care of itself."

Kaczynski's plan for revenge began with a prank, but then escalated rapidly into violence. He wrote of pouring sugar into the gas tanks of snowmobiles to disable them, but soon the plan took a darker turn. He fantasized about becoming a sniper, concealed in the woods, prepared to shoot trail bikers.

Yet over time he realized that his attacks on random individuals would accomplish little. Gradually, he devised a scheme that would be noticed, by targeting those who were agents of the technological plague. "Considering technological civilization is a monstrous octopus, the motorcyclist, jeep drivers, and other intruders into the forest are only the tips of the tentacles." He wanted to strike at the heart.

Kaczynski's rage was nothing new. There were times during adolescence when he would become very angry, a feeling intensified by an inability to express himself openly: "However, I never attempted to put any such fantasies into effect because I was too strongly conditioned . . . against any defiance of authority. To be more precise, I could not have committed a crime of revenge, even a relatively minor crime, because my fear of being caught and punished was all out of proportion to the actual danger of being caught."

Kaczynski embraced his anger only after a period of intense personal confusion. In graduate school, he had a dramatic and pivotal confrontation with the ambiguity of his sexual drives. After several weeks of intense arousal, he began fantasizing that he was a woman, to the point where he considered a sex change. He set up an appointment for medical advice, but while waiting to see the psychiatrist, he lost his nerve. As a cover, during his session, he discussed being depressed about the possibility of being drafted. As he left the building, he felt rage at the doctor who had consulted with him, and within this rage, he felt the possibility of release and renewal, as he realized he could find a new mission in life through violence.

> Just then there came a major turning point in my life. Like a Phoenix, I burst from the ashes of my despair to a glorious new hope. I thought I wanted to kill that psychiatrist because the future looked utterly empty to me. I felt I wouldn't care if I died. And so I said to myself why not really kill the psychiatrist and anyone else whom I hate? What is important is not the words that ran through my mind but the way I felt about them. What was entirely new was the fact that I really felt I could kill someone. My very hopelessness had liberated me because I no longer cared about death. I no longer cared about consequences, and I said to myself that I really could break out of my rut in life and do things that were daring, irresponsible or criminal.

Several years after acknowledging his violent preoccupation, on June 3, 1980, Kaczynski mailed a bomb to United Airlines president Percy Wood. "I feel better," he wrote. "I am still plenty angry. I'm now able to strike back! I can't strike back to anything like the extent I wish to, but I no longer feel totally helpless, and the anger duzzent gnaw at my guts as it used to."

Before he was apprehended, Kaczynski would mail twelve more bombs, killing three and wounding twenty-two others.

When we read the work of criminologists who have studied the mind-set of violent perpetrators, we encounter many of the characteristics of capture: first there is an attentional cueing; then a preoccupation with the provoking cue; and finally, a change, often extreme, in emotion, resulting in feelings of intense anger, offense, or injustice. Dr. James Blair, chief of the National Institute of Mental Health's Unit on Affective Cognitive Neuroscience, and other researchers have identified a neurological pattern common to many violent offenders in which an underlying hypervigilance or reactivity to specific cues generates intense and overwhelming feelings of shame, humiliation, and anger. These feelings, in turn, intensify the individual's sensitivity and response to the cues.

If your threat circuitry is overly responsive, you become hypervigilant. The neural circuitry becomes sensitized, so perceived threats generate larger responses.

When we are captured by an emotional stimulus, Dr. Blair reminds us, the brain's amygdala is activated. Because the amygdala is connected with the prefrontal cortex, the region of the brain involved in executive functions such as decision making, the amygdala's fight-or-flight impulses will narrow attention to perceived threats, thus warping our ability to interpret actual threats in the environment. Ultimately this process erodes our ability to make rational choices; the brain becomes a single-purpose sentinel.

"If you're in a situation where you're hyperresponsive to threat," said Dr. Blair, "you're going to be very irritable. But if your decision making is relatively intact, you should still be able to select out, reasonably efficiently, an appropriate response. On the other hand, if the emotional connotations of the future consequences are obscured, then we might lash out or punch the person."

By exploring the role of capture in violence, I do not mean to offer an excuse for harmful or criminal conduct; intense feelings of hurt, fear, rage, or offense do not prevent most people from modulating their behavior. Many of us have ways to defuse negative feelings, or at least suppress them until a violent impulse passes. We bite our tongues. We walk away. We hit a punching bag or go for a long walk. We may act out momentarily, but we have the ability to stop ourselves before we go too far. In other words, most of us recover our balance after we lose it. Some of those who are captured, however, do not recover this balance. Capture short-circuits the brain's ability to link a complex range of emotions to actions and consequences.

When considering the role of capture in violent behavior, one of the most difficult questions to answer is why people exposed to the same stimulus can respond in dramatically different ways—some violent, others benign. One person may suffer discrimination and not only resist the urge to retaliate but also channel that anger into a meaningful pursuit; another person becomes fixated on his or her pain, until thoughts of revenge feel impossible to resist. Past exposure to similar stimuli has some role in shaping this response. Having been wronged, ridiculed, or judged by others increases an individual's sensitivity to perceived slights and injustices. This sensitivity, in turn, predisposes the individual to attentional biases, overwhelming feelings, and in certain cases, an excruciating urge to act out violently. Prior exposure to particular cues, as well as our past responses to such cues, begets the narrowed focus and obsessive thinking that lie behind some forms of aggression and violence.

Our reaction to salient stimuli is also influenced by experiences and ideas that capture us, including ethical and moral values. While one person's moral conviction may prevent him from committing a violent act, another person may come to believe that violent acts are justified.

THE ASSASSINATION OF ROBERT KENNEDY

In May 1968, CBS aired a documentary entitled *The Story of Robert Kennedy*, which records the senator's support for Israel. Kennedy's assassin, Sirhan Sirhan, would later testify that the broadcast overturned everything he'd once believed about Kennedy. Sirhan said at his trial that he had long thought Kennedy was "for the underdog and also . . . for the disadvantaged and for the scum of society, that he wanted to help the poorest people . . . and the weakest." Yet the pictures of RFK in Israel, "helping to celebrate the Israelis . . . and the establishment of the state of Israel," struck a nerve: "It just bugged me . . . burned me up, and up until that time I had loved Robert Kennedy." Galvanized by these images, Sirhan found a new and passionate sense of purpose. "I thought to fight the minute that I saw that television program."

Born in Jerusalem in 1944 to Christian parents in a predominantly Muslim area, Sirhan was forced to leave his childhood home in the Old City during the Arab-Israeli conflict in 1947–48. "The Jews kicked us out," he would later testify. Shortly thereafter he would witness the death of his brother, who was killed by a truck as it swerved to avoid sniper fire. Horrified by the daily bombings and the devastation surrounding him, Sirhan became anxious and withdrawn, developing a rigid ethical code by which he judged his own and others' behavior.

At the age of twelve, Sirhan came with his family to the United States and resolved to put the trauma of war behind him. Like many new immigrants, however, he struggled to succeed in his adoptive country. He failed in many endeavors, both in school—he was ultimately dismissed by Pasadena City College—and in a string of dead-end jobs. He became socially isolated, bitter, and angry, believing that he was better than others and nurturing an increasingly elaborate fantasy of worldly success that was completely at odds with his everyday reality. While unemployed, Sirhan read voraciously about the history of Palestine and the Arab-Israeli conflict, devouring publications from the Arab Information Center and a volume on Zionist influences on U.S. foreign policy. He came to see himself as the victim of discrimination and identified strongly with minorities and the poor.

As Kennedy edged closer to becoming his party's nominee for president, Sirhan grew angrier. A diary entry from the morning of May 18, written at 9:45, reads like an incantation: "My determination to eliminate RFK is becoming more the more of an unshakable obsession . . . RFK must die. RFK must be killed . . . Robert F. Kennedy must be assassinated before 5 June 1968."

The date marked the first anniversary of the beginning of the Six-Day War between Israel and the Arab world. To Sirhan, it signified the "beginning of the Israeli assault, the Israeli aggression against the Arab people. . . . It [evoked] in me something that I can't describe—I have the same feeling about Zionists as you do about Communism." On May 26, 1968, he came across a column syndicated in the Pasadena *Independent Star-News* by editorialist David Lawrence. The piece criticized RFK's foreign policy, pointing out that Kennedy favored "engagement" in Israel but "disengagement" in Vietnam. "Presidential candidates are out to get votes," Lawrence wrote, "and some of them do not realize their own inconsistencies. Just the other day, Sen. Robert F. Kennedy of New York made a speech in Los Angeles which certainly was received with favor by Protestant, Catholic and Jewish groups which have been staunchly supporting the cause of Israel against Egypt and the Arab countries." The next day, the newspaper printed a photograph of Robert Kennedy speaking at Temple Neveh Shalom in Portland, Oregon. The accompanying article noted Kennedy's commitment to send fighter jets to Israel.

On June 1, 1968, during a visit to Temple Isaiah in Beverly Hills, Kennedy repeated his pledge to send Israel Phantom jets. Sirhan heard reports of the speech on the radio. He became furious and disoriented. Of the senator, Sirhan said, "He bugged me to the point where I was concentrating in the mirror, instead of seeing my own face, [there] was Robert Kennedy's."

The following day, just two days before the California primary, Sirhan went to a Kennedy rally at the Ambassador Hotel in Los Angeles. When he saw Kennedy's smiling face at the rally, Sirhan's resolve briefly wavered—"He looked like a saint to me"—but the sense of betrayal soon overwhelmed this view.

On June 5, during dinner at a restaurant in Pasadena, Sirhan came across a newspaper ad for the Miracle Mile March for Israel to commemorate Israel's victory in the Six-Day War. "A fire started burning inside of me at seeing [how] Zionists, these Jews, these Israelis . . . were supporting this ad, and the fact that they had beat the hell out of the Arabs one year before."

Sirhan claims that, in his rage, he mistakenly understood that the parade was scheduled for that night. He set out to see it. A gun that he had used for target practice earlier in the day was in the backseat of his car. Stopping at a local Kennedy campaign office, he heard about a "big party" at the Ambassador Hotel and decided to follow some partygoers there. As he walked toward the hotel, with his gun still in the car, Sirhan passed a sign for "some Jewish organization," which only "boiled him up again."

Hours later, at 12:15 a.m., Robert Kennedy declared victory in the California primary. Once he had delivered his remarks, he was escorted away from the lectern to meet with press in the Colonial Room. Kennedy was ushered through the hotel's pantry area, where Sirhan lay in wait.

With a .22-caliber revolver, Sirhan shot at Senator Kennedy and into the crowd around him. Kennedy died nearly twenty-six hours later. Five others suffered bullet wounds but recovered. In Sirhan's pocket was a clipping from the Pasadena newspaper about Kennedy's support for Israel.

When the police searched Sirhan's house in Pasadena, they discovered rambling denunciations of Kennedy in Sirhan's notebooks: "Kennedy must fall. Kennedy must fall. . . . We believe that Robert F. Kennedy must be sacrificed for the cause of the poor exploited people . . . Robert Fitzgerald Kennedy must soon die."

THE COLUMBINE
SCHOOL SHOOTINGS

Psychopathy has traditionally been understood as a lack of conscience, an absence of anxiety or guilt, and a failure of empathy, but these focus only on what is absent in the mind of the psychopath. What, then, one might ask, is present?

Along with Dylan Klebold, Eric Harris murdered thirteen people at Columbine High School in April 1999. Harris deliberately left behind a trail of journals, essays, videos, and a Web browser history that allowed investigators to track his studies. His writings provide a window into the mind of an exceedingly self-conscious killer, one captured by the urge to commit violence. Eric was determined to avenge the perceived injustices of a world that consistently made him feel inferior. At the same time, he was watching himself do it, "making a statement," leaving behind a testament to his motivations and ideas.

His journal begins a year before the massacre and continues up to a couple of weeks prior to the event. The document at first has a monotone quality. It sounds like one long, continuous scream of inarticulate rage. Yet, upon closer inspection, we can discern a modulation of tone, at times, in his reactions to the world. There is also a strong element of self-awareness, as if he's looking in a mirror to monitor the role he's playing. And there are moments of searing candor, even vulnerability. This impending act of terror is his moment on the stage, his bid for recognition, and will be what he imagines as his greatest achievement. It's clear that he's writing not simply to himself, but for anyone who might find his journal after his death and write about him. He's both venting his anger and posing as a monster, showing off for posterity so that his mark on the world will seem that much more important and indelible. He's captured by the urge to commit murder as a way of asserting his power but also as a way of establishing that he's superior to everyone who has persecuted him.

As he says on December 17, 1998, when he was seventeen years old, "I wonder if anyone will write a book on me, sure is a ton of symbolism, double meanings, themes, appearance vs. reality shit going on here, oh well, it better be fuckin good if it is writtin." It sounds like a badly phrased gloss on something he's reading for his English class, and it also suggests that he's playing with his reader's mind, manipulating what he presents in order to construct a persona people will respect and remember. The reader soon sees that Harris's admission that he's a compulsive liar is compounded by his self-conscious desire for fame. His ravings in sympathy with Nazi Germany, his phrases in German to emphasize this, are an embrace of white supremacy but also a mask he wears to provoke the reader's outrage and disgust and fear. Mostly, it makes him appear more formidable, which is the goal. The most prominent dimension of his journal is the engine of extreme emotion in every word. There is no lack of affect here. It pulses through every syllable.

Throughout the journals, Harris poses as someone who has a grudge against his peer group, who resents the social injustices of a high school culture that has excluded him. Yet he widens the significance of his revenge by making it a gesture of more than vengeance; he turns it into a protest against the inauthenticity of human society. He extrapolates from his personal pain an entire philosophy to justify murder. It's a common element of the capture that governs the nihilistic mood of the killers in these stories— how personal pain begets an emotional repulsion at everything. Just as Ted Kaczynski came to see the slightest telltale evidence of modern technological society as evil, Eric turns his anger over a personal wound into an indictment of human existence. Grandiosity, that black mirror reflection of deep-seated insecurity, is a potent delusion.

The recurrent spasms of revulsion in this yearlong account of Harris's rage recur like building waves of nausea. He begins in April 1998 by condemning everyone and everything, dismissing compassion in favor of social Darwinism: anyone who can't carry his own weight deserves to die. It's a generalized condemnation, not yet focused on anyone in particular. Along with this condemnation comes a feeling of immense power. He is free now to do whatever he wants: "I know what all you fuckers are thinking and what to do to piss you off and make you feel bad . . . I feel like GOD and I wish I was having everyone being OFFICIALLY lower than me. I already know that I am higher than almost anyone in the fucking welt." He has asserted his supremacy, but not yet committed himself to killing classmates. He ratchets up his rage month by month by fixating on cues that arouse his sense of being excluded, left behind by the more popular and socially adept students around him.

The next month, his school yearbook arrives, and he reacts to the gallery of happy kids by dismissing popularity as a way of melting into the herd. The rant that runs throughout his journal recurs: the rest of the world is full of imitators and cowering conformists; only Harris has the strength to recognize society as it is. Therefore, what sets him apart is a virtue rather than a failing. "It has been confirmed, after getting my yearbook and watching people . . . the human race isn't worth fighting for, only worth killing . . . nothing means anything [any]more . . . I don't want to be like you or anyone which is almost impossible this day w/ all the little shits trying to be 'original-copycats,' I expect shits like you to criticize anyone who isn't one of your social words; 'normal' or 'civilized' . . ."

Two months later, Harris begins to write specifically about murder. Again, it isn't just revenge; he imagines himself a lone Nietzschean figure who has overcome the world, taking responsibility for his actions and fully identifying with his absolute, amoral freedom. He boasts of how his crimes will define him: "I know I could get shot by a cop after only killing a single person, but . . . I chose to kill that one person so get over it! It's MY fault! Not my parents, not my brothers, not my friends, not my favorite bands, not computer games, not the media. IT is MINE!"

With the passing months, Eric's hunger to kill grows more and more encompassing. As the journal inches closer to the date of the killing, he writes about turning off his feelings in order to kill all but "about 5 people." By October, his plan has escalated from guns to bombs, and he imagines "half of denver on fire . . . napalm on sides of skyscrapers and car garages blowing up . . ."

Eric tries to suppress any feelings of disgust at his increasingly lurid fantasies. He resolves not to "be sidetracked by my feelings of sympathy, mercy, or any of that. I will force myself to believe that everyone is just another monster from Doom like FH or FS or demons, so it's either me or them. I have to turn off my feelings," he writes. Before carrying out his horrific plan, Eric apologizes to his mother: "I am really sorry about this, but war's war." He adds, on a jarringly affectionate note, "my mother, she's so thoughtful. She's helping out in so many ways." He needs, he realizes, to distance himself from his parents. "I don't want to spend any more time with them. . . . I wish they were out of town so I didn't have to look at them and bond more."

In November, he writes, "If I could nuke the world I would," yet this is balanced by the most humanly recognizable expression of emotional pain in the journals. It's his most vulnerable admission of how all this rage originates in the hallways of his high school: "Everyone is always making fun of me because of how I look, how fucking weak I am and shit, well I will get you all back: ultimate fucking revenge here, you people could have shown more respect, treated me better . . . treated me more like a senior and maybe I wouldn't have been as ready to tear your fucking heads off. That's where a lot of my hate grows from. The fact that I have practically no self-esteem, especially concerning girls and looks and such . . ."

From there, it's a direct line to the killing in April 1999.

His final entry, only days before the murders, is heartrending for how much he sounds like an ordinary ostracized kid, a putative loser who just doesn't have what it takes to be popular: "I hate you people for leaving me out of so many fun things . . . no, no, no, don't let the weird looking Eric KID come along, ohh fucking noooo."

Eric Harris went from shunned teenager to mass murderer. His inability to handle his torment made the high school he shunned a household word.

THE MURDER OF JOHN LENNON

It took patience to get a glimpse of John Lennon. An entire day could pass without any sign of his emerging from the Dakota, the nineteenth-century luxury apartment building overlooking Central Park where he lived with his wife, Yoko Ono, and their son. But Mark David Chapman was on a mission, and nothing—not the long wait or the chilly December air—would force him to abandon his watch. Now and then he fingered the gun in his pocket. Under his arm, as a ploy to look like a fan waiting for an autograph, he held Lennon's latest album, *Double Fantasy*; in his pocket he carried a copy of *The Catcher in the Rye*. His entire life, he later told journalist Jack Jones, had reached its culmination in this very moment. He was consumed by one goal: "to kill the phony."

He hadn't always hated Lennon. As a child, Chapman, like so many members of his generation, had idolized the Beatles. He had even staged impromptu concerts in his garage, where he'd lip-synch "She Loves You" for an audience of neighborhood kids. Chapman identified most with Lennon, of all the Beatles, because he and John both had troubled upbringings. Like Lennon, Chapman adored his mother, and although his father hadn't abandoned him as Lennon's had, the man was so emotionally distant—and, on occasion, violent— that in Chapman's view, he might as well have left. Lennon's lyrics revealed to the world a complex young man, both deeply sensitive and blisteringly angry. Chapman saw himself as similarly in conflict: brimming over with compassion for others, yet full of rage. He rarely lashed out physically, though. Only once had he been so angry that he grabbed a kitchen knife and lifted it toward his father, but the larger man easily disarmed his son.

Chapman flirted with drugs as a teenager, but this phase passed quickly. Soon he had discovered a new fixation. Growing up in Georgia, he had a number of religious acquaintances, and after spending some time at a youth group retreat, he decided to become a Christian himself. Born-again Chapman denounced the Beatles as evil as zealously as he'd once worshipped them because in his mind they pushed an anti-Christian agenda. He told his best friend's sister he didn't like the band anymore "because John Lennon had said that they were more popular than God." Around this same time he read *The Catcher in the Rye*, identifying with its antihero, Holden Caulfield, to an extent that would grow delusional as the years passed.

Chapman, in his last years of high school, found work as a YMCA counselor, a job at which he excelled. In 1975, the organization sent him to Lebanon. Because of the ongoing civil war there, the Lebanese Y was forced to close, so Chapman accepted an offer to transfer to a YMCA-run refugee camp in Fort Chaffee, Arkansas. There he became instrumental in processing tens of thousands of "boat people" escaping to the United States after the fall of South Vietnam.

It was as if he'd found his calling. He worked sixteen hours a day with the refugee children; he started a band and a softball team for them. On an official visit to the refugee camp, President Gerald Ford congratulated Chapman on his stellar performance. "I just could never say enough good about the Mark Chapman I knew for those five or six years in Georgia," his former supervisor later said.

But then the camp closed, and Chapman left to join his fiancée, Jessica Blankenship, a childhood sweetheart, at a Christian college in rural Tennessee. The routine life of a college student sent Chapman into a decline. "With the YMCA in Lebanon . . . I had been a somebody. . . . [Afterward] I rose to an even greater position of importance at Fort Chaffee. Then, when the job ended . . . I became a nobody. I was a regular college student with regular responsibilities and that was it. That was all," Chapman told Jones. A course on the history of war particularly depressed him—it taught him that "all of human history was nothing but a history of great battles," as he recalled for Jones, "and all those monuments and books were created just to celebrate death." His darkening mood hindered his ability to study, and he began to think of himself as inferior to his classmates. He left at the end of one semester and returned to the YMCA at Fort Chaffee, hoping to re-create his best days. But there were no boat people this time, no children eager to play with him—just a few spoiled campers and weary counselors. Disappointed, he returned to Atlanta and found work as a night-shift security guard—a job that brought long hours of solitude. Being a security guard was a job for a nobody like him, he thought. Spending so much time in isolation from others only compounded his sense of worthlessness. He talked increasingly of suicide. Faced with worrisome changes in her fiancé's personality, Jessica broke off the engagement.

Rejected and despondent, Chapman booked a spur-of-the-moment trip to Hawaii, a destination he'd long fantasized about visiting. His idea was to have one final adventure before he killed himself—but his last hurrah evolved into an extended party as he lounged at the upscale Moana Resort, drinking and fraternizing with other tourists. By the spring of 1977, the sheen of island life finally wore off. Chapman drove to the beach, affixed a hose to the tailpipe of his car and fed it into the window, and turned on his car's ignition, hoping to asphyxiate himself. But a fisherman rapped on the window the next morning, and Chapman awoke, startled to find himself alive. The hose he'd attached to the tailpipe had melted, saving his life. Chapman believed God had intervened, and he resolved at that moment to live. Later that day, he walked into a mental health clinic in Honolulu and told a counselor what he'd done. She immediately drove him to Castle Hospital, where he was admitted to the psychiatric unit. He quickly became the most popular patient on the unit, chatting with the staff, playing his guitar, and even becoming a kind of ersatz counselor to the other patients. The staff liked him so much, in fact, that they offered him a maintenance job at the hospital after his release. By 1978, Chapman seemed to have gotten his life on track. He had a close-knit circle of friends at work who regarded him as one of the family, and he could afford a nice apartment in Honolulu. He had a girlfriend, Gloria Abe, whom he married a year into their courtship in June 1979.

Then, just as quickly as things had come together, they began to fall apart. Chapman had been promoted to a new job in public relations and printing. He didn't mind the nature of the work, but the solitude of the hospital's print shop wasn't good for him—Chapman thrived on the attention of others. He was alone for long stretches of the day. His depression sparked his anger and led to disputes with the coworkers he had considered friends. Eventually, he was asked to resign.

Chapman's behavior became increasingly erratic after he was let go from his job—he went on spending sprees and alcohol-fueled benders. He isolated himself at home or holed up in the local library, rereading *The Catcher in the Rye* obsessively. It was as if Holden Caulfield was voicing Chapman's disgust at the world's "phoniness." The line between Mark David Chapman and J. D. Salinger's bitter young man began to blur to the extent that Chapman told his wife he was considering changing his name to Holden. At the library, he also discovered a coffee-table book entitled *John Lennon: One Day at a Time*. A photograph in the book hit Chapman hard—Lennon on the roof of his famous building, smugly posing for the camera. He had praised the virtue of poverty in his lyrics, but he had become one of the idle rich. Lennon was so egomaniacal he'd had the gall to compare the Beatles to Jesus Christ!

"He told us to imagine no possessions, and there he was, with millions of dollars and yachts and farms and country estates, laughing at people like me who had believed the lies and bought the records and built a big part of our lives around his music."

Chapman's outrage—what he deemed "the tornado"—started pulling him in. He checked the book out of the library and brought it home, where he could turn its pages while listening to his wife's Beatles records. He worked himself into a frenzy thinking about Lennon the liar, Lennon the big shot, Lennon the popular rock star. And then, Chapman had a revelation. "It was almost like I was handed something . . . here was the solution: *Kill John Lennon.*" Lennon's death was a fait accompli at that moment, the historic act that would tie together the loose threads of Chapman's life. "After [the revelation] happened, there was no power on earth that could save John Lennon's life."

And now here he was, less than two months later, a marksman casing his target, standing in front of Lennon's apartment building. When the limo pulled up, he knew who was inside. He'd had his first chance earlier in the day, but he'd been so starstruck he'd bungled it. Now he was ready. Lennon and his wife got out of the car and Chapman moved in. As the famous couple walked through the archway that led to the building's inner courtyard, Chapman aimed his revolver and fired five times into Lennon's back.

Chaos ensued, but Chapman froze in place. He had nowhere to go now. Part of him had believed that when he succeeded in killing Lennon, he would finally become Holden Caulfield, and he was dazed when that didn't happen. The doorman screamed at him to leave. But he was done. Chapman let the doorman kick the gun away. He paced, staring at the pages of *The Catcher in the Rye*, and waited for the police to arrive. When they did, he took off his coat to show them he had no weapons. He wanted the police to arrest him, not shoot him.

THE MURDERS AT SANDY HOOK ELEMENTARY SCHOOL

Hour after hour, Adam Lanza kept dancing. He lost himself in the energetic music and flashing images of *DanceDanceRevolution*, a heart-pumping interactive video game that he played at home and at an arcade nearby in Newtown, Connecticut. An American version of a Japanese game, it prompted players to execute dance moves, directing them to strike their feet in various places on a floor pad, over and over and over. Adam would play almost without pause for as many as ten frenetic hours a day, sweat pouring down his body. He'd created his own private rave.

This trance quelled his anxieties and resentments, and he sought out this sheltering absorption for almost a decade, which was also almost half his life.

On December 14, 2012, he shot his way into the Sandy Hook Elementary School with a Bushmaster model XM15-E2S rifle. Within a matter of minutes, six adults and twenty first-grade children lay dead. He then turned a Glock 20 10-millimeter pistol on himself, ending his tormented struggle at the age of twenty. The body of his mother, Nancy Lanza, was discovered shortly afterward, in the two-story colonial house they shared, where she had been killed with a Savage Mark II rifle.

The mass murder stunned a nation already confounded by similar tragedies at Columbine and Virginia Tech. In response to that collective public anguish, local, state, and federal law enforcement authorities poured thousands of hours into a subsequent investigation, as if they were on a massive manhunt for the actual culprit. In a way, they were. They wanted to know why.

They gathered physical evidence and sorted reports, statements, interviews, lab tests, photographs, and search warrants into seven hundred files. They calculated how much ammunition Lanza had carried with him on that terrible day—more than five hundred rounds, weighing about fourteen pounds. Their records include photos of the Lanza home and its affluent setting, lists of the video games found in the basement (*Left for Dead*, *Doom*, *Team Fortress*, and *Dynasty Warriors* among them), and mention of the Christmas check Adam's mother had given him to purchase a new pistol.

The Connecticut State Office of the Child Advocate joined in the vast information-gathering effort, seeking opportunities to improve the public health system so that it could better respond to the red flags in the behavior of kids like Lanza. Coupled with media investigations, the documentation about the Newtown massacre came to seem encyclopedic, yet the murder defied understanding. Something had taken hold of Adam Lanza that grew into a single-minded fixation on adding his mark to the history of mass killings. Adam knew what he was doing that wintry morning and had planned it well in advance, in deliberate detail, and with great care. No one can fully explain why.

Those who knew the young man recognized that he was deeply troubled. He "presented with significant developmental challenges from earliest childhood," according to the Office of the Child Advocate's report, "including communication and sensory difficulties, socialization delays, and repetitive behaviors." Adam's parents, who were divorced, at first misunderstood the extent of his difficulties. At various times, they asked for help from school officials and the mental health system, ignored warning signs, changed his surroundings in the hope of changing his behavior, tried to impose discipline, and gave in to his demands, thinking that capitulation might curb his tantrums. None of it worked.

At the age of thirteen, Adam was diagnosed with Asperger's syndrome, an autism spectrum disorder that impairs social interaction. Common features include extreme adherence to repetitive routine, an inability to interact appropriately with peers, and hypersensitivity to slights or rejection. A tendency to fixate on minute details leads many people with Asperger's to dedicate vast amounts of time to a single potent interest. Some people learn everything there is to know about the Civil War or trains or the world's tallest mountain ranges. Some become consumed with firearms and war. Adam's refuge became mass murder.

For many years, experts believed that people with Asperger's syndrome lacked empathy and were repelled by close human connections, but that theory has shifted in recent years to its polar opposite. Current thinking now holds that those with the diagnosis are so extremely sensitive to their environment that they cannot cope with the powerful tide of emotion it evokes. They shrink from touch and are easily overwhelmed by sensory experiences because they cannot filter any of them out. Barraged by color, sight, and sound, they withdraw in order to cope.

Adam received only intermittent supportive services over the years. School records tell the story of his childhood development. Early in elementary school, he was given speech and occupational therapy as part of a special education plan, but they were discontinued by fourth grade, when his academic and social performance was deemed age appropriate.

In 2002, when he was in fifth grade, Adam produced the "Big Book of Granny" as part of a class project. The spiral-bound book features a character with a gun in her cane that she uses to shoot people. It includes a section labeled "Granny's Clubhouse of Happy Children," which contains dialogue for an imaginary television show. In one episode, Granny punches a young boy, sets off an explosion, and threatens to kill a group of children. In another, a character describes a game called "Hide and Go Die." It was a book of extreme violence, according to the Child Advocate's report, "not the sort of creation that most children would even know to invent."

By the age of ten, Adam was already socially withdrawn and fearful. A report by the Office of the State's Attorney in Connecticut indicated that "he did not think highly of himself and believed everyone else in the world deserved more than he did." He had also become so compulsive about washing his hands that he developed an extreme skin irritation.

The hints of psychological distress in his elementary school years became pronounced in middle school. As a seventh-grader in a new setting, he was expected to function more independently and to forge social connections on his own. Yet, at Newtown Middle

School, Adam seemed terrified by stimuli as common and ordinary as the noise and confusion of hallway traffic. He clung to the walls, avoiding eye contact, making his way from math to social studies.

In this new environment, "his social, emotional and communicative struggles appeared to have become increasingly intense." The frantic outbursts that had long been a feature of his behavior at home took on new rancor as well, and eight months into the school year, his mother moved him to a local Catholic school, evidently hoping he would benefit from more personal attention and the school's more structured and uniform routines.

The transition did nothing to help him, and his violent thoughts became more frequent. Responding to an assignment to write a page or two on any topic, Adam turned in ten pages so graphic in their depiction of battles, destruction, and war that a teacher brought them to the principal's attention. The twelve-year-old also drew scenes of death and printed out images of violence from his computer. He began to suffer panic attacks. In June 2005, just eight weeks after he had enrolled, Adam's mother pulled him from his latest school.

Over the next year, homebound, Adam never received from the school district the tutoring and other support services to which he was entitled. Either the school failed to follow through on its obligations or his mother declined the services. Inside his bubble, Adam grew more anxious and socially isolated, and he never returned to an eighth-grade classroom. Symptoms of obsessive-compulsive behavior became more apparent as he worried about dirt and contamination, developed extreme food rituals, and refused to touch doorknobs.

Adam visited at least two mental health providers during this time, but was very reluctant to take the recommended medication. An expert in developmental delays at the Yale Child Study Center emphasized to Adam how important drug therapy was, telling him he was "living in a box right now, and the box will only get smaller over time" unless he had further treatment. The family did not insist that Adam comply with the doctors' suggested treatments, accepting his argument that the drugs caused all sorts of unpleasant side effects. Ultimately, Adam and his mother parted ways with Yale.

In ninth grade, Adam was able to ease back into public school classes, although he was spending much of his at-home time behind a closed bedroom door, playing his video games. Along with *Dance-DanceRevolution*, *World of Warcraft* was a particular favorite. School staff communicated with his mother regularly, and they formed a sort of support network for Adam. These signs of improvement were followed by something of a breakthrough at the start of tenth grade, when he joined the Technology Club.

Those with Asperger's syndrome often have an affinity for technology, perhaps because computers can harness the hyperfocus and attention to detail that are defining elements of the disorder—and because electronic technology is designed to be entirely predictable, a tool under the user's total control. In any case, Adam participated actively in the club and made a small number of social connections there. He developed a good relationship with the faculty adviser, who kept a protective watch on him, and even held a party for club members at the Lanza house.

But the panic attacks Adam had suffered over the years never diminished, and he did not return to mainstream high school classes in his junior year, instead receiving individual tutoring. A year later, he was able to enroll in a community college and graduated early. By 2009, at age seventeen, now an avid gamer, he had assumed the online persona of "Kaynbred" and was increasingly pulled into a fantasy world. His computer logs show more than five hundred hours playing *Combat Arms*, where players kill other players, rise through military ranks, and earn Gear Points, which can be used to purchase electronic weapons and accessories. It was little more than a standard online shooter game of the sort played by millions now, but it's hard not to imagine that, for Lanza, it may have felt like boot camp.

Adam also joined a gun enthusiasts' message board, where he shared detailed information about weaponry, ballistics, and gun laws. He bookmarked hundreds of websites about firearms, killing sprees, and mass murders, filed away a video dramatization of children being shot, and collected electronic images of himself holding various guns to his head. His basement game room at home was decorated with war and weaponry posters, and he created an indoor shooting range where he could use a pellet gun.

His emerging obsession with mass killings intensified as Adam studied shootings that had occurred in schools, workplaces, shopping malls, a teen prayer rally, and a cafeteria. The shootings at Columbine High School especially fascinated him, and he had a complete copy of the Columbine investigative report, along with several video clips related to the young gunmen. He collected surveillance videos from the Westroads Mall shooting in Nebraska and the Stockton schoolyard shooting in Cleveland. In his bedroom, he kept a photocopied newspaper article describing a school shooting in 1981, and a copy of *Amish Grace: How Forgiveness Transcended Tragedy*, a book about the 2006 murder of five young Amish girls in a one-room schoolhouse. On the Wikipedia website, where contributors can edit and alter entries, Adam corrected numerous details about mass murders. To track all past atrocities, he created a seven-by-four-foot chart, a spreadsheet of sorts, with the names, number of victims, and types of weapons used.

In the last years of his life, Adam's field of attention narrowed to his frequent immersion into *DanceDanceRevolution* and exchanges with an online community that shared his obsession with mass murders. As his mind shrank to a single focus, his body shriveled. He became sickly; his weight dropped to 112 pounds on a six-foot frame. His temper tantrums, always ferocious, grew more frequent, and he became increasingly withdrawn.

By 2010, Adam had quit communicating with his father, and his mother had grown desperate at her inability to help him. She continued trying to appease her son, washing his laundry daily, arranging the food on his plate the way he preferred it, serving it on the dishware he specified. But she also began spending more time away from home, often staying late at a local bar and restaurant. If she was aware of his cyber life, she did not talk to anyone about it, although she did tell a few friends that she feared Adam didn't care about her.

As he became more and more immersed in past horrors and the violent fantasies of cyberspace, the real world essentially became imaginary, like the setting of a game. Sequestered in a bedroom whose windows were taped over with black trash bags, he had disassociated from the world.

Yet he was sufficiently in touch with reality to plan one of the most horrific acts in American history. He plotted meticulously, studying security procedures at Sandy Hook Elementary School and examining its student handbook. Three days before the shootings, he had e-mail discussions about the "aesthetics" of killing. "The inexplicable mystery to me isn't how there are massacres, but rather how there aren't 100,000 of them every year," he wrote. "While granting that modus operandi really isn't that important, I just can't get into vehicular slaughterers. It seem[s] too mediated, like using remote explosives (too hot). And knives stray too far from the whole 'mass' aspect (too cold). The aesthetic of pistols tends to be just right." It's a chilling appraisal of weapons as if they were customizable adjustments to the flow of a video game.

Shortly before heading to the elementary school, Adam killed his mother in her bed. Over the preceding few months, he had become as disconnected from her as he had from the rest of the world. Killing her was, in a sense, an act of suicide: she was his only genuine lifeline. There was no turning back.

Adam Lanza, Mark David Chapman, Eric Harris, Sirhan Sirhan, and Ted Kaczynski were all captured by a swirling sense of humiliation, injustice, and worthlessness. As a result, they became hypervigilant—to perceived wrongs, to threats in their environments, to intrusive thoughts, or merely to annoying traits in others. Each of these men turned to violence to be free from the grip of obsession, to regain a sense of control over an otherwise disappointing, confusing, or painful reality.

A THEORY OF HUMAN CAPITAL

James Holmes thought he had a "broken brain." That was the phrase he first used as a young teenager, and that's what he scrawled, many years later, in a notebook he began keeping in June 2012. All those years, he struggled mightily to fix that broken brain, searching for insights in neuroscience, seeking out psychiatric care, and committing his analytical thinking to paper.

None of it worked. "That's my mind," Holmes wrote. "It is broken. I tried to fix it. I made it my sole conviction but using something that's broken to fix itself proved insurmountable." In the end, Holmes saw only one way out: "Life's fallback solution to all problems—Death."

On July 19, 2012, the twenty-four-year-old man mailed his notebook, with its mix of incoherent ramblings, clear commentary, and cold logic, to his psychiatrist, Dr. Lynne Fenton. A few hours later, at the midnight premiere of the Batman film *The Dark Knight Rises*, he tossed a canister of tear gas into the audience, and began shooting. Twelve people died in the Aurora, Colorado, attack, and seventy others were injured. Holmes called the injured "collateral damage." To him, fatalities were all that counted.

At the trial that took place almost three years later, four court-appointed psychiatrists agreed that James Holmes fell somewhere on the schizophrenic spectrum. But when "Jimmy" was a little boy in Castroville, California, no one had thought he had a broken brain. Former neighbors described him as sweet, bright, mellow, and fun to be around. Something changed around the age of eleven, when his family moved to San Diego: thoughts of suicide entered his mind. In his middle school years, Holmes made what he called a "para-suicide" attempt, slashing his wrist with cardboard with enough force to break the skin. He told his parents that it was a paper cut, because he thought he would be punished for showing any signs of weakness.

Over the next few years, he grew more socially isolated and withdrawn, preferring to stay indoors, sleep long hours, and play video games rather than socialize. Although he joined a track team, his coach called him a "shadow figure" who had to be told sternly to be part of a group photograph.

Yet Holmes was bright, finishing high school with a 4.0 grade-point average. And his social anxieties did not extend to his home life, where he talked comfortably with his parents and sister, shared hugs, and had "loving" relationships.

On the surface, his college experiences seemed relatively unremarkable. He attended the University of California, Riverside, and lived in an honors dormitory. Classmates remembered him as very shy, never initiating social interaction, but a willing participant when they drew him out. He earned his undergraduate degree in neurobiology with a 3.9 GPA.

For most of those years, Holmes was keeping a lot hidden. After the move to San Diego, he participated briefly in family therapy with a licensed clinical social worker, but wrote that he "revealed nothing as to not appear weak amongst family." Believing that his mind was not like that of his peers, by age fourteen he took an interest in neuroscience and wanted to learn how ordinary brains worked and whether his could be repaired. In high school, random, violent thoughts began to haunt him; he imagined people having their heads cut off.

In 2011, Holmes was offered one of six coveted spots in the graduate neuroscience program at the University of Colorado, outshining most of the seventy other applicants. On paper (college transcripts, test scores, letters of recommendation, research experience) Holmes looked impressive. He never volunteered answers in graduate seminars but usually knew them when called upon. He completed his assignments on time and earned an A on one midterm. Still, he remained a loner, outside the group even when he was part of one. "When I'm around people I'm kind of anxious," he said. "I know they expect you to say something." Holmes often froze when forced into conversation, seemingly unable to get words out of his mouth. One professor recalled him as a poor communicator who could not interact with others, a particular obstacle to effective lab work, and several of his professors felt that he was insufficiently motivated.

Christmas of that year seemed to mark the beginning of a downward spiral. "My mind was kind of falling apart," Holmes recalled. He shared nothing of his deteriorating mental state with his parents or sister. "I didn't want my family to know that I was sick," he said. Ever since adolescence, he had believed that such an admission would be "bad" and meant "that you're a burden." Nonetheless, he sustained an intimate relationship with a girlfriend, Gargi Datta, for some months. On Valentine's Day 2012, Holmes cooked her an onion-crusted chicken, serving ice cream for dessert. They lit candles and watched *Saturday Night Live* and music videos. But they broke up not long afterward, only intensifying Holmes's already "depressed" and "pessimistic" outlook. Increasingly, he seemed to lack the drive to finish anything.

He continued to have intrusive thoughts of suicide and homicide, which began to narrow to a single focus. Mass murder, what he called "the last escape," was becoming increasingly salient. Holmes became persuaded that the value of his victims' lives would somehow accrue to him. His theory held that every individual was worth one point. "If you attribute value to killing people, you become more valuable if you [kill]," he explained.

Holmes, it seemed, had journeyed from obsession to delusion: from "intrusive thoughts, images or impulses [that] are experienced as excessive, unreasonable, and thus distressful" to "false beliefs based on incorrect inference about external reality that are firmly sustained despite what almost everyone else believes."

For the first time, Holmes sought psychiatric help, arriving at the office of Lynne Fenton on March 21. At that first meeting, according to Fenton's notes, Holmes said his obsessive-compulsive symptoms had worsened and that he saw homicide as the only solution to his suffering. At the same time, he remarked, "You can't kill everyone, so that's not an effective solution." Fenton took his comment as a sign that he was unlikely to act out his fantasies. She prescribed clonazepam, used to treat panic disorders, and sertraline, a selective serotonin reuptake inhibitor designed to treat depression, social anxiety, and obsessive-compulsive disorders.

Holmes agreed to a follow-up appointment a week later. Meanwhile, he shared some of his violent thoughts with Datta, writing in a Gmail chat that he felt like doing something evil. Asked what that would be, he responded, "Kill people of course."

Datta suggested it was not worth the trouble. "Most people are not worth what might happen to you coz of the attack."

"That's why you kill many people," Holmes declared.

During the next several months, he met regularly with Fenton and told her he was having thoughts of homicide three to four times a day. Although she believed he had a "schizoid personality disorder" and "paranoid delusions," Fenton did not perceive that his thinking had changed significantly over time, and she did not think he was an imminent threat, partly because he never described a specific plan. She adjusted his medications, increasing his dose of sertraline, and brought in another psychiatrist for a second opinion. In early June, Holmes saw Fenton for the last time and told her that he had failed his oral exams and was withdrawing from graduate school. Soon after, she contacted the university's threat assessment team to ask for a background check. The team reported that Holmes had not been in the military and did not have a concealed weapons permit. Fenton's notes indicate "a clean bill of health."

Meanwhile, he was stockpiling an arsenal. Holmes made his first purchases (a Smith and Wesson folding knife and a stun gun that resembled a cell phone) on Amazon. He then acquired a .40-caliber Glock 22 at the Gander Mountain store in Aurora and paid for a shotgun at Bass Pro Shops. Purchases seemed to drive more purchases—"going from compulsion to compulsion," he said. Soon enough, Holmes had added a Glock 23, an assault rifle, laser sights, explosives, tear gas grenades, a gas mask, handcuffs, and thousands of rounds of ammunition to his cache. After researching firearms laws and mental illness, Holmes had expected that his psychiatric encounters might prevent him from buying weapons, but he met with no such barriers. He bought a full suit of body armor and photographed himself wearing it, a firearm at his side. He started target practice at a local gun range.

"It just kept escalating," Holmes said. His intent in buying the first pistol had been self-defense, but "then it went towards other purposes. Offensively."

Holmes recounted that he had "transferred my suicidal thoughts into homicidal," as he clarified his goals and defined his mission: "to go to the theater and shoot as many people as possible." He was no longer thinking unrealistically about the destruction of mankind, an image fueled by the generalized discomfort he always felt around people. No longer was his mission to obliterate the world with nuclear bombs or biological agents "that destroy the mind." Instead, it had become a set of "realistic thoughts about a group of people getting killed."

In the end, Holmes was driven by what he called his "theory of human capital": the idea that taking someone else's life would increase the value of his own. "It just took hold of my mind," he explained. "I don't know why it dominated. But it did." Despite his "hatred of mankind," he was not spurred forward by blinding emotion. "To me hatred is kind of like hating broccoli or something. Not a fiery, angry passionate hate." Rather, he acted because he believed he had no choice but to execute his all-consuming mission: "I was almost, like, catatonic. I didn't have any drive to do anything other than the mission. . . . I had a purpose to fulfill."

The psychiatric drugs prescribed by Fenton may have intensified his responses. Soon after starting on medication, he reported that the "first appearance of mania occurs, not good mania. Anxiety and fear disappears. No more fear, no more fear of failure. . . . No fear of consequences. . . . Intense aversion to people, cause unknown. . . . Love gone, motivation directed to hate and obsessions. . . . No consequences, no fear, alone, isolated, no work for distractions, no reason to seek self-actualization. Embraced the hatred, a dark knight rises."

It remains unclear whether Holmes truly wanted to act on his delusion. There were always two sides to his story. One was about a young man desperate to get well, who had sought out mental health services, kept every appointment, and later told a court-appointed psychiatrist that he wanted someone to stop him from the shooting: "I wanted to be fixed. To be normal."

Holmes had labeled a page of his notebook "self-diagnosis of a broken mind," and presented a bulleted list of thirteen illnesses, including dysphoric mania, generalized anxiety disorder, Asperger's syndrome/autism, schizophrenia, obsessive-compulsive personality disorder, psychosis, and chronic insomnia. In five more pages, he had detailed the many symptoms attributed to those diagnoses, from catatonia, excessive fatigue, self-isolation, hair-pulling, and an inability to communicate to "brief periods of invincibility," "concern with cock . . . excessive stimulation," and "the obsession to kill."

He had also searched for alternatives to murder, even as the possibility assumed an ever more prominent role in his thinking. Among the options he considered, and their outcomes: "Ignore the problem. If the problem or question doesn't exist, then the solution is irrelevant. Didn't work . . . Delay the problem. Live in the moment without concern for answering the problem at present. Didn't work . . . Pawn the problem. If one can't answer the question themselves, get someone else to answer it. Didn't work."

The other side of Holmes's story was of a man who meticulously planned carnage. A man who caused grief and pain. He knew his actions were illegal, and he knew they were wrong.

As Holmes collected weapons, he thought carefully about the most suitable method and venue. Bombs wouldn't work, he wrote, because they are "too regulated" and "suspicious"; he put an X by that idea. Serial murder, too, was ruled out: "too personal, too much evidence, easily caught, few kills." Another X. Finally, there was a check mark on the words "mass murder/spree," which had the advantages of "maximum casualties, easily performed with firearms," despite being "primitive in nature."

He briefly weighed an airport as the site of his shooting, but rejected the idea, partly because of the substantial security in the terminals and the legacy of terrorism associated with the setting. "Terrorism isn't the message. The message is, there is no message," Holmes wrote.

So he decided on the sixteen-screen Century Aurora Multiplex, which was "isolated, proximal and large." Where better to "case the place" than "an inconspicuous entertainment facility"? Shortly after he dyed his hair a shocking reddish-orange in late June, Holmes began taking photos outside the movie house. He purchased tickets there three times between July 7 and 17. He devoted four pages of his notebook to sketches of various theaters within the complex and made notes about the pros and cons of each layout. For example, the ability to lock the double doors in one theater provided the opportunity for "increasing casualties," while another had the disadvantage of offering many escape routes to the audience. He also created a profile on AdultFriendFinder, an online sex hookup site, and updated it on July 18 with the headline "Will you visit me in prison?"

After months of planning, with the theater selected and his weaponry assembled, Holmes was ready. He rigged his apartment with incendiary devices and set a timer on his computer for midnight so that techno music would blare at the highest possible volume, which he hoped would summon and distract law enforcement authorities. After driving to the multiplex, Holmes entered Theatre 9, where Batman was on the screen, but stepped out shortly afterward, pretending to answer a cell phone call. Leaving the rear exit door propped open behind him, he returned to his car to don his protective armor and gather his weapons. Before heading back inside, he made one last call, to the University of Colorado mental health hotline, but hung up without talking to anyone. "Last chance to turn back," he explained. "Doubts I guess."

Eighteen minutes after the movie began, 911 dispatchers received their first phone call. Screams and continued gunfire could be heard in the background. As blood poured down the aisles, Holmes's gun jammed, and he finally stopped shooting. He walked at a normal pace through the emergency exit, rifle in hand, pausing to glance back and take notice of the survivors still crouched on the floor. He thought one man smiled at him as he departed.

Holmes had no plans for the aftermath. "I would go to jail or I would die. Things would take their own course. There was the third possibility of getting away but I didn't consider that." Still dressed in his protective gear, he stood by his car until the police arrived. "I considered the mission over," he said.

On July 16, 2015, a twelve-person jury convicted James Holmes on 165 counts of murder and attempted murder, rejecting the defense's insanity plea. Three weeks later, the jurors failed to reach the unanimous decision required to impose the death penalty. Nine favored it, two were uncertain, and one was adamantly opposed. Holmes was sentenced instead to life in prison without the possibility of parole.

6

CAPTURE
AND IDEOLOGY

Acts of terror are perpetrated by people who are captured by an idea. At some point in their development, terrorists often become enthralled by the belief that they are fighting for a cause larger than they are, a truth that transcends the self. Such ideologies, whether political or religious in nature, are all-encompassing systems of belief, potent stories that render a frustrating and complicated world seductively simple.

To the faithful, ideologies promise a kind of immortality, beyond the gates of heaven or in the chronicles of world history. In the case of radical Islamism, this promise is made all the more powerful by a profound sense of political injustice. Since the dawn of colonialism, devout Muslims have been under attack, economically and politically oppressed, or forbidden from living in accordance with their own laws and customs. Part of the appeal of radical Islamism is the promise of a return to an idealized past, a golden age when the faithful lived in harmony with God and with one another.

But al-Qaeda and the Islamic State of Iraq and al-Sham, or ISIS, are by no means the only groups to yoke utopia to violence. By their very definition, ideological narratives such as Hitler's theory of Aryan supremacy, the political paranoia of Stalinism, and the genocidal fervor of the Khmer Rouge promise to deliver the faithful from the confusing, complicated, ambiguous everyday if only they pledge themselves to fighting the oppressor. It is a seductive and powerful promise, one that preys on the individual's desire to escape the confines of the self, all in the name of a greater path, to connect to something more meaningful.

The most horrific acts of violence or terrorism are, then, not merely expressions of sadism or depravity. Often, at the heart of seemingly inexplicable violence lies capture, or what political scientist George Kateb calls "ideological half-thinking."

Kateb has devoted his career to the role of ethical deliberation in liberal democracies. He is, in this sense, a direct descendant of Hannah Arendt, perhaps the West's most trenchant theorist of political evil. For Arendt, the atrocities of twentieth-century history—Stalin's gulag and Hitler's death camps—could not be explained through the traditional vocabulary of the vices. Behind such seemingly inexplicable events was not some demonic creativity. Rather, these full-scale atrocities were perpetrated by people who were captured by ideological half-thinking or, as Kateb puts it, "ideas that are not carefully thought through but that are so attractive that they get us to act as if we were beside ourselves, indeed not ourselves."

Ideology can lure us into a kind of trance: we are, by our very nature, susceptible to ideas that allow us to make sense of the world in all its teeming diversity. As Kateb explains, these fictions are so seductive because they "lend the world a coherence, a kind of power or beauty, that it did not otherwise have." These narratives depart from everyday reality in order to invest in "a completion, a structure, a magnificence" that reality otherwise lacks. All too often humans have acted on such fictions, hoping to realize that magnificence, and in so doing, have become willing to destroy everything that exposes the story as untrue.

Arendt identifies this pattern of capture in the great tragedies of modern history. Ideology, in all its many guises (jihadist martyrdom or anticommunist hysteria or the desire for *Lebensraum*), promises nothing short of spiritual redemption, albeit by worldly measures. Many people who initially know such an ideology to be fictitious, or invented, eventually come to see it as real, even inevitable. "You begin by telling a story," Kateb explains, "and the longer you tell it, the louder you say it, the more you're taken in by the deception that you thought you were putting over on someone else. It's now being put over on you by yourself. . . . If a story begins in contrivance but nonetheless affects people's actions, if it seems to be making things happen, then, *Why, look, it's working. It has to be true. It couldn't be otherwise.*"

We know that in the process of capture, attention becomes progressively automated and focused; judgment and critical thought recede, replaced by a "pre-reflective" mental state. Kateb explains how ideologies appeal to this biological mechanism: "We all have susceptibilities, vulnerabilities. Things that sweep us up, exert an almost impossibly strong influence on us." Once we are within their grasp, "we find ourselves committed before we know just what has happened to us."

These susceptibilities extend the radius of capture beyond the private struggles of the addict, the suicidal patient, the obsessive artist, and the school shooter. What was once personal becomes social, even societal. Still, the quest for elusive and illusory control revolves around the seduction of "if only": If only I had a drink, a hit of cocaine. If only the entire world recognized this truth, lived according to these rules.

In the twenty-first century, one ideological force that has confounded Western understanding has been the pull of radical Islamism.

THE AMERICA I HAVE SEEN

"The American," wrote Sayyid Qutb in 1951, "is primitive in his appreciation of muscular strength and the strength of matter in general."

For the Egyptian religious scholar, this primitiveness was nowhere more evident than on the football field. Unlike its European counterpart, the "rough American style" had "nothing to do with its name, for the foot does not take part in the game." Rather, Qutb explained to his readers, each player tries "to catch the ball with his hands and run with it toward the goal, while the players of the opposing team attempt to tackle him by any means necessary, whether this be a blow to his stomach, or crushing his arms and legs."

Qutb was just as struck by the sight of American fans as they cheered on the grunting quarterbacks, or whooped at smoky boxing championships and "bloody, monstrous wrestling matches." Their ardor was pure "animal excitement born of their love for hardcore violence": "Enthralled with the flowing blood and crushed limbs," these creatures delighted not in athleticism but in displays of savagery: "Destroy his head. Crush his ribs. Beat him to a pulp." In Qutb's mind, lurking in the grandstands of suburban America was the senseless barbarism that defined a barbaric nation.

Sayyid Qutb was an unusual spectator in the football stadium of Greeley, Colorado. Born in 1906 to a respectable but struggling family in Upper Egypt, he spent his childhood in the rural village of Musha. His early education was primarily religious: by the time he enrolled at a recently opened government school, he had memorized the entire Qu'ran. A quick-witted and diligent student, Qutb managed to secure a highly sought-after place at a teacher training college in Cairo after graduation.

In the 1920s, the Egyptian capital was a bustling, chaotic city of cinemas, cafés, and grand boulevards adjacent to medieval slums. To the pious Qutb, Cairo was at once alluring and repellent: its cosmopolitan elite disdained local customs and sought to shed Islamic traditions in favor of Western cultural values. By the 1930s, Qutb had found a spiritual home for himself among the city's leftist intelligentsia, penning reviews and poetry in Egypt's fledgling literary journals.

With the outbreak of World War II, Qutb's journalism became increasingly politicized: he decried the abuses of Allied troops, who "ran over Egyptians in their cars like dogs." The Americans, he declared in 1946, are "no better than the British, and the British no better than the French." All Westerners were "sons of a single loathsome material civilization without heart or conscience." Around the same time, Qutb wrote a sharp denunciation of Egyptian radio stations for broadcasting morally debased popular songs; only spiritually exalted music should be aired, he argued.

Perhaps because of his increasingly strident political journalism, Qutb received a grant from the Egyptian Ministry of Education to travel to the United States to study the American education system. Perhaps the Egyptian government wanted to contain the unruly intellectual's radicalization, or simply keep him out of Egypt for a time. In any case, in 1949 Qutb set off for the Colorado State College of Education.

It was in Greeley, Colorado, that Qutb first encountered football—and the equally perplexing institution of the American church. Qutb observed that despite Americans' fervor for building churches, they had little interest in "the spirituality of religion and respect for its sacraments," and "nothing was farther from religion than the American's thinking and his feelings and manners." Americans, Qutb concluded, go to church not for spiritual uplift but for "carousal and enjoyment, or, as they call it in their language, 'fun.'"

In his travelogue, Qutb records the landscape of 1950s suburban America: "Each church races to advertise itself with lit, colored signs on the doors and walls to attract attention, and by presenting delightful programs to attract the people much in the same way as merchants or showmen or actors." To the Muslim visitor, even more disturbing than neon church signs was the free commingling of religion and sex: ministers felt "no compunction about using the most beautiful and graceful girls of the town" to attract parishioners.

For Qutb, this spectacle represented the contradictions at the heart of American society. Repulsed by its secularism, materialism, and moral laxity, he decided to return to Egypt early. His frustration with Western culture, however, followed him back to Cairo: he soon resigned from the civil service and joined the Muslim Brotherhood, a grassroots organization dedicated to the revival of traditional Islamic values throughout the Arab world. In the political maelstrom of midcentury Egypt, the Brotherhood represented a powerful rebuke to the incursion of Western culture and to the rise of secular ideologies throughout the Islamic lands. The creation of a Jewish state in Israel and the resulting displacement of Palestinian Arabs only added a sense of urgency to the Brotherhood's call for action against the forces of the secular West.

In 1954, Qutb was arrested and charged with orchestrating an attempt to assassinate Egypt's secularist president, Gamal Abdel Nasser. Over the course of his imprisonment, other members of the Brotherhood were regularly arrested and tortured by Egyptian authorities: guards suspended the prisoners with their arms tied behind their backs, beat them with clubs, and subjected them to near drownings. One particularly sadistic guard killed twenty-one Muslim Brothers and wounded nearly twice as many when they refused to break stones in a local quarry.

As Qutb witnessed the maimed corpses of his comrades being carried through the corridors of the prison, his sense of injustice transformed into indignation. For Qutb, the line separating *Hizb Allah*, the "party of God," from *Hizb al-Shaytan*, the "party of Satan," could not have been starker.

Qutb was incredibly prolific while imprisoned; during his twelve years of confinement, he wrote his most widely read works. These volumes, including *Milestones*, were smuggled out of Cairo's Tura Prison, copied by hand, and circulated throughout the Islamic world. In *Milestones*, Qutb's worldview became increasingly Manichean; he called for the "extermination of all Satanic forces and their ways of life." Qutb blamed "Zionist Jews" and "Christian crusaders" for all the ills of modern society, from prostitution and drug abuse to capitalistic greed and spiritual anomie. "Humanity today is living in a large brothel!" he proclaimed. "One has only to glance at its press, films, fashion shows, beauty contests, ballrooms, wine bars, and broadcasting stations! Or observe its mad lust for naked flesh, provocative postures, and sick, suggestive statements in literature, the arts and the mass media."

In the face of this moral and spiritual corruption, he argued, Muslims were morally obligated to wage war against the forces of *jahiliyya*, or pre-Islamic ignorance and barbarism. Only a literal interpretation of the faith (a return to the *salaf*, or "fundamentals") would bestow on mankind true freedom. In every other system, men serve other men, Qutb explained; in Islam, men serve only Allah. "Islam is a universal truth," he concluded, "acceptance of which is binding on the entire humanity. . . . If anyone adopts the attitude of resistance, it would then be obligatory on Islam to fight against him until he is killed or he declares his loyalty and submission."

Qutb urged his readers to view the Qu'ran not as a theological tract but as a manual for action, "as a soldier on the battlefield reads his daily bulletin so that he knows what is to be done." The controversial "Sword Verse" of the holy text, which commands the faithful to take up arms against paganism, was to be interpreted not as broad spiritual guidance but as an operations manual for jihad. In this sense, Qutb is the intellectual who most immediately shaped the thinking of today's radical Islamist leaders, and his works continue to occupy a central place in the canon of militant groups, including al-Qaeda and Islamic jihad. It is no coincidence that Qutb's younger brother, Muhammad, taught the young Osama bin Laden.

Qutb returned again and again to the myth of a golden age. He was enthralled by nostalgia for a bygone era (albeit an imagined one) defined by spiritual clarity, social cohesion, and moral simplicity: "Mankind today is on the brink of a precipice, not because of the danger of complete annihilation which is hanging over its head—this being just a symptom and not the real disease—but because humanity is devoid of those vital values which are necessary not only for its healthy development but also for its real progress." These virtues, he believed, could be found in the early Islamic era, long before the tides of colonialism eroded the social and political values of the Muslim world. Only by reclaiming Islam as a way of life, an all-encompassing social and political system, could Muslims restore their civilization to its former glory. Qutb dedicated *Social Justice in Islam* to "the youth whom I behold in my imagination coming to restore this religion as it was when it began . . . striving in the way of God, killing and being killed, believing profoundly that glory belongs to God, to His Apostle and to the believers."

In August 1965, Qutb was rearrested and sentenced to death for plotting to overthrow Nasser's secular government. When Nasser offered him the chance to avoid execution, Qutb staunchly refused to negotiate with his enemy and remained stoic in the face of his impending death. In a June 1966 letter to the Saudi Arabian author 'Abd al-Ghaffar 'Attar, he described himself as having undergone a spiritual epiphany in prison: "I have been able to discover God in a wonderful new way. I understand His path and way more clearly and perfectly than before. My confidence in His protection and promise to the believers is stronger than ever before."

After his execution in 1966, Qutb was hailed as a martyr by not only fellow members of the Brotherhood but Muslims worldwide who had found in his writings an escape from spiritual confusion. "Death," Qutb insisted, "does not represent the end. Life on earth is not the best thing God bestows on people. There are other values and nobler considerations." Here, Qutb quoted from the Qu'ran: "If you should be slain or die in God's cause surely forgiveness by God and His grace are better than all the riches that [others] amass. If you shall die or be slain, it is to God that you should be gathered." In a world that seemed hell-bent on silencing the faithful, few words could have provided greater comfort to the spiritually and politically dispossessed.

"Believers," Qutb enjoined, "fight those of the unbelievers who are near you, and let them find you tough; and know that God is with those who are God-fearing." He described the heavenly paradise that awaited those Muslims willing to embrace martyrdom; there, they would "rejoice in what Allah has bestowed upon them of His Bounty." Those who die in the service of Allah, Qutb proclaimed, "are alive, with their Lord, and they have his provision."

Qutb's belief in a golden era not only captured him, but also provided the ideology to capture future generations of followers.

THE OBLIGATION OF OUR TIME

During his eighteen months in a Yemeni jail cell, much of it in solitary confinement, the radical cleric and jihadist Anwar al-Awlaki found comfort in Qutb's writings. "Because of the flowing style of

Sayyid I would read between 100 and 150 pages a day," he later re-counted. "I would be so immersed with the author I would feel Sayyid was with me in my cell speaking to me directly."

When Awlaki finally emerged from prison, he was unrecogniz-able to those who knew him. "Prison had hardened him," explained his sometime student Morton Storm. "I could see it in his eyes. They'd danced before; now they were steel. There was also a hint of paranoia; he saw spies everywhere." Just as prison had served to fur-ther radicalize Sayyid Qutb half a century earlier, months of captiv-ity only strengthened Awlaki's conviction that violent jihad was inevitable. In an interview broadcast over the Internet, Awlaki echoed Qutb's indictment of the West: *"Jahiliyya* is the ignorance of the pre-Islamic era," Awlaki explains. "You will find that Sayyid Qutb uses this word a lot in reference to the times that we are living in. . . . *Jahiliyya* comes from the root word *jahal*, which is ignorance. So it is the time of ignorance. The absence of the message."

Into this absence Awlaki projected a vision of the modern West as inimical to Islamic values. In a series of YouTube sermons, he argued that *jahiliyya* wasn't merely the historical period before the birth of Islam. Rather, it was the state we all live in, a world that has not yet accepted the revelation of Muhammad or the wisdom of shari'a, or Islamic moral code. For Awlaki, this stubborn refusal to recognize and abide by the moral authority of Islam called for unrestrained violence; jihad became a source of meaning in a world otherwise de-fined by passive ignorance. "We will implement the rule of God on earth by the tip of the sword," he declared.

In advocating unrestrained jihad, Awlaki blithely dismissed the Qu'ranic injunction against wanton violence. America, he argued, was responsible for the deaths of millions of Muslims, including women and children; the blood of the victims of 9/11 represented but "a drop in the ocean in comparison." The faithful need "not consult anyone in the matter of killing Americans," he explained. "Combat-ting the devil does not require a fatwa, nor consultation, nor does it require prayer to Allah." Awlaki again echoed the words of Qutb: "They are the party of Satan, and fighting them is the obligation of our time."

Born in Las Cruces, New Mexico, Anwar al-Awlaki moved to Yemen with his family when he was seven years old. There, he attended a secular private school, along with children from other elite families. After graduating from high school, Awlaki returned to America in 1991 to study engineering at Colorado State College, where he served as president of the Muslim Students Association.

At a small mosque a few blocks from campus in Fort Collins, Awlaki cultivated a talent for preaching. His signature blend of folksy American cultural references and learned Qu'ranic exegesis quickly endeared him to the local community. Shortly after graduation, he married a Yemeni cousin and decided to leave engineering behind: he resolved to become an imam.

Subsequent stints took Awlaki to the Arribat al-Islami mosque in San Diego and to Washington, DC, where he served as imam at the Dar al-Hijrah mosque and as Muslim chaplain at George Washington University. Over the course of these travels, he assumed multiple, often contradictory, roles: the representative of a tolerant, cosmopolitan Islam on national television; a grassroots community organizer and legal activist; and an increasingly paranoid conspiracy theorist who publicly questioned the role of Muslims in the 9/11 attacks.

These contradictions extended to Awlaki's personal life. Though he exhorted his congregations to lead lives of moral and spiritual purity, he was far from puritanical: he was twice arrested for soliciting prostitutes and spoke regularly, with a certain intimate knowledge, of the dangerous "temptations of American life," in which, "especially in Western societies, every *haram* [forbidden pleasure] is available." To his neighbors, he was an assimilated American who enjoyed deep-sea fishing and dreamed of entrepreneurial success. He eventually recorded a very popular series of lectures, available as a boxed set, on the life of Muhammad and the lesser prophets of Islam.

Immediately after the attacks of September 11, 2001, Awlaki became a de facto spokesperson for the American Muslim community, appearing on national media and denouncing the hijackers as wayward fanatics: "There is no way that the people who did this could be Muslim, and if they claim to be Muslim, then they have perverted their religion." "We came here to build," he insisted, "not to destroy." But when the American authorities began to uncover a web of connections that linked him to at least three of the hijackers, Awlaki became indignant. (In fact, as early as 1999, he was known to the FBI, which was concerned about his connections to a small Islamic charity that funneled money to al-Qaeda.) Awlaki left the United States for Britain in 2002, in a state of spiritual and financial distress, and he returned to Yemen in 2004.

Over the course of the subsequent years, his sermons, broadcast over the Web to a global audience, became more stridently anti-American. As Morton Storm explained, "His eloquent and authoritative tone was pitch-perfect; he made the radical sound reasonable." Behind Awlaki's colloquial self-styling and references to "Joe Six-pack" was an urgent call for a return to *salaf*, the fundamentals of Islam. His approach to the Qu'ran was infused with Qutb's perspective, treating the holy text as a set of literal prescriptions to be applied to the present day.

In August 2006, Awlaki was arrested by the Yemeni authorities. Though no charges were ever brought, he would spend the next eighteen months in jail. After his release, he moved to a remote mountain hideout, where he broadcast video messages to Sunnis around the world, calling for a global jihad against the West. When Storm lingered after a study group led by Awlaki, his teacher shed any pretensions to liberalism: "9/11 was justified," he proclaimed.

In his Internet sermons, Awlaki argued that the American people, not simply their government, had declared war on Islam: "They are participants as they voted for this administration. They are the people who are financing this war." In a moral universe defined in such stark terms, the only hope for redemption lay in martyrdom: "America as a whole has turned into a nation of evil. I eventually came to the conclusion that jihad against America is binding upon myself, just as it is binding on every other able Muslim."

Awlaki was captured by this image of violent redemption infused with Qutb's messianic zeal. Awlaki's version of jihad against America captured, in turn, many Internet followers.

In December 2008, Awlaki sent a message of congratulations to the Somali Islamist group al-Shabaab. He thanked them for "giving us a living example of how we as Muslims should proceed to change our situation. The ballot has failed us, but the bullet has not."

I'M GOING TRAVELING

In December 2008, Major Nidal Hasan, a psychiatrist in the U.S. Army, began a mostly one-way e-mail correspondence with Awlaki. Over a period of six months, Hasan asked him about the teachings of the Qu'ran and the spiritual responsibilities of Muslims in the American military.

In one e-mail, Hasan discussed Muslim American soldiers who "appear to have internal conflicts and have even killed or tried to kill other [U.S.] soldiers." He referenced the actions of Sergeant Hasan Akbar, who had thrown a grenade into a tent of U.S. troops, killing two officers. If such people die while engaged in jihad, Hasan asked, "would you consider them shaheeds [martyrs]?"

Awlaki responded briefly to the e-mails, but never answered Hasan's questions about the tenets of Islam. He did promise to "keep an eye for a sister" because Hasan had mentioned that he was "looking for a wife that is willing to strive with me to please Allah."

Though they met only briefly, Hasan regarded the spiritual leader with great respect and even attempted to launch an essay contest that would award a five-thousand-dollar scholarship for the best answer to the question "Why is Anwar al-Awlaki a great activist and leader?" To Hasan's disappointment, the contest never took place, because Muslims in the community were "petrified by potential repercussions," he later wrote. Nonetheless, Hasan assured Awlaki that nothing would stand in the way of his devotion. Allah, he wrote, had "lifted the veil from my eyes" just before the attacks of September 11: "I have been striving for Jannat Firdaus ever since." Jannat-ul-Firdous is the highest place in Paradise.

The e-mail exchanges went silent in June 2009. But five months later, Awlaki had something to say publicly about his acolyte. "He is a man of conscience who could not bear living the contradiction of being a Muslim and serving in an army that is fighting against his own people," Awlaki posted on his website. "He did the right thing." The "right thing" was Major Hasan's November 5, 2009, rampage in Fort Hood, which began with the shout "Allahu Akbar!" and ended ten minutes later with thirteen people dead and thirty others wounded. Almost four years later, twenty-three senior military officers voted unanimously to sentence Hasan to death.

Born in 1970 to Palestinian parents who had immigrated to Virginia from the West Bank, Hasan joined the U.S. Army as an infantryman at age seventeen. From that point on, his life unfolded entirely within the world of the military. While serving as an enlisted soldier, he earned a medical degree and did his residency in psychiatry at Walter Reed Army Medical Center.

Though his childhood had been largely secular, and even included Christmas celebrations, Hasan grew increasingly religious after his mother's death in 2001. An imam at the Muslim Community Center in Silver Spring, Maryland, where Hasan attended daily prayers, often in army fatigues, did not hesitate to describe him as devout. At the same time, Hasan became increasingly discontented with the military, and complained that he was being harassed because of his religion. One soldier scratched his car with a key, apparently irritated by a bumper sticker that read "Allah Is Love." Another left a diaper in his car with the message, "That's your headdress." Hasan explored the possibility of leaving the service, but because the army had paid for his education, his commitment would have been difficult to break.

Even as Hasan's ideas became increasingly radical, they generated no official response. Though the FBI's Joint Terrorism Task Force did intercept Hasan's e-mails to Awlaki, investigators concluded that the exchange was "fairly benign." Family members did not recognize the transformation that was taking place; they regarded Nidal as a gentle, sensitive man who was still mourning the loss of his beloved parents.

But there were many signs of a growing obsession. Hasan was consumed by what he saw as the incompatibility of his faith and his military allegiance: he was to become a Muslim soldier fighting wars against other Muslims. In June 2007 he gave a presentation at Walter Reed on "Koranic World View as It Relates to Muslims in the U.S. Military." Striking an academic tone, he spoke in the third person on the topic of why Muslims might feel conflicted about service in Iraq and Afghanistan. He sounded mild mannered, but quoted sources that belied his even tone.

During the lecture, Hasan explained that many Islamic scholars had issued religious rulings, known as fatwas, forbidding American Muslims from joining the U.S. military. He also read verses from the Qu'ran to indicate, albeit by implication, where the holy text stood on that question: "Whoever kills a believer intentionally, his punishment is hell, he shall abide in it, and Allah will send His wrath on him and curse him and prepare for him a painful chastisement."

Hasan quoted one Muslim-American soldier's brother: "It's getting harder and harder for Muslims in the service to morally justify being in a military that seems constantly engaged against fellow Muslims." Further, Hasan claimed that thousands of U.S. soldiers had converted to Islam while stationed in the Persian Gulf. He even claimed that such soldiers were regularly joining the Taliban and al-Qaeda after being discharged. "There is something out there with these groups that is really resonating as Islamic," he concluded. "These guys are willing to go over there and fight."

Hasan described in detail a series of "adverse events" involving Muslim-American soldiers. One soldier faked his own kidnapping so that he could desert the military; a sergeant refused to deploy to Iraq because of his religious beliefs; Hasan Akbar threw grenades at fellow servicemen. Nidal Hasan suggested that his colleagues exercise particular caution around soldiers with obviously Muslim names or those who read the Qu'ran devoutly, as such soldiers might be "more predisposed to having conflicts."

To put these events in context to the group, he explained that obeying and fearing God is a core obligation of Islam, and that failure to submit brings grave punishment: in one Qu'ranic story, the children of Israel are turned into apes and swine after they persist in working on the Sabbath. But, Hasan insisted, the rewards of doing good are just as important to the Islamic faith: "And their recompense shall be Paradise, and silken garments, because they were patient. . . . Reclining therein on raised thrones, they will see there neither the excessive heat of the sun nor the excessive bitter cold. . . . And amongst them will be passed round vessels of silver and cups of crystal."

Because the Qu'ran is a "progressive revelation," Hasan explained in his lecture, earlier verses were sometimes abrogated by later ones. In one example of shifting attitudes, the text initially denies Muslims the right to defend themselves, then permits them to act in self-defense, and ultimately allows them to strike the first blow against their enemies. "The verses of defensive jihad start abrogating the peaceful verses," he claimed. "This becomes important because you'll see sometimes people cherry-picking verses, trying to show Islam in a more peaceful light where, indeed, you could make a great theological argument that those verses were actually abrogated."

Hasan even gave a nod to Osama bin Laden, noting that he had given up the comforts of a multimillionaire's life to embody "the spirit of jihad." For Hasan, that willingness to sacrifice status in pursuit of a just cause required careful psychological analysis: "Instead of just labeling someone as a terrorist, we have to really understand why these Muslims really identify with that and how do we fix that. Is there really injustice that is going on, or is it truly just some aberration, which I doubt?"

Hasan's gradual descent into radical Islam became more apparent during the two-year fellowship that followed his residency. In August 2007 he gave a presentation to classmates entitled "Is the War on Terror a War on Islam? An Islamic Perspective," which was so inflammatory that his instructor ended it early. In class, he justified suicide bombings and told peers that religious beliefs took precedence over allegiance to the U.S. Constitution. All the while, Hasan was studying jihad on the Web. At least two military colleagues called him a "ticking time bomb."

Yet Hasan was also rising up through the army ranks: he was promoted to major in May 2009, even as he was corresponding with

Awlaki. Hasan's deployment to Afghanistan was scheduled for late November; as the date drew near, his anger about the wars in Iraq and Afghanistan only intensified. As a military psychiatrist charged with evaluating returning soldiers, he was exposed daily to troubling war stories, including one about the killing of a civilian and another about the deliberate dumping of fifty gallons of fuel into the Iraqi water supply. He asked supervisors for guidance about reporting such atrocities, closing his e-mail request with the phrase "All praises and thanks go to Allah, The Cherisher and Sustainer of all the worlds."

Hasan had not yet received a response on the November morning when he attended prayers at his Fort Hood mosque. That day, he told a fellow worshiper, "I'm going traveling." He had already given away most of his possessions. A few hours later, he opened fire.

@SLAVEOFALLAH

The video opens with a low boom and a red-splattered logo with the words "YOUTH TALK DAWA"—*dawa* means "proselytizing"— underneath. Anwar al-Awlaki's echoing voice plays over a plaintive melody. The subject of the speech is "Women of Islam"; according to its YouTube page, it has been viewed more than forty thousand times.

Awlaki narrates the life of Asiya, wife of the pharaoh who reigned in the time of Moses. Asiya's "willingness to give up all of these worldly aspects for the sake of Allah," he explains, "is a reflection of the depth of her faith." Later Awlaki would write, "Jihad must be practiced by the child even if the parents refuse, by the wife even if the husband objects."

Teenage sisters Fatima and Amal Farah and their friend and neighbor Anisa Ibrahim found in Awlaki's words a highly personal spiritual message. Convinced that their faith was under threat, the three girls resolved to flee the United States for the Islamic caliphate proclaimed by the Islamic State of Iraq and al-Sham, where they would marry soldiers fighting on behalf of Islam. Thanks to personal blogs such as *Bird of Jannah—Jannah* means "paradise" or "heaven"—penned by a woman who had traveled to Syria to support ISIS, they knew that they would be expected to take care of the homes and babies of jihadists. They felt prepared to do anything—abandoning their families and the bounties of America, even risking their lives—in the name of *tawheed*, or "the singularity of Allah."

The three young women hadn't always been so fervently religious. A year earlier, they were, in many respects, typical American teenagers. They were responsible students who never gave their parents or teachers any indication that they were unhappy. They had long felt somewhat distant from their non-Muslim peers, though they had found a sense of belonging in the area's sizable Somali and Sudanese community. Both families lived in apartment complexes just a few miles from the mosque where they occasionally prayed, the Colorado Muslim Society, which advertises itself as "the biggest Muslim community in Colorado." The young women attended one of the most diverse high schools in Colorado, where they were part of an African American majority. In March 2014, just seven months before the three girls left for Syria, the school's girls' soccer team put on hijabs to protest a referee who had barred a Muslim teammate from wearing her headscarf.

All three teens were active on social media, but Fatima was particularly prolific; she posted primarily on an unmonitored forum called Ask.fm, where mostly teenage users asked one another questions via the anonymity of a user name. She told her online correspondents that she felt proudest of herself when she passed a math test, and that she hated being woken up in the morning. "I have a lot of friends," she wrote. "I love them all, they are amazing." She wrote about her hobbies (tennis, swimming, and listening to music) and said that she hoped to work one day at "a fashion business!! Duh."

A single post hinted that Fatima was heading into the digital universe of radical Islamism. In late October 2013 she told a correspondent on Ask.fm that during bouts of insomnia she would "watch lectures on youtube and stay on twitter." Fatima and her friends were attracted by what some scholars have started calling the "wireless caliphate," the online sphere of Islamic proselytizers who have inundated Tumblr, Twitter, Pinterest, and other social media sites with propaganda in support of the Islamic State. On their personal pages, they post photos of soldiers in black face masks, machine guns pointing at the sky, with captions from the Qu'ran; they express admiration for those who heed the call to "drop the nationalist flags and raise the banners of *tawheed*," and decry what they see as rampant Islamophobia. There is also a subculture of women who call themselves *Umm*, or "Mother," and actively recruit young females to the cause. The blogger who created *Bird of Jannah* was one of these women; she provided tips on traveling to Syria undetected, including what to pack and what to expect on arrival. "Sisters, please drop your whatsapp contact via message and I will contact you . . . [this service] is the easiest app for me altho I'm aware it's not safe. Please forgive me," she tells her readers. "I will only reply [to] those who I trust in sha Allah" (i.e., "with Allah's help or permission").

What made these young women such easy targets for seduction? Why did they trade comfort for fanaticism, poverty, and violence?

In the wireless caliphate, the girls found instant meaning, a heritage, and a future all for the taking. Whatever doubts they had about their purpose were assuaged by Qu'ranic verse; whatever frustrations they had about their surroundings were validated by Internet friends who assured them they were living in a modern-day Babylon. The social media accounts of the three young girls began to reflect their burgeoning faith. In April 2014, just a few months after she wrote that the greatest lesson she had ever learned was to "do [her] homework," Fatima told a correspondent who asked about her new zeal that she had "realized [her] purpose in life.

"Islam makes things a lot easier," she said.

By April, the girls' social media threads were dominated by Islamic themes. The Farah sisters created second accounts under their *kunyas*, the Arabic pseudonyms used among some radicals as noms de guerre. Fatima posted as Umm Sufyan and @carrierofsins, while her sister, Amal, appeared as Umm Suleiman @_SlaveOfAllah_. Amal's bio was summed up in a single quote about modesty from Umar, a compatriot of the Prophet Muhammad: "Lower your gaze from the world and turn your heart away from it." The three girls quickly became active in the world of "Muslim Twitter," exchanging Qu'ranic verses with members of ISIS and occasionally expressing political views that most would consider extreme. Anisa, for example, tweeted in early June that "those who identify as 'gay' and 'Muslim' at the same time deserve death"; a few months later, she wrote that Muslims who "hand out apologizes [sic] because of 9/11 are a disgrace." To the question "What are you doing tomorrow?" Fatima responded, "Going to the masjid [mosque] in sha Allah." Later, when asked when she'd like to get married, seventeen-year-old Fatima replied, "As soon as possible in sha Allah."

October 17, 2014, was a mild day in Aurora, and began seemingly inauspiciously for the parents of the three girls. The Farah sisters stayed home from school, feigning illness. Anisa left her house and headed toward school, but around ten o'clock her father received word that she hadn't arrived. When he called her on her cell phone, she claimed that she was late for class; in fact, she and the Farah sisters were already on their way to Denver International Airport, armed with their passports and two thousand dollars they had stolen from the Farahs. As they waited to board a plane to Frankfurt, they asked their nearly two thousand Twitter followers to pray for them. By the time they arrived in Frankfurt, their parents had reported them missing and the German authorities were waiting for them.

Once the girls had been returned to the care of their parents, Anisa's father told the *Denver Post* that the three "realized they [had] made a mistake." But he may have been speaking wishfully. Less than three weeks after returning from Germany, Fatima took to her Tumblr account again, this time to post a screenshot of an exchange on Ask.fm. "Where are you from?" the writer asked.

"I live in some far away land," she wrote, "where the *Ghurabah* [righteous loners] are few."

PART III

7

CAPTURE
AND SPIRITUALITY

In my study of capture, I returned again and again to a fundamental paradox: although it often causes great pain, capture also underlies exalted experiences. Having witnessed the suffering that capture can sow, we might well ask how we can exchange dangerous forms of capture for benign or even productive ones.

This sort of transformation may occur suddenly: we can be seized by a spiritual insight seemingly without warning. But we can also consciously try to make ourselves more receptive to such experiences. Hence William James's belief in the power of free will to redirect our attention. If we seek out positive forms of capture, we may be able to guide ourselves out of suffering.

While religious experience offers us one opportunity to do just this, perception of the divine is unique, just as the individual perceiving it is.

Spiritual experiences have been described as a feeling of absolute dependence, of being grasped by an ultimate concern. They may involve moments of release from ordinary perception. The catalyst may be spiritual or aesthetic in nature—a poem, or a landscape, or a moment of quiet meditation.

"The feeling of it may at times come sweeping like a gentle tide, pervading the mind with a tranquil mood of deepest worship," the German theologian Rudolf Otto once wrote. "It may pass over into a more set and lasting attitude of the soul. . . . It may burst in sudden eruption up from the depths of the soul with spasms and convulsions or lead to the strangest excitements, to intoxicated frenzy, to transport and to ecstasy."

What is this all-powerful *it*? Philippe Borgeaud, professor of ancient history and religion at the University of Geneva, explained it to me this way: "God is a value for the best or the worst. But he is not an allegory. A god is not something signifying another thing. Aphrodite *is* beauty, love, desire, sex. She *is* that. . . . The god is recognized by the emotion, but the emotion is the god, also."

David Foster Wallace understood this basic fact about the relationship between the human and the divine. Wallace came to believe "there is no such thing as not worshipping. Everybody worships." Our only choice, then, is *what* to worship. "And the compelling reason for maybe choosing some sort of god or spiritual-type thing to worship—be it JC or Allah, be it YHWH or the Wiccan Mother Goddess, or the Four Noble Truths, or some inviolable set of ethical principles—is that pretty much anything else you worship will eat you alive."

To see how religious scholars would respond to the idea of capture, I paid a visit to two Amherst professors, Susan Niditch and her husband, Rob Doran.

"It's ancient," explained Niditch, an expert on Judaic studies. "In the Hebrew Bible, Jeremiah describes being seduced by God and filled with the spirit."

Both she and Doran, a scholar of early Christianity, see capture at work in religious history.

"Being a prophet of God meant Jeremiah was a vessel for God's message," said Niditch. "He believed that he could not escape this line of work: a force had acted on him."

Paul, too, was utterly transformed by what he described as a force outside himself. In his letters to the Galatians, Paul writes, "You revealed yourself to me." Raised as a Pharisee, Paul was sent by the leaders in Jerusalem to quiet a group of renegade Jews who were proclaiming that the end was nigh and that Jesus was the Lord. Indeed, Paul believed that Jesus was a common criminal who had been rightfully executed by the Romans.

Yet when God revealed himself to Paul, he was struck dumb and began to preach that Jesus is the Lord. Such stories of conversion often portray the experience as "something that comes out of the blue," Doran explained. Paul suddenly began to embrace views that we would likely describe today as delusional: he believed that Jesus was alive, even though he knew that he had been crucified. For Paul, Jesus was the new Adam, the harbinger of a new world order. Paul recognized that we either accepted this new order or perished with the old one. That sense of a radical break with the past remains a central element within Christianity. "We have to be captured," Doran said. "The same is true of Muhammad: he has a vision, and the vision transforms him."

That said, Doran warned that we must understand this sort of conversion experience within the context of the culture in which it takes place. In the first century, for instance, there was a widespread notion that God could appear to you. Niditch described a social world in which everybody believed in God, and in which the divine had enormous power over human affairs. This sort of transformative experience, in other words, existed within the larger cultural vocabulary.

Still, Doran believes that capture plays a central role in the history of almost every world religion. Time and time again, captured individuals display a kind of charisma that allows them to draw others into the faith, to evoke strong emotional responses in their followers. "Though it begins as a private, individual experience," Doran explained, "capture soon becomes public."

CAPTURE BY THE DIVINE

Edith Stein was not looking forward to her visit with the newly widowed Anna Reinach. Stein feared Anna's raw response to the death of her husband on the battlefield in Flanders in World War I, but more difficult to face was Stein's own grief about his tragic death. Adolf Reinach, only thirty-four when his life was taken, had been a generous friend to her, a kind teacher and mentor who had opened many doors. Along with Anna, he had welcomed Stein into their lively intellectual circle at the university in Göttingen when she'd arrived as a doctoral student in philosophy just two years earlier. Stein had come to study with Edmund Husserl, a philosopher and the father of phenomenology—as well as a colleague of Reinach's. Now Anna had written asking Edith to come visit and help put her husband's papers in order. Though it was not possible to decline, Stein felt unprepared for the emotional toll she expected the meeting to take.

When Stein arrived, she found that Anna was not in need of consoling. The widow's tranquility quickly transformed Stein's despair at her friend's untimely death. Stein knew that Anna and her husband had recently converted to Christianity; they'd been baptized together during one of Adolf's furloughs. Yet it was clear that Anna's brave acceptance of fate had little to do with religion in the abstract, or the dogma of the church. Hers was, as Stein immediately recognized, a dynamic faith in God—warm, alive, and personal.

"It was my first encounter with the Cross and the divine power that it bestows on those who carry it," Stein would later recount of the experience. "For the first time I was seeing with my very eyes the church, born from its Redeemer's sufferings, triumphant over the sting of death. That was the moment my unbelief collapsed."

Stein's father died before she was two years old, leaving her, along with her six siblings, to be raised by her mother alone. In the years that followed, Stein's mother transformed her husband's debt-saddled lumberyard into a thriving business. A devout Jew, she managed to maintain her family's morale in the midst of tragedy. Stein, her youngest child, was born on the Day of Atonement; as Stein wrote in her autobiography, "I think, more than anything else, it made [me] especially dear to her."

It was particularly painful to her mother, then, when Stein announced, at the age of thirteen, that she was an atheist. For the first time, Stein had questioned the existence of God and, in an early flexing of her powerful intellect, concluded that she saw nothing there. Though Stein was never one to doubt her own mind, she felt her mother's heartache with every step she took away from her childhood religion.

Years later, inspired by Anna Reinach's quiet religiosity, Stein came to see religious faith in a new light, as a way of "resting in God"—"of being exempted from all anxiety and responsibility and duty to act." In the wake of this epiphany, Stein approached her work as a philosopher with a renewed sense of vitality and purpose. Her writings soon became infused with a spiritual vocabulary as she trained her focus on the palpable gap between sensory perception and the experience of the divine. A perceptual experience, of a landscape or a face, "is marked by an immediacy, in the sense that what is present itself is said to be experienced immediately, not what is merely grasped through its effects or made present through its messengers." That is, we can actually see the landscape or the face—not only its effect in the world. God, however, is not "immediately intuited" in this way; the divine is experienced at a remove, through his effects on worldly beings.

Though Stein now believed in God, she had yet to encounter him again as she had during her visit with Anna Reinach. Four years later, in 1921, however, when she was thirty years old, she pulled the autobiography of St. Teresa of Avila from a bookshelf while visiting philosopher friends at their farm. She did not put it down until dawn. "As I closed the book," Stein recalled, "I said, 'This is truth.'"

That same morning, she went into town to buy a Catholic catechism and missal, and soon after she found herself at Mass at the local parish church, taking the first steps toward baptism. The transformation seemed natural, like falling in love; every encounter seemed to point in one direction.

Though Stein traveled a more winding route than most, from Judaism to atheism to Christianity, she was steadfast in her decision to convert. From that point on, she knew she wanted to become a Carmelite nun, as St. Teresa had been, but was dissuaded for some time by her family and spiritual mentors. In 1934, however, as Hitler began to seize power, Stein returned home to visit her mother for the last time. They went to synagogue together. The next day, she took a train to Cologne, where she joined the Carmelite convent.

While there, Stein did not fully retreat from the world. She had witnessed the rise of the Nazi Party with a clear-eyed awareness and knew better than most where the prejudice against Jews was leading. The same year she joined the convent, she also wrote her memoir, *Life in a Jewish Family.* So at the point when she could have, understandably, relinquished her previous life, taking on a new name in a new family, Stein chose to stand and claim her Jewish identity. Her purpose was clear: to demonstrate that Jews were not monsters but fellow human beings. She believed that only empathy could possibly forestall the approaching terror.

She also wrote to Pope Pius XI, imploring him to condemn the "deeds perpetrated in Germany which mock any sense of justice and humanity." And she issued a prescient warning to the pontificate, one that the Vatican managed to hide from the public eye for half a century: "The responsibility must fall, after all, on those who brought them to this point and it also falls on those who keep silent in the face of such happenings."

Stein now had an undeniable commitment to God, but she had also come to understand that this did not mean she could look away. "During the time immediately before and quite some time after my conversion I . . . thought that leading a religious life meant giving up all earthly things and having one's mind fixed on divine things only," she wrote. "Gradually, however, I learnt that other things are expected of us in this world. . . . I even believe that the deeper someone is drawn to God, the more he has to 'get beyond himself' in this sense, that is, go into the world and carry divine life into it."

In the dedicated life of the convent, in addition to the liturgical cycle of prayers, two hours daily were given to silent prayer. Stein could be seen in those hours through the window of her cell, perfectly still, on her knees with arms outstretched. In a poem, she expressed the wordless content of those prayers:

> Who are you, kindly light who fills me now,
> And brightens all the darkness of my heart?
> You guide me forward, like a mother's hand,
> And if you let me go,
> I could not take a simple step alone.
> You are the space,
> Embracing all my being, hidden in it.
> Loosened from you, I fall into the abyss
> Of nothingness from which you draw my life.
> Nearer to me than I myself am,
> And more within me than my innermost self,
> You are outside my grasp, beyond my reach,
> And what name can contain you?
> You, Holy Spirit, you, eternal Love!

As the Holocaust consumed Germany, Stein was transferred from Cologne to the safer reach of a convent in the Netherlands. Her arrival coincided with a strong statement from the Dutch church in defense of the Jews, which infuriated the Nazis. Any hope of protection for converts of Jewish origin vanished, and Stein was arrested at the convent in August 1942. Within a week, she was deported to Auschwitz, where she would soon be murdered in a gas chamber.

PAYING ATTENTION

Spiritual awakenings are often not consciously orchestrated or willed into existence; rather, they descend upon us unbidden. Might we somehow coax ourselves into a state of spiritual capture? Simone

Weil spent much of her philosophical career reflecting on the act of turning toward the divine.

Weil always paid close attention to the world. At the age of six, she gave up sugar on behalf of the French soldiers in World War I, whom she'd been told had to go without it.

"At fourteen . . . I seriously thought of dying because of the mediocrity of my natural faculties," Weil later wrote. "I did not mind having no visible successes, but what did grieve me was the idea of being excluded from that transcendent kingdom to which only the truly great have access. . . . I preferred to die rather than live without that truth." It was only after working through this internal struggle, however, that Weil came to the more egalitarian conclusion that anyone can enter such a kingdom after all—"if only he longs for truth and perpetually concentrates all his attention upon its attainment."

This notion—that attentiveness offered spiritual opportunity to everyone—drove Weil ever after. It led her as a teacher and an activist, when her altruism was a kind of instinctive spiritual endeavor, and culminated in her work for a year, beginning in 1934, as a manual laborer in factories in Paris, in order better to know the conditions of the working class. These factories were her first contact with adversity outside her privileged life—and where she found "the affliction of others entered into my flesh and my soul."

As an adolescent, Weil had determined there was no way to know whether there was a God—"the data could not be obtained here below"—and she'd put the matter aside. Yet, as she described it, Christian motives (kindness, acceptance, generosity, justice) always guided her nonetheless. "That is why it never occurred to me that I could enter the Christian community. I had the idea that I was born inside," Weil once professed—despite the fact that she'd been raised by Jewish parents. "But to add dogma to this conception of life, without being forced to do so by indisputable evidence, would have seemed to me like a lack of honesty."

So she held herself back from organized religion, unwilling even to visit churches, in an attempt to keep her spiritual truth from being corrupted. Ironically, it was this kind of intellectual rigor and conscientiousness that in the end became her unwitting preparation for accepting God.

At the age of twenty-seven, after her year of factory work, feeling emotionally pummeled and physically frail, Weil traveled with her parents to Portugal. There she went to a little village on a day it was celebrating a patronal festival. "The wives of the fishermen were going in procession to make a tour of all the ships, carrying candles and singing what must certainly be very ancient hymns of a heart-rending sadness," Weil recalled. "I have never heard anything so poignant unless it were the song of the boatmen on the Volga." She was moved in a way that felt different to her.

The following year, Weil went alone for two days to Assisi, Italy. Breaking her vow not to enter churches, she visited the chapel of Santa Maria degli Angeli, where St. Francis, whom she greatly admired, used to pray. On that day, she felt an invisible force compel her to drop to her knees.

The next year, Weil experienced the last in her triumvirate of powerful revelations when she chose to spend ten days going to religious services from Palm Sunday to Easter Tuesday. For much of this time, she was suffering from a terrible headache—Weil frequently had migraines—and in the church "each sound hurt me like a blow." In making the extraordinary effort required to see past her pain, she felt herself rising up, leaving her suffering body behind to find "pure and perfect joy in the beauty of the chanting." At this time, Weil believed that the "thought of the Passion of Christ entered into my being once and for all."

In a letter Weil wrote to Joseph-Marie Perrin, a Dominican priest who was a friend and deep influence in her life, she depicted her life after religious conversion, making it clear that she conducted it with the same effort and discipline she'd earlier brought to her humanitarianism.

She began, for example, reciting the Lord's Prayer every morning. Having initially memorized it as an intellectual exercise, she soon found herself so taken by it that she was compelled to repeat it as a daily ritual. "If during the recitation my attention wanders or goes to sleep, in the minutest degree, I begin again until I have once succeeded in going through it with absolutely pure attention," Weil recounted. "Sometimes it comes about that I say it again out of sheer pleasure, but I only do it if I really feel the impulse. The effect of this practice is extraordinary and surprises me every time, for, although I experience it each day, it exceeds my expectation at each repetition."

Weil offered a description of the abandon she felt during this exercise: "At times the very first words tear my thoughts from my body and transport it to a place outside space where there is neither perspective nor point of view." Often she felt the presence of Christ during the recitation, "infinitely more real, more moving, more clear than on that first occasion when he took possession of me."

Weil, like William James, believed that attention was key. Yet where James argued that it was the method by which we narrated our lives, choosing only what interested us to be part of the story, Weil stressed that it was the way in which we became poised to receive God.

Weil felt that when we make a sustained effort, even if it is only at something as mundane as a school exercise "which we have failed through sheer stupidity," we are faced with the irrefutable fact of our smallness, even mediocrity, and the consequent need for further reflection and absorption. "Attention consists of suspending our thought," Weil wrote, "leaving it detached, empty and ready to be penetrated." Such cultivation sparked Weil's spiritual awakening. The capacity to attend in this way, she concluded, "is a very rare and difficult thing; it is almost a miracle; it *is* a miracle."

CAPTURED BY A MESSAGE

American theologian Howard Thurman's four-month visit to India was nearing its end, and he was beginning to get anxious. The

Howard University dean had lectured endlessly on African American religious culture; seen the sun rise over Kinchinjunga, in the Khyber Pass; sat under the legendary banyan tree in Shantiniketan—but he had yet to meet Mahatma Gandhi.

Some months earlier, a British devotee of the Mahatma, a young woman named Miriam Slade, had come to Thurman's office at his request to speak on the topic of *ahimsa*, or "nonviolence." Thurman was eager to learn more, so Slade agreed to arrange for Thurman to meet with Gandhi when the former traveled to India. When Thurman arrived in Ceylon in early October 1935, a telegram from Gandhi was waiting for him. "I shall be delighted to have you and your three friends whenever you can come before the end of this year," Gandhi wrote.

Yet now it was February 1936, and the meeting had failed to materialize. After two full days of lecturing at Bombay University, Thurman turned to his wife, Sue Bailey Thurman, a lecturer, author, and the traveling secretary for the YWCA, and told her he couldn't wait any longer; he would send Gandhi a message that afternoon. On his way to the post office, Thurman passed a young man in a white Gandhi cap. The two stopped and grinned at each other before the young boy handed Thurman a note. It was an invitation from Gandhi to meet at nearby Bardoli.

The next day, the Thurmans and another delegate from the YWCA found themselves sitting on the floor of Gandhi's tent, answering the Mahatma's questions about every possible aspect of black American life. When it was time for Gandhi's guests to pose questions in return, Thurman asked why Gandhi's movement of nonviolent civil disobedience had failed to expel the British from India. Gandhi told the visitors that his fellow citizens, whose dedication to the cause was paramount to the success of the movement, had "lacked vitality" to hold such a lofty ideal in their minds for a sufficient amount of time. The first reason they lacked vitality, he explained, was hunger. (India at the time was in the throes of its own Great Depression, partly brought on by the British government's high taxation and fees and its refusal to let Indian citizens export

goods.) The second was that they lacked self-respect—and not because of British rule, but as a result of their own unethical behavior.

"We have lost our self-respect," Gandhi told them, "because of
the presence of untouchability in Hinduism. . . . If the shadow of an
untouchable falls on the Hindu temple or, in some instances, on the
street on which the Hindu temple is located, the temple is considered
to be contaminated."

Thurman sensed something significant in this message. "How on
earth did you attack such a thing as that?" he asked.

"The first thing I did as a caste Hindu," Gandhi replied, "was to
adopt into my family an outcaste and make that person a member of
my family, legally, and in all other ways. This announced to other
caste Hindus, 'This is what I mean by what I am saying.' Then I
changed the name from outcaste to 'Harijan,' a word that means
'Child of God.'"

Hearing this phrase, Thurman was transported back to his childhood in Daytona, Florida. His father died when he was seven years
old, so Thurman and his two sisters were raised by their mother and
grandmother, an experienced midwife. His grandmother had spent
the first twenty-two years of her life as a slave. She couldn't read, and
often asked her grandson to read aloud from her Bible—though he
was not permitted to read from Paul, who in his Epistle to the Ephe-
sians justified slavery. His grandmother told him that a few times a
year, the slave master where she grew up would allow a local black
preacher to speak to his slaves. The preacher was theatrical in style,
bombastic in tone. As the sermon reached a high point, the preacher,
dripping with sweat, would move his eyes from one face in the crowd
to another.

"You are not niggers!" he would bellow. "You are not slaves! You
are God's children!"

Throughout his years in college and later at seminary, Thurman
pondered this concept. How was it that he, the grandchild of a slave,
an African American growing up in the Jim Crow South, had so
wholly absorbed this message—and how had that belief affected
him? He knew that his profound conviction that his life was important stemmed from his family's abiding faith in Jesus and the way his

community, the black neighborhood of Waycross and the Baptist Church he joined at age twelve, watched over each child. It was his faith in God and his belief that he was a child of the divine that had motivated him to excel in school and had offered a shield from the institutionalized bigotry and everyday cruelty of Jim Crow. When a white man would demean him, addressing him as "boy," Thurman's resolve and self-respect were steeled by the knowledge that in the eyes of God all men were equal. Jesus himself had been part of the persecuted minority of Jews in Palestine during the Roman occupation, but he had prevailed because he knew that to consider himself less than the Romans would be to allow resentment into his heart—resentment that would surely destroy him from inside.

Even before his visit to India, where many Hindus had asked him how blacks could associate with the faith of their conquerors, Thurman had been wrestling with the split between the "religion of Jesus," as he called it, and the religion of Paul. In the first version of Christianity, each soul was beloved, and righteousness, charity, and simplicity were valued. The second version, as he saw it, had been used to justify imperialism, slavery, and violence for centuries throughout the world. It had fostered a system of profound inequality, one in which the majority of people lived "with their backs against the wall." How could those distressed people find any solace in Christianity when it had been used as a weapon against them for so long? In 1935, before he set sail for India, Thurman wrote a paper on that very topic. In "Good News for the Underprivileged," published in the journal *Religion in Life*, Thurman argued that the Pauline ethos was a perversion of authentic Christianity. True Christianity, he said, could be found in the words of the poor Jew fighting, theologically if not literally, the militant Roman forces in his land. What path did Jesus offer the occupied Jews? Should they attempt to assimilate or should they fight back? Should they isolate themselves? Should they allow their resentment to fester inside? No, Jesus argued: they should be humble and meek, all the while remembering that God's kingdom was within them. They needn't look to the Romans, or to their hatred of the Romans, to find a solution. They needed only to look within themselves.

While touring the subcontinent, Thurman saw a racial divide that was in some ways deeper and more brutalizing than the one he experienced every day back home. Christianity was an unattractive option for Indians, most of whom were Hindu or Muslim, because converting would be akin to joining the ranks, in name if not in color, of the British ruling class. As an observer, Thurman watched the interactions between black and white people, both at home and abroad, and asked himself if religion could break the color bar. Thirteen years after his conversation with Gandhi, Thurman would write a short book on this topic, *Jesus and the Disinherited*. In it, Thurman recalls Gandhi's assertion that a human needs a solid core of self-esteem before he can commit himself to personal, let alone social, change. "It is quite possible for a man to have no sense of personal inferiority as such," Thurman wrote, "but at the same time to be dogged by a sense of social inferiority. The awareness of being a child of God tends to stabilize the ego and results in a new courage, fearlessness, and power." A man who knows that he is a child of God does not fear his captors, nor does he fear death. There is no hate in his heart toward those who oppress him, because he holds a deep sense of "moral responsibility" toward his maker. He cannot be pulled asunder by the limitations of society.

Sitting on the floor of that sparsely furnished tent in India, Thurman saw how faith in God, self-esteem, and the ability to dedicate oneself to social change were inextricably bound together; when one of these precepts failed, the others would follow suit. Thurman was lucky: he had always known he was special in the eyes of God. But for a majority of his peers, this was not the case. They had grown up believing they were less than full citizens, that they couldn't achieve equality even if they wanted to. How could Thurman ensure that they come to know, with the same urgency that he did, that they were God's children, and that they all had the power within them to overcome their obstacles?

Before the group left to catch a train back to Bombay, the "little brown man of India," as he was affectionately known throughout the world, surveyed them one last time. "Well," Gandhi said, "it may be through the Negroes that the unadulterated message of non-violence will be delivered to the world." Decades later Thurman's message had become widespread, and influenced the work of more than a few civil rights activists. A young Martin Luther King Jr. was known to sit and read from *Jesus and the Disinherited* while preparing for a protest or a march. In the final lines of the book, Thurman writes, "When men look into [Jesus's] face, they see etched the glory of their own possibilities, and their hearts whisper, 'Thank you and thank God!' "

THE REVELATION OF NATURE

Spiritual capture does not necessarily depend on belief in the divine. Other sources of wonder can play a similar role in our psychic lives.

> *Five years have passed; five summers, with the length*
> *Of five long winters! And again I hear*
> *These waters, rolling from their mountain springs*
> *With a sweet inland murmur.*

Returning after a long absence to the banks of the River Wye, William Wordsworth finds a landscape blissfully indifferent to the passing of years. For the poet, the intervening five years were defined by violent change, both political and personal, but the familiar scenery around Tintern Abbey proves immune to the erosions of time. From the "steep and lofty cliffs" to the "plots of cottage-ground" and "orchard-tufts," these reassuring sights remain steadfastly the same.

Wordsworth finds in the passing of years a consolation for the loss of youthful vitality. The abbey's unchanging natural beauty becomes a source of spiritual solace, a temporary escape from the inevitable disappointments of daily life, from the "lonely rooms" and "the din / Of towns and cities." As the poet's attention gradually broadens to encompass the entirety of nature, he is lifted, if only momentarily, out of the temporal realm, released from the confines of the self and into a union with his surrounding world:

> *A presence that disturbs me with the joy*
> *Of elevated thoughts, a sense sublime,*
> *Of something far more deeply interfused,*
> *Whose dwelling is the light of setting suns,*
> *And the round ocean, and the living air,*
> *And the blue sky, and in the mind of man—*
> *A motion and a spirit that impels*
> *All thinking things, all objects of all thought.*

Upon reading Wordsworth's "Tintern Abbey," Samuel Coleridge struggled to contain his elation. "We see into the *Life* of [the] Things," he wrote in a feverish journal entry. "By deep feeling we make our *Ideas dim*—& this is what we mean by our Life—ourselves. . . . (let me) think of *myself*—of the thinking Being—the Idea becomes dim whatever it be—so dim that I know not what it is—but the Feeling is deep & steady—and this I call *I*—identifying the Percipient & the Perceived."

Intense aesthetic experience (marveling at the immensity of the ocean, or seeing it through the eyes of a poet) fuses subject and object, perceiver and perceived. For many poets, musicians, and artists, and for their audiences, this momentary experience of selflessness offers a private consolation against the disappointments of living. For Wordsworth in particular, spiritual experience did not depend on institutionalized religion, or even on the existence of God. Aesthetic experience, too, could temporarily halt the flux of time, blurring the boundary between the solitary self and the surrounding world.

The ancient Greek philosopher Longinus called this facet of aesthetic experience the sublime. The carefully wrought artifice of rhetoric and poetry, Longinus observed, often inspires a particular form of awe, one that is not merely a reaction to the force of its content.

Indeed, the sublime accounted for the singular power of language to persuade by its very majesty.

For Wordsworth, the sublime had an inherently therapeutic quality, a "renovating virtue" whereby "our minds / Are nourished and invisibly repaired." In *The Prelude* a wintry memory from the poet's adolescence gradually acquires a numinous aura: "All shod with steel / We hissed along the polished ice in games / Confederate." As the skaters gave "their bodies to the wind,"

> all the shadowy banks, on either side,
> Came sweeping through the darkness, spinning still
> The rapid line of motion; then at once
> Have I, reclining back upon my heels,
> Stopp'd short, yet still the solitary Cliffs
> Wheeled by me, even as if the earth had roll'd
> With visible motion her diurnal round.

Here, the normally disparate realms of space and time dissolve into the single fact of motion: the wheeling cliffs and even the rolling earth are indistinguishable from the circling skaters. The young Wordsworth loses himself, if only momentarily, in the glittering, icy scene.

This form of crystalline perception, which seems to halt the march of time, was for the Romantics the essence of poetic achievement. "If the doors of perception were cleansed," insisted William Blake, "everything would appear as it is, infinite." Only then would we "see a World in a Grain of Sand / And a Heaven in a Wild Flower."

Over the two millennia following Longinus's death, philosophers and artists have tried to make sense of the sublime. Why, and how, does beauty reconfigure our relationship to the world around us? In the eighteenth century, a number of European philosophers (including Immanuel Kant, Joseph Addison, and Edmund Burke) returned to the idea of the sublime, arguing that sublimity was distinct from mere beauty: it depended on the experience of terror, on the "idea of pain and danger," albeit of a particular sort. The sublime

was, in Addison's words, "an agreeable kind of horror," often inspired by profound works of art or music or, alternatively, by the
sheer grandeur of nature: alpine landscapes, raging storms, endless
seas. Such experiences awaken our fear but, at the same time, evoke
within us a sense of unity with the cosmos.

There is, of course, a tension inherent in this kind of experience.
By its very nature, the sublime fuses, and perhaps even obliterates,
seeming opposites: self and world; change and permanence; danger
and ecstasy. It is at once elevating and threatening, momentary and
everlasting; it draws us to it with an irresistible force, even as it
promises to annihilate the world as we know it.

8

CAPTURE
AND CHANGE

Capture allows us to focus, to be moved, and to act with purpose. The mechanism does not, in and of itself, give meaning to our lives. Rather, it allows us to search for and experience meaning.

In addition to causing significant mental distress, capture can also enable growth and recovery. Many people do have moments of clarity and changes in perception. We can arrive at a point where we no longer care about what has been salient in the past. An understanding of the process of capture can lead to a more balanced life. At the very least, such insight can allow us to discern, and perhaps even influence, what captures us. Still, I have struggled with a basic question: is the only way out of an unhealthy form of capture another, more positive form of capture?

Just as we have explored how capture can shape experience and take control of our lives, it is important to delve into stories about people who, having realized they were on a destructive path, were able to change course. Before looking at these stories, however, I want to step back and look at capture from a different vantage point.

Throughout the book, we have seen individual, dramatic outcomes of capture, a spectrum of experience that charts the progression of capture's influence. We have taken an intimate look at a continuum of mental afflictions and obsessions, beginning with inner torment and ending with devastating suicides. Capture can lead to behavior that is harmful to the self, and can even destroy that very self, when we most desperately need to quiet a storm of psychic pain.

The progression of capture can become equally, if not more, dangerous when its object is external. When a person becomes captured by an abiding sense of rage, the burden is no longer his alone to bear. That mental suffering can endanger entire societies. Violence against others is an extreme consequence of capture, but one that commands our collective attention.

Capture has even more significance when the salient object is ideological. Ideologies have a powerful allure to the disenfranchised: the individual becomes dedicated to a higher cause, which promises to give meaning to his life, to connect him to something greater than the self. Of course, conversion, or dedication to a greater cause, can also result in the opposite of terrorism: benevolence that positively influences the lives of others.

The most heartening aspect of capture is the possibility of being released from suffering by escaping the orbit of a feedback loop that causes profound distress. We find stability and self-awareness by exchanging one capture for another. The stories that follow show how capture can also be a profoundly positive experience, charting a way out of mental anguish.

MARTIN LUTHER'S *ANFECHTUNGEN*

When Martin Luther prayed from the modest quarters of his monastery cell in Wittenberg, a wave of dread would rise up. He had struggled with this inner turmoil for more than a decade. Year after year, he prayed for insight, but it rarely seemed to help; so he had learned to be patient. These moments of *anfechtung* (the equivalent of nausea) overpowered him, immobilizing his mind with despair. The young Catholic monk desperately sought assurance that he was a genuinely good Christian worthy of God's love, but all he recognized were his sins.

Those who knew him would have been bewildered by Luther's vision of himself. The behavior of this monk, now in his early thirties, hardly justified his profound sense of inadequacy. He was a brilliant student of theology and philosophy, sharp-witted, good company—even charming. He was also a humble, obedient monk, assiduous in his observance of rituals. He'd come to Wittenberg to earn his doctorate and he'd proven himself a quick study. He could master any academic subject. When I visited Wittenberg and spoke with a Luther scholar, Silvio Reichelt, he told me Luther had gotten a "free ride" at the University of Wittenberg, an opportunity to earn his PhD without a fee. "He came here to start from scratch."

Living conditions in the monastery were harsh. Like other monks, Luther lived in poverty; he had only a small cell, frigid on cold winter nights and hot in the summer. He would pray up to seven times a day, with maybe five hours of sleep between one day's evening prayer and the next morning's first church service. It was a strenuously quiet life, with an emphasis on reading, study, and contemplation. When he arrived at the monastery, he would have been loaned only two blankets and two sets of clothing. His bed consisted of a narrow stone platform with a bit of straw for a mattress. He may have had a desk and a chair, possibly a picture of the Virgin Mary or one of the saints on his wall. His only personal possessions, though, were a crucifix and a begging bowl.

Luther's father had wanted his son to become a lawyer. When Martin opted instead for the monastic life, the elder Luther voiced his doubts: "Let's hope it's not all deception and delusion." That response took root in Luther's mind as the beginning of preoccupations that would, as he put it, drive him toward his world-changing insights. Yet almost a decade later, he was still in the dark.

As the years at the monastery progressed, Luther became increasingly concerned with what would have seemed a dry and abstract theological query. For him it was a matter of more than life or death: a tormenting anguish about the fate of his soul. He sought to understand how he, as a "fallen" individual, could save himself by doing good deeds. How could anything he undertook be of any avail, since it was, by definition, the act of a soul cut off from God by sin? And if he couldn't do anything to earn salvation, how could he ever know he was saved?

For a dozen years, Luther wrestled, in a harrowing, private way, with this fundamental uncertainty, his studies and prayers offering little peace. During Mass, his anxieties burned at him to the point where he felt Jesus Christ loomed over him as an angry judge, not a savior. His dread over his inability to do what he believed God wanted him to do (to live a life of perfect moral integrity, perfect righteousness) grew so intense that during one sermon, Luther collapsed onto the floor shouting cryptically, "I am not! I am not!" As he told his adviser later in life, "God is impelling me, driving me on, rather than leading me. I cannot master myself; I want to be calm, yet I am driven."

Why, with his tireless and devoted study and adherence to all the ritual, had this *anfechtung* not abated? Regardless of what he did, he felt he would never earn God's forgiveness and love. He was a man not only captured by harrowing self-doubt, but also compelled to behavioral extremes:

> I often accumulated my appointed prayers for a whole week, or even two or three weeks. Then I would take a Saturday off, or shut myself in for as long as three days without food or drink, until I had said the prescribed prayers. This made my head split and as a consequence I couldn't close my eyes for five nights, lay sick unto death, and went out of my senses. Even after I had quickly recovered and I tried again to read, my head went round and round.

Despite this inner turmoil, amazingly, he remained brilliant, charismatic, and funny. At his first monastery south of Wittenberg, in Erfurt, Luther's talents and intensity caught the eye of his superior, Johann von Staupitz, who had been appointed as a reformer with a mandate to steer monastic life closer to its roots in Augustin-ian theology. He saw in Luther a potential ally, but he also felt a deep respect for, and connection with, the young man's integrity and innocence.

It was a fortuitous meeting of mentor and protégé. Staupitz became a spiritual father figure for Luther, one of the few people who could provide the young monk the solace he craved. Luther depended deeply on Staupitz for encouragement and acceptance, and once spent six hours continuously confessing his sins to the older man. During one of these sessions, seeing how harrowing Luther's self-doubt had become, Staupitz suggested that Luther abandon abstract questions about moral judgment and simply contemplate the humble and compassionate sacrifice Jesus made on the cross. He steered Luther into a deeper study of theology, possibly as a way of distracting the monk's thoughts, but also as a means of preparing Luther for a career that Staupitz may already have had in mind for his student.

All this helped wean Luther from his sense of Jesus as wrathful enforcer and opened him up to a vision of Jesus as a fellow sufferer whose life and death brought forgiveness, not justice. Yet moments of *anfechtung* kept recurring for Luther, overpowering him with anxiety about his spiritual peril. He wrote about these hellish immersions in vivid detail:

> I myself knew a man who claimed that he had often suffered these punishments . . . they were so great and so much like hell that no tongue could adequately express them, no pen could describe them, and one who had not himself experienced them could not believe them. At such a time God seems so terribly angry, and with him the whole creation. At such a time there is no flight, no comfort, within or without, but all things accuse . . . in this moment, it is strange to say, the soul cannot believe that it can ever be redeemed.

This was not the experience of a scholar or theologian debating questions of sin and grace. His was existential torment, an all-consuming descent from which Luther couldn't imagine the possibility of rescue. In response to his student's distress, Staupitz advised Luther to read the German mystics, such as Meister Eckhart, with his emphasis on the terrible emptiness (what another mystic, St. John of the Cross, referred to as the dark night of the soul) that precedes spiritual insight. It didn't immediately help; Luther was no mystic. Yet Staupitz offered the idea that Luther's sense of abandonment, the sense that he was lost, could be a sign that he was closer to God than he suspected. Staupitz hinted that the awareness of his immense separation from God could have been, in effect, progress. It was a prolonged awakening to the reality of sin, which was one of the final steps for Luther to then recognize the reality of God's love. Staupitz offered Luther a way to see these periods of obsessive self-doubt as genuine, involuntary penitence, a prelude to forgiveness and understanding. In response, Luther began to regard his suffering as practice, a path of learning how to get closer to God. Yet reframing his bouts with *anfechtung* as necessary didn't make them any easier to endure, and Luther still craved a different understanding of God.

Staupitz suggested that seeing penitence as the first step toward God was putting things in the wrong order. First came love, then penitence. As he listened to Staupitz's words, something opened up within Luther. Suddenly it made sense. Penitence followed from a love of God, not the other way around. Penitence, his painful awareness of his sinfulness, was a sign that he loved God—which was all God wanted. His anguish was to be welcomed. How simple! He wrote:

This your word stuck me like some sharp and mighty arrow . . . and then, what a game began. The words came up to me on every side, jostling one another and smiling in agreement, so that, where before, there was hardly any word in the whole of Scripture more bitter than [penitence] . . . now nothing sounds sweeter or more gracious to me.

In 1516 he looked to the Bible, hoping again to see the truth it contained in a new light. And he did. What he found was a new understanding of faith as the first and, in a way, final step: faith, and not good works, was what brought righteousness into a person's life. With this insight, his personal doubts and despair dissipated. His eyes fell on Romans 1:17: "In the Gospel, the righteousness of God is revealed, a righteousness by faith from first to last, as it is written, 'the righteous will live by faith.' " With these words, everything fell into place. Luther later wrote:

> If God was committed . . . to condemning all those who failed to achieve moral perfection, then the central act of the Christian religion, which was re-enacted at every mass, the death of Christ on the cross, became pointless. I began to understand the justice of God as that by which the just lives by the gift of God, namely by faith. *The just lives by faith.* This straightaway made me feel as though reborn, and as though I had entered through open gates into paradise itself.

Luther had not only successfully broken free from the anguish of his capture, but he was now able to channel that same focus and intensity into meaningful action. The very question that had plagued him for more than a decade, the efficacy of good works, would become the basis for his rejection of the Catholic Church and its various practices and rituals that he viewed as corrupt or an abuse of power. The realization he'd been struggling toward for a decade not only dispelled his *anfechtungen*, but also drove him to work for change in the world, setting in motion the Reformation and the division between Protestant and Catholic.

MEANINGFUL ASSOCIATION

In his memoir, *Darkness Visible*, William Styron wrote of the moment he decided he would take his own life: "I felt my heart pounding wildly, like that of a man facing a firing squad, and knew I had made an irreversible decision."

As he soberly prepared for his own demise—disposing of his journal, visiting his lawyer to draw up his final will, attempting and failing to write a suicide letter ("reduced to an exhausted stutter of inadequate apologies and self-serving explanations")—Styron observed his actions from an unbridgeable distance.

In *Darkness Visible*, Styron cannot fully explain his depression. "I was sixty when the illness struck for the first time, in the 'unipolar' form, which leads straight down," he wrote. "I shall never learn what 'caused' my depression, as no one will ever learn about their own." Instead, he offers a host of probable reasons: he had recently quit drinking, his sleep was restless enough to make him feel he'd been sentenced to a lifetime of insomnia, he was fearful about having recently turned sixty—"that hulking milestone of mortality"—and he was experiencing unprecedented difficulty with his writing.

"He gave up drinking because, as he wrote, he felt alcohol was no longer his friend. And then he went into a kind of withdrawal, as is natural, but he didn't come out if it," Rose Styron recalls of her husband during this period. "It became clear to me that a lot of his buoyancy and ability to write and write and write every day—he never drank when he wrote, but he drank for a couple of hours in the evening afterward and that allowed him to relax and think of what he was going to do the next day. Suddenly he didn't have that."

Styron was exquisitely aware of himself and the surrounding world—this was, after all, the force behind his finely wrought fiction. "Self-loathing was part of both his depressions in 1985 and in 2000," Rose explains. (Styron died of pneumonia in 2006, at eighty-one years old, after another struggle with depression.) "He was uniquely sensitive to every change in the wind and the moon, general or personal. He was extremely sensitive to anything he perceived as a criticism or a slight. . . . Every big writer gets good and bad reviews. He pretended he'd never read any of the bad reviews, but it wasn't true. He internalized and fretted about them."

Styron also speaks of these emotions in his memoir: "Of the many dreadful manifestations of the disease, both physical and psychological, a sense of self-hatred—or, put less categorically, a failure of self-esteem—is one of the most universally experienced symptoms, and I had suffered more and more from a general feeling of worthlessness as the malady had progressed."

In 2000, after being "absolutely depression-free," as Rose describes her husband, this feeling returned, like a wrecking ball abruptly collapsing the fifteen years in between, and Styron sank into a depression once more. "He said the most interesting thing: the worst thing about him is that he'd been unproductive and fallow," his daughter Susanna noted about a visit to him in the hospital during this time. "He hadn't written a novel in 20 years, hadn't done the work he was supposed to do. I realized how painful that was for him and that there were no reassurances, about his having written three masterpieces, which is more than most people ever do, that could possibly soothe him, though I tried." She recounted that he had told her sister Polly that "besides the great novels, there were many smaller things he wanted to write, that they were like little beasts and he had watched them turn their backs on him and walk away."

As was true for David Foster Wallace and Ernest Hemingway, these thoughts were predicated on an accumulation of experiences and triggers over the course of a lifetime. "Depression, when it finally came to me, was in fact no stranger, not even a visitor totally unannounced," he conveys in *Darkness Visible*. "It had been tapping at my door for decades."

Styron's own father, with whom he always had a close relationship, suffered from depression throughout his life and was hospitalized during an emotional collapse that, as Styron came to recognize in his adulthood, greatly resembled his own. Yet, as described by Styron himself, the event that had the most dramatic influence, the one that would continue to color the rest of his life, was the loss of his mother to cancer when he was thirteen years old.

Not long after he realized he was hurtling toward a death of his own making, and while he was still in the process of making the logistical arrangements for it, Styron found himself wandering his house on a bitterly cold night, bundled up and restless. With Rose asleep in bed, he decided to settle in the living room by the fireplace and watch a videotape of the film *The Bostonians*. There was a young actress in the cast who had also played a small role in an off-Broadway play Styron collaborated on.

At a certain point in the film, characters walked "down the hallway of a music conservatory, beyond the walls of which, from unseen musicians came a contralto voice," Styron writes, "a sudden soaring passage from the Brahms *Alto Rhapsody*." The powerful voice of a woman intoning lyrics Brahms had fashioned from a Goethe poem—lamenting the pain of a lonely wanderer and praying that he might become alive to the world again—shot like an arrow through the thick fog that had enveloped Styron for months. "In a flood of swift recollection," he recalls, "I thought of all the joys the house had known: the children who had rushed through its rooms, the festivals, the love and work, the honestly earned slumber, the voices and the nimble commotion, the perennial tribe of cats and dogs and birds."

Yet something else may have come to him even more swiftly than these memories, a flash of remembrance that occurred before he was even able to register it.

"Brahms's *Alto Rhapsody* was what Bill's mother used to sing," Rose Styron explains. "He had a very loving and complicated relationship with her because she became sick after he was born. . . . I realized, after living with him all the time that I lived with him, for fifty-three years, that he had had premonitions always that his mother would die, and he carried that through in his fear that all of us, his wife and children, would suffer pain and die. That was certainly part of his eternal pessimism for those he loved. He was incredibly fearful if I was home a half hour late. He imagined the worst because he always thought his mother was going to die and then, when he was thirteen, she did. So he had this incredible emotional pull, and his best-remembered happy moments were of his mother singing and playing the piano. . . . I can only feel that when he heard the *Alto Rhapsody* again, having watched this young actress in *The Bostonians*, that he was suddenly filled with a different kind of emotion because he had survived and we had all survived. Perhaps he thought that if he killed himself, the pain would be as intolerable for all of us whom he loved and cherished . . . that he couldn't do that to us, that his mother had left him and his father completely bereft. But that was not her fault, and this would be his fault, and he felt he couldn't do it."

For that moment, if not forever, the association for Styron between Brahms's *Alto Rhapsody* and the memory of his mother was so strong, so deeply embedded in him, that it had the power to reverse the tide of his savage emotions. "I drew upon some last gleam of sanity to perceive the terrifying dimensions of the mortal predicament I had fallen into," Styron declares. "I woke up my wife and soon telephone calls were made. The next day I was admitted to the hospital."

MOMENTS OF CLARITY

In my years of studying addiction, I have always been intrigued by Alcoholics Anonymous (AA). It is a powerful and effective program for many people, and its twelve-step system is credited with saving lives. But how does AA actually work?

AA was founded in 1935 when Bill Wilson, now better known as Bill W., reached out to local ministers while tempted to drink on a business trip in Akron, Ohio. Wilson had long struggled with his alcoholism and knew that in order to stay sober, he needed to speak with a fellow sufferer—someone who knew the fight from the inside. He was directed to Henrietta Seiberling, a member of a religious organization called the Oxford Group, which advocated for "the power of God to change lives." Seiberling, in turn, introduced Wilson to Dr. Bob Smith, a physician who had turned to the Oxford Group to help him with his own struggle with alcoholism. Upon meeting, the two men spoke for six hours; in fact, Smith became the first alcoholic Wilson successfully brought to sobriety. Soon thereafter, they partnered to create Alcoholics Anonymous, a program strongly influenced by the Oxford Group.

Prior to meeting Seiberling and Smith, Wilson had already heard of the Oxford Group. His friend Ebby Thacher, who became sober after joining the organization, had told Wilson of his high regard for the group. Wilson recalled that Thacher had told him the group "emphasized getting honest with one another, with making a survey of their defects, admitting that they were powerless without the help of God." This was what had finally released Thacher from his, as he put it, "awful compulsion to drink."

At the time, Thacher's sentiments sounded clichéd to Wilson, but his story of recovery made a deep impression. Wilson felt he could follow most of the principles that Thacher described—except for the last: he had no faith in God. "This admitting you were hopeless, well, that wasn't too hard," Wilson explained in an interview, reflecting on his attitude at the time. "This getting honest with yourself and another person, well, one could do that. This making restitution for harms done to other people, a tough job, but certainly one would try that. Working with others without any demand for money or acclaim, well, that would be just wonderful. But when he came to the God part, again, I remember a terrible balking."

As Wilson's depression and drinking worsened, however, he landed in the hospital for the last time. "I still gagged badly on the notion of a Power greater than myself, but finally, just for the moment, the last vestige of my proud obstinacy was crushed." Wilson later said:

> All at once I found myself crying out, "If there is a God, let Him show Himself! I am ready to do anything, anything!" Suddenly the room lit up with a great white light. I was caught up into an ecstasy which there are no words to describe. It seemed to me, in my mind's eye, that I was on a mountain and that a wind not of air but of spirit was blowing. And then it burst upon me that I was a free man. Slowly the ecstasy subsided. I lay on the bed, but now for a time I was in another world, a new world of consciousness. All about me and through me there was a wonderful feeling of Presence, and I thought to myself, "So this is the God of the preachers!" A great peace stole over me and I thought, "No matter how wrong things seem to be, they are still all right. Things are all right with God and His world."

There have been many interpretations of Wilson's spiritual experience, with some dismissing it as the product of hallucinations brought on by the injections of belladonna he received in the hospital. Regardless of the source, the experience triggered in Wilson a spiritual awakening and an evangelical zeal to help other alcoholics.

Many years later, in an interview, Wilson offered a profound insight into how he envisioned AA working. "The secret would be deflation at depth, [forcing the alcoholic to face the enormity of the problem] using the tools given us by science, the compulsion plus the allergy," he explained. "Then, when one alcoholic carried this message to the next, he could reach the newcomer where he lived, even below consciousness. Then, bringing about the deflation, he would create a dependent of the newcomer upon a sponsor, upon a group, if there were any, and, finally, upon God." In other words, AA offers a substitute form of capture, exchanging dependency on alcohol for reliance on community and spirituality. AA's actual success rate remains unclear. There is no doubt, however, that it allows some people who previously felt hopeless to reclaim their lives. Some participants experience what they describe as a "moment of clarity," wherein the mind becomes sufficiently detached from its thought pattern long enough to see the pattern—this moment itself is freedom. The difficulty is maintaining that freedom.

A CREATIVE LIFE

"I just want to fall asleep and never wake up again."

These words occupy the center of a diagram that spreads over the inner cover of a slim hardcover book, Chris Ware's graphic novel *Building Stories*. The diagram itself is a flowchart of the author's endless cycle of tortured rumination, a metaphorical perpetual motion machine. Its cogs are small icons, including a pistol, a noose, and pills.

The book is one of fourteen items, including folded strips of paper, pamphlets, broadsheets, and board-mounted panels, that together re-

semble the makings of an elaborate board game. The boxed assem-
blage pushes the boundaries of what constitutes a novel, graphic or
otherwise, yet it remains firmly rooted in storytelling. The novel's
protagonist is an unnamed young woman with a prosthetic leg whose
life seems framed by unremitting loneliness and bleak horizons. With
no set entry point, the story, unfolding on various pieces of the whole,
offers an accumulation of mundane detail and layered memories that
create the subtle dimensions of a lived world. A flowchart depicts life
paths that include existing alone in front of a TV or existing in front
of an identical TV next to a faceless partner.

> How do I do it how do I do it how do I do it?
> But who will find my disgusting bloated body?
> (I really should lose some weight . . .)
> My parents?
> The landlady?
> The cat?
> I can't leave the cat behind . . .

The woman's imagined corpse, discovered again and again in al-
ternate scenarios, is detailed with flies. The cat nibbles on toes.

> Is it possible to *hate* yourself to death? If it is, I'm trying . . .

In my conversations with the author, Ware describes his experience
with depression as physical and emotional: "It sets in the bones, in a
way. There's something of a heaviness, or a leadenness in the limbs
and in the joints. It's almost like a hopelessness that spreads from
your toes upward. You'd think it would be the other way around.
This sounds really embarrassing, but the phrase 'sad wall' comes to
my mind. I don't understand why. It's like I've hit a wall, literally,
going at a certain speed. It overwhelms me to the point where I
almost can't move."

It can last, he says, anywhere from a day to weeks at a time. But
that understanding of a finite duration, a prognosis that experience

knows to be limited, is not accessible to the mind in the midst of it. You can't see over the sad wall. "When you're in the bubble of despair, you just feel like, 'Oh yeah, this is it. How am I going to find my way out of it? I never will. I'll be here forever.' And then when you're out of it you're like, 'That was too bad. Well, hopefully that won't ever happen again.'" It's a lesson that he never fully learns, for reasons that he doesn't understand. He jokes that he should print a laminated card, inscribed, "Get ready. It's going to come back and you will forget how horrible it is."

Chris acknowledges that his work serves as both the most common trigger for his episodes of depression and the readiest antidote. His resolve to continue making art is a mark of inner resources that have carried him safely through past episodes of depression. Whether putting himself through the public paces that his career prescribes or mining his most troubling memories for material, he exhibits a resilience that belies his trumpeted self-doubts. Chris notes that he has never been hospitalized for depression or taken medication for it, despite the recommendations of doctors, and says that his condition has never been as profoundly debilitating as that of others he has observed.

"Part of the solipsistic ridiculousness of it is that I've spent most of my life trying to figure it out within my own artwork, and of course it brings it on at the same time, but I've thought a lot about my childhood and my upbringing and my experiences." He describes his early life as confounded by uncertainties: his father's abandonment even before his memory registers and the ensuing open questions of self-blame; the subsequent paternal figures as initially unknown quantities; being uprooted from his childhood home at a critical point in adolescence; the casual cruelty of inconsistent friends. Much of his emotional turmoil arose in the aftermath of failed relationships, "where you wake up in the morning for a second or two not remembering what had happened the night before and then suddenly remembering—oh yeah, that girl broke up with me— and then all of a sudden that sinkhole opening up in your chest and everything caving in on itself."

The swings in self-judgment that plague creative work are familiar to many artists and writers. "You can sit at a table, start working," Chris says, "which is essentially tantamount to thinking, it's another kind of thinking . . . and in a matter of sometimes just an hour you can come to the conclusion that you're a worthless human being with absolutely no accurate read on reality or yourself and that you should just end it all." But on reaching those worst of all possible moments, "The only thing to do is to try to keep working beyond that. When I was younger, there were a lot of temper tantrums, etcetera, a lot of things thrown out the window. I've gotten beyond that now."

What makes that shift in perception possible? Chris describes having an early sense of vocation. At the age of eleven or twelve, he began studying cartoon art from instructional books, and within a couple of years he was submitting work to publishers, though he received no response. He describes a childhood where the turmoil of uncertainty was balanced by hours of happy solitude spent drawing in his attic bedroom and the warm encouragement of his mother and grandmother. He notes that he has tried to re-create the environment of that attic room for his current work space, and that he felt a similar "embryonic sense of individual efficacy" while working in his dorm room at college. If the process of the work provides its own pleasure, an easeful forgetting of self in the focus it requires, it also creates a sense of purpose and validation. He has a calling, one that necessarily involves summoning the courage to face a blank page and the painful memories that will be transmuted to fill it.

To the extent that he has realized his ambition, his art is an antidote for his depression. If unruly emotions and thoughts can be molded into a pleasing new form, if the artist's hand can intimate some sense of larger purpose in the composition of his story, there is hope that meaning might exist even beyond the dominion of the pages he has created.

In 2005, Chris's daughter, Clara, was born. Chris says that the birth of his daughter and the unforeseen transformation that parenthood brought him was akin to a religious experience, or the closest to such a thing that he's ever come: "It really is the most miraculous thing to ever happen to anyone, and it happens to everyone. Women can turn the universe inside out, literally. It's astonishing when you think about it. It's starting time over again, in a way. It's almost making time meaningless." From the day of his daughter's birth, he says, a veil lifted. Everything "that had been hanging over my head, the leaden dangling criticisms floating around me just seemed to evanesce. They just disappeared.

"It's probably very chemically explicable." He laughs. "I think probably pretty much all new parents go through it." As if to demonstrate, Chris has placed a life-size drawing of a newborn baby girl at the center of a broadsheet in *Building Stories*. The heroine of his story has become a mother, and her experience is as unexpectedly affirming as Chris's.

Chris admits that since becoming a parent, thoughts of suicide still occasionally surface, but it is no longer a real option. In the light of parental responsibility, to oneself as well as to the mental health of a child, the longing for extinction somehow fades to a "shell-like sense of the rhetorical." Some of his friends survived the suicide of a parent in childhood. "I wouldn't want to do that to anybody. It's simply not worth it." His "prime directive" now is learning to parent in ways that avoid the repetition of the dysfunctional patterns that have haunted his life.

"All of a sudden you're no longer the protagonist. The movie has a new cast. All of a sudden you're a supporting actor and you suddenly realize that's what you've been all along, and that's the way every human being should be, and that's the way you really should be living your life." In the light of new priorities, it becomes clear to him that, at least in his case, his earlier social anxiety and hypersensitivity are "really just a very self-indulgent kind of melancholy and egotism on the part of the bearer, and it's something you really just need to keep to yourself."

For Chris, becoming a father shone a light on a simple truth: depression is an exercise in torment that is entirely focused on the self. Any lasting solution requires a redirection of attention elsewhere.

COMPELLED TO BE DIFFERENT

Growing up in Iowa farm country, Margaret knew her home life was somewhat chaotic. Both her parents made it clear they adored her and her younger brother, but neither of them, she knew intuitively, was a responsible, healthy adult. Her father was by trade an odd-job worker (a trucker, a door-to-door potato salesman), but more often than not he made his money by conning people. Uneducated but handsome and charming, he sweet-talked the man next to him at the coffee shop counter into accepting a bad check or buying a piece of stolen farm equipment; often, he gambled away any money he made from the deal. He was frequently absent from the home, either securing farm equipment or drinking and philandering. Whereas her father was charismatic and confident to the point of self-delusion, her mother was a person almost without a self. Though she was an intelligent woman who had once dreamed of going to medical school, she had succumbed to familial expectations and moved back to the small farming town where she grew up, only to find herself trapped with an unfaithful husband and a crushing sense of disappointment. Margaret's mother was often depressed and rarely interacted with people outside the immediate family, sometimes lying in bed for days. As a young girl, Margaret surveyed the scene around her and thought, "I don't want to be like this. I don't want this to be my life."

As many intelligent children do, she dealt with the instability by achieving. Starting in grade school and on through high school, she stuck to a rigorous schedule of studying and activities. "Class got out around three in the afternoon," Margaret, now nearing fifty and a college professor in Colorado, remembers. "I would jump in the car, change into my dance clothing, and my mom would drive the twenty miles to the next town, where there was a dance teacher. I would go to dance class for two hours or three hours, I'd get back in the car, we'd drive back another forty minutes home, and then I'd do homework. On the weekends I'd have dance class all day Saturday. I'd go to dance recitals, and then in high school there were speech contests and theater and band concerts and so on." Both she and her brother knew that succeeding in school was a means of getting out of their provincial hometown, where the teachers often read directly from textbooks as lessons and most people would look at you cock-eyed if you wanted to do anything other than be a farmer or marry one. Right around the time the family moved to a modern split-level house—where they remained for six years, the longest the family would be able to afford to stay in one place—Margaret began running track in school and fell immediately in love with the sport. She had known since age seven, when she began dancing, that individual sports were an effective outlet for her—tossed balls were not her friends—but running was more liberating than dancing, which involved a painstaking focus on perfection. "Running was a free, calm, isolating, *isolated* event, and I really enjoyed that," she says, seeming almost soothed just by talking about it, even now. She wasn't a fast sprinter, so the track coach started her out in two-mile races, and even though she didn't win her first one, she still felt blissful from the experience. "Whatever that feeling was, that was really good. I liked that." From then on she ran constantly—with friends, solo, in snowstorms or oppressive midwestern heat. She ran around her tiny town's one paved road and out on the smaller gravel roads in the farmland. Running, she felt unencumbered by her family life, in some ways depleting herself of energy so she didn't have the strength to worry. "I could put myself into a mode where I would master something and *that* would be the thing that I would think about, not about who was going to pay the rent or whether my father was going to come home."

Running has been so potent a balm for Margaret that she has continued doing it for more than thirty-five years. She's run marathons and adventure events, and has gotten increasingly into triathlons as her joints have aged and she's had to cut back on her weekly mileage. Now her average run during the week is five to seven miles, and on weekends it's eight to ten, and she supplements her running with cycling and swimming. "I don't know anything, really, about Buddhism, but it seems to me that whatever I'm worried about, I start to disassociate from in a way. Sometimes I actually say to myself, 'I'm going to work on that on my run.'" She also lapses into a state of "flow," during which she says she feels "attentionless, as opposed to very focused." At the end of the run, she feels "smoothed out," she says, and is able to face a long day dealing with students and administrators with energy, enthusiasm, and flexibility.

For all the positives running has brought to Margaret's life, I ask her what would happen if she weren't able to run. How would she feel? She immediately starts laughing. "Very, very anxious. My anxiety level would rise pretty rapidly within a couple of days. And I wouldn't feel . . . like myself," she says, reiterating how important running is to her identity. It's so integral to her sense of who she is that she has continued running even after undergoing several foot surgeries a few years ago. "I can't really do a full marathon run anymore, but I could do a half marathon run. I think that the hardest thing is trying to get the bump in the mood when you can't go as hard as you used to go."

"Have you felt the need to take it up a notch over the years, and have you been able to control that?" I ask.

"I fantasize all the time about quitting my job so I can train full time, and that's not healthy," she says, before pausing. "I have a little bit of an all-or-nothing view of it. The parameters can't change too much or I don't feel like I've actually met my goal. I'm attracted to opportunities to set goals and then work intensively to achieve them." This is something Margaret clearly both loves and fears about herself: her ability to ignore doubts and withstand discomfort in service of the goal. "I think that there's some quality in artists and entrepreneurs and criminals and cheats that's similar, which is that they really do have to believe in something and they have to convince other people that it's true, too." This is a talent she believes she adapted from her father, the man who called her, at twelve, a "lying bitch" for telling her mother he was seeing another woman, the man who promised to mail in her deposit check for her freshman-year dormitory but pocketed the money instead. "I just didn't want to be a person who couldn't master my urges, I guess," Margaret says. She stops for a moment, presumably connecting the dots in her head. "You find ways to create meaning and order in your life, and having this physical routine gives you all those benefits, plus mental mojo. It's a survival technique. It's not just like 'I feel better.' It's survival."

BEING IN THE RIGHT PLACE

"It's not shelter or nourishment, but it feels like a need," Jerome says to me.

He goes to more than three hundred concerts a year. At thirty, he is unmarried and works for a not-for-profit. Though music is not his only avocation—he knowledgeably, and often passionately, follows sports and politics—his house is filled with more than four thousand vinyl records.

The reason for his prodigious concertgoing, he says, is best summarized by a conversation he had with Trey Anastasio, the guitarist and lead singer for Phish.

"I remember explaining to Trey that the first time I had heard the word 'flow' in a certain context was in a Phish song called 'The Lizards.' And then later, I learned in therapy what the word means in that context. I told him the reason I go see Phish is to achieve that experience. When I am seeing and hearing the band, nothing else matters. He said that's why they play, to feel that level of focus."

Jerome describes the effect of music on him—what's in front of him is the only thing that is part of his thought process. He doesn't drink or do drugs.

It's simply that "*this* is all that matters at this very moment." He goes to see music in order to discover and to be engaged. "A good live music experience can bring energy and a positive feeling, and if I've had a long or a negative workday, that experience can bring distance from that. When I return to work from a trip to see music, I often get asked whether I had 'fun.' I struggle more than I should with that question. Fun? Sure. But it doesn't matter. In instances in which I'm getting on a plane or driving for hours, I hardly have any choice in the matter. It's where I'm supposed to be."

When he was four, Jerome had a brown Fisher-Price tape player. He used to play the B-52's album *Cosmic Thing* while walking around the house. A few years later he heard Pearl Jam for the first time; they became his favorite band. He's now seen them seventy times. "They're still my favorite," he says.

Jerome's parents encouraged this type of behavior. His mother in particular nourished his interest in music: "Before I was old enough to ride in the front seat, the quizzes would begin. She would turn on the local oldies station and test me on different Motown acts. Understanding what made the Four Seasons bad and the Four Tops good was as important as any life lesson. I don't know how old I was when I completely understood, but I know now—she was teaching me about authenticity."

What's most important to Jerome in music is honesty and earnestness. "Even if it's a sad song or an excruciatingly painful song, if you're a fan, you see the achievement there, and that can be a positive experience, too. You don't walk out of a concert after an hour of sad songs necessarily feeling sad. You might, but there's the potential to understand that you were part of something great, that you witnessed something special." There are times when music can move Jerome to sadness. Angel Olsen's *Burn Your Fire for No Witness*, his favorite album to come out in 2014, with its melancholic character, evokes a feeling of empathy with the artist's expression. Vic Chesnutt's song "Flirted With You All My Life" can trigger pain.

"Because of my depression, it was difficult for me to seek out positive experiences," Jerome explained. Going to concerts was not always easy, so he began setting a schedule and buying tickets well in advance. There were times when he bought tickets only to throw them in the trash the night of the show.

One such moment still bothers him. St. Vincent was playing a headlining show for the first time in DC. The sense of regret when he folded up that ticket was instantaneous, but the negative inertia that kept him from going was even stronger.

While Jerome had a strong relationship with music since childhood, in later years, during his depression, he challenged himself to go to concerts even when it was difficult. His efforts proved worth the investment. "People who want to die don't make plans.

"I've spent the majority of my life wanting nothing more than to just be in the room with the music I loved," he explains. "I've worked hard to make sure nothing takes that away from me again."

Why push himself to go? "There is a sense of belonging—not belonging in the sense that you belong to a community, but that you belong, that you're in the right place."

What grips him most is the search for new music, the promise of discovery: "I'm drawn to finding more things that are good. I don't care what anyone doesn't like, but I'm curious about what people like.

"It's the idea that there's something out there that's good that I would take pleasure in. There's more out there that I want to find. I don't like the idea that there's something out there that I don't know about."

How might this thrill of discovery help to assuage the pain of mental illness? This path to relief is not unique. Creating incentives to seek out positive experiences can prove to be lifesaving, literally vital. On our own, we are capable of short-circuiting the endless loop of rumination and self-defeating thoughts.

DISTRACTING THE BLACK DOG

In 1911, during his term as home secretary, Winston Churchill wrote a letter to his wife, Clementine, about his "Black Dog." Those two words have since become one of the most famous figures of speech in political leadership. Churchill's letter was infused with a spirit of hope; he had just learned of the recovery of an acquaintance who had been treated for depression: "Alice interested me a great deal in her talk about her doctor in Germany, who completely cured her depression. I think this man might be useful to me—if my black dog returns."

Ever since then, scholars and historians have asked, what exactly was Churchill's canine shadow? Throughout his career, he shouldered the burden of sending many thousands of people to their deaths. As head of the Home Office, he oversaw dozens of cases of capital punishment. In his next role, as first lord of the Admiralty during World War I, he was saddled with the responsibility for casualties of war. The same held true when he became prime minister in World War II, sending men into battle and functioning as the leader of a nation under daily assault from Germany. His was a life that would inspire a certain level of melancholy in nearly anyone.

Yet Churchill never revealed the depths of his dark moods. Over these years, he published three dozen books, and in 1953 he earned a Nobel Prize in Literature. He never lost his remarkable energy and mental vitality. It wasn't enough to be running the world's greatest empire; he felt the need to publish a small library of history, biography, and fiction. The burdens of his life seemed to spur his productivity.

While his letter speaks to his yearning for a cure, there's no evidence that Churchill was ever incapacitated by depression. The evidence does suggest that during his time at the Home Office his responsibility for the fates of other human beings caused him great anguish. Yet he was so effective in his role that he was promoted to the Admiralty. Even when his Black Dog loomed most threateningly, he was somehow able to overcome its grip.

Painting was central to his admitted self-therapy, though his reputation as a prodigious drinker suggests he had an ancillary one as well. (Observers reported that while he was never visibly drunk, he often sipped a weak concoction of Johnnie Walker and water, closer to mouthwash than a mixed drink, throughout the morning.) He first applied his brush to canvas in 1915 to pass the time as he waited to be sent into battle in France. He'd resigned from the Admiralty in disgrace after a major defeat in the Dardanelles, and he joined the British army to risk his life in the comparatively humble rank of colonel. He fell in love with painting, feeling that he'd freed himself not only from the pressures of the world around him, but also from the ongoing war between his light and dark moods. It was his hobby, his recreation, but mostly his escape—his distraction from the burdens of living.

Churchill's book *Painting as a Pastime* reveals how he fought the Black Dog with art. With a mix of statesmanlike lucidity, self-awareness, and gratitude, he describes the inner tranquility his hobby brought him. Painting seemed not only to quiet his mind, but also to teach him the value of meditative detachment. "The tired parts of the mind can be rested and strengthened," he wrote, "not merely by rest, but by using other parts. It is not enough to merely switch off the lights which play upon the main and ordinary field of interest; a new field . . . must be illuminated."

With painting, nothing was irrevocable. A badly painted stretch of canvas can be scraped clean. As he described it: "You can build it layer upon layer if you like. You can keep experimenting. You can change your plan to meet the exigencies of time or weather. And always remember to scrape it all away."

Early on in the book he writes, "One can only gently insinuate something else into [the mind's] convulsive grasp." It's a telling turn of phrase. Churchill's gloom had a sort of manic momentum and insistence, echoing Styron's characterization of his depression as a brainstorm. His mind was nearly impossible to slow down once it had begun to race. After a long day, painting allowed him to throttle this relentless storm, to focus on a harmless canvas where no lives were at stake:

Painting is complete as a distraction. I know of nothing which, without exhausting the body, more entirely absorbs the mind. Whatever the worries of the hour or the threats of the future, once a picture has begun to flow along, there is no room for them in the mental screen. They pass out into shadow and darkness. All one's mental light, such as it is, becomes concentrated on the task. Time stands respectfully aside, and it is only after many hesitations that luncheon knocks gruffly at the door.

BELIEF

Reverend Rachel Rivers's understanding of God is not that of most pastors you will meet. As a minister of the Swedenborgian Church, a Christian denomination, and as a psychotherapist, Rivers believes that God cannot be defined by an institutionally determined set of qualities, let alone contained in a three-letter word. In fact, when parishioners confess to Rivers that they've lost their faith, she encourages them to describe the God they no longer believe in. Almost invariably, the God they depict is not one she believes in, either.

God, says Rivers, is best understood as the fact of love. To be at one with God is to recognize "all the way through your bones . . . not just intellectually, but emotionally . . . that you're known and loved for who you are." This sort of radical acceptance does not depend on adherence to a particular set of behaviors or faith in a predetermined divine.

Rivers has long been interested in the nature of belief. How does belief in God offer comfort to the faithful? Is faith an entirely internal process, or might it originate outside the mind? During her training as a psychologist, Rivers found herself drawn to these metaphysical questions about spiritual experience. She embarked on an exhaustive study of psychological texts, but the answers proved elusive; it only was when she turned to the philosophy of the eighteenth-century renegade theologian Emanuel Swedenborg that she began to find satisfying answers.

A scientist by training, Swedenborg was by no means a conventional scholar of Christian theology. He recognized that religious belief is necessarily personal, indeed idiosyncratic; no single religious dogma could possibly speak to every soul, and no description of faith could capture the manifold experience of the faithful. Swedenborg therefore encouraged believers to chart their own ways to God—"different paths up the same mountain"—even as he insisted that we are not merely solitary, wandering beings: we belong to a collective united by God's love.

An active pastor and psychotherapist, Rivers has drawn freely from both Swedenborgian theology and contemporary psychology in her ministry, both in counseling sessions and in sermons. When we met in San Francisco, Rivers said she found in my theory of capture a confirmation of her own experience of spiritual anguish. I asked her if the solace of religion might be yet another instance of capture. Is the addict or the depressive or the criminal who "finds God" merely substituting one form of capture for another?

"That's what it feels like," Rivers responded, "but only when we are looking at it from the outside—imagining what it would feel like to be captured by the divine." Whereas other forms of capture erode our autonomy, she said, being captured by God is an experience of radical freedom. To illustrate her point, Rivers contrasted the problem of faith to the quandary of the overeater. While pondering a pint of ice cream in the freezer, we may feel that depriving ourselves of the sensory pleasure of just one scoop would involve an abnegation of freedom. We deserve that ice cream, after all, and it's our right to claim it. Only later, when we are no longer gripped by blinding desire, do we realize that the exact opposite is true: choosing not to eat the ice cream represents the purest exercise of the will.

Turning toward God, Rivers explains, is structurally different from turning to addictive substances or immersing ourselves in piles of work. Religious belief in its truest sense involves fostering a reciprocal relationship with God, whereas other objects of capture will never return our affection or investment. When we are captured by the divine, says Rivers, we are not "hijacked," but rather are freed from the everyday distractions and quandaries that otherwise obscure our most basic needs. For the faithful, making decisions still requires effort, and this, too, is an important facet of religious experience. In our embrace of God there is a "feeling of partnership" that nonetheless preserves our autonomy. We are not simply at the mercy of our neural wiring or our brain chemistry; rather, we must, in Rivers's words, be "recaptured over and over and over again."

For Rivers, to be free is to feel oneself recognized and loved, not in a vacuum but in the context of an organic, evolving relationship to the world. The believer is no longer "a prisoner to a small idea, a more limiting idea of who she is," but understands herself as part of an infinitely expansive whole.

GOOD-BYE TO ALL THAT

The vivid flashbacks and intrusive memories that characterize post-traumatic stress disorder would seem very clear examples of capture. The poet and novelist Robert Graves, who fought as a soldier in World War I, brings a precise self-awareness to his own story of wartime trauma and recovery. There is much to be learned about both the torment and the potential healing power of memory from his memoir, *Good-Bye to All That*, written in an intense stretch of four months more than a decade after the end of the war.

In it, Graves recounts the odds against surviving in the trenches on the Western Front. Within three months of enlisting, one in five younger officers would be killed, one would be seriously wounded, and three would sustain wounds minor enough for them to return to the front within weeks. Not surprisingly, this first modern war also saw the first officially recorded descriptions of what we now call posttraumatic stress disorder (then known as shell shock or neurasthenia) and its first study by the medical and psychiatric professions.

Graves writes in his memoir that an exceptionally difficult spell of combat, along with the death of a close friend, pushed him to the edge: "My breaking point was near now, unless something happened to stave it off. Not that I felt frightened. I had never yet lost my head and turned tail through fright, and knew that I never would. Nor would the breakdown come as insanity; I did not have it in me. It would be a general nervous collapse, with tears and twitchings and dirtied trousers; I had seen cases like that." Only after another hundred pages of wartime stories does Graves open up more fully about how he is feeling. Long after the peace is celebrated, he writes, "I was still mentally and nervously organized for War. Shells used to come bursting on my bed at midnight, even though Nancy [his first wife] shared it with me; strangers in daytime would assume the faces of friends who had been killed." The flashbacks would continue to haunt him for more than a decade after the war.

At first Graves did not seek help for his neurasthenia. He did, however, organize an intervention for his close friend and fellow poet Siegfried Sassoon, whose own breakdown had led him to protest the war publicly and to desert. Graves got Sassoon admitted to Craiglockhart War Hospital in Edinburgh, Scotland, where the psychiatrist W. H. R. Rivers was experimenting with psychoanalysis in the treatment of soldiers' nervous symptoms. Rivers was especially interested in the poet-soldiers under his care, both Sassoon and the English poet Wilfred Owen. Through his letters and conversations with Sassoon, Graves was exposed to Rivers's ideas, which influenced many of the choices he made as he managed his own recovery.

In late 1917, the same year he began working with the war poets, Rivers delivered a paper to the Royal Society of Medicine on "The Repression of War Experience." In it, he argues for the constructive role of memory in balance with forgetting, then a new concept. The accepted wisdom of the time prescribed the deliberate repression of traumatic memories, including the avoidance of any mention of the war in conversation. Rivers noted how such a strict regimen did not prevent nightmares, anxiety, and psychosomatic ailments from surfacing, but that these symptoms subsided when the individual relaxed his efforts at repression. Only by communicating these traumatic memories, Rivers argued, could veterans achieve true catharsis.

Graves struggled to write a novel that he hoped would "rid [him] of the poison of war memories," but he "had to abandon it—ashamed at having distorted my material with a plot, and yet not sure enough of myself to turn it back into undisguised history." He would not find a true voice for the work until 1929, when he wrote his memoir. Graves's recovery, and the integration of painful memories, had a timetable that could not be rushed.

Graves found some solace in daily life. Marriage brought children and an unforeseen advantage well ahead of the times: his wife was an ardent feminist, and his days became filled with housekeeping duties and child care, which suited him well. Graves and his wife were active socialists, too, which distanced him further from the Britain of rigid class barriers and imperial power that prevailed before the war. Much of the black humor running throughout his memoir turns on the absurdity of military rituals contrived not only to enforce discipline and inspire loyalty, but also to preserve the rigid divisions between officers and men in the lower ranks that prevailed in the civilian world. Any intellectual tool, be it socialism or feminism, that put space between him and the military mind-set could be useful.

Graves also put miles between himself and the war, moving to Cairo for a teaching job. Eventually, at the age of thirty-three, having just completed *Good-Bye to All That*, he moved permanently to Majorca, where he would spend the rest of his life. A full six years later, he could finally write with the ease earned through long effort and hindsight:

> *Entrance and exit wounds are silvered clean,*
> *The track aches only when the rain reminds.*
> *The one-legged man forgets his leg of wood,*
> *The one-armed man his jointed wooden arm.*
> *The blinded man sees with his ears and hands*
> *As much or more than once with both his eyes.*
> *Their war was fought these twenty years ago*
> *And now assumes the nature-look of time,*
> *As when the morning traveler turns and views*
> *His wild night-stumbling carved into a hill.*

RECONCILIATION AND FORGIVENESS

Though the injustices of South African apartheid swarmed around her, Thandi Shezi had always been a vibrant young woman. The cheerful mother of two was raising her son, Ayanda, and daughter, Mbalizethu, in a large and loving multigenerational family, surrounded by aunts, uncles, and grandparents.

But there were few opportunities for schooling in her Soweto township. So, in the late 1970s, her father took Shezi to the Zulu capital of Ulundi so she could complete her education. As a student, she says, she was talkative and loud, though rebelliousness was not built into her character—she didn't smoke or drink, and was dutiful about her schoolwork. Shezi loved to socialize with her classmates, and was known for her willingness to lend a hand to anyone who needed help. It was that generosity of spirit, combined with her outgoing nature, that got her elected as a student representative to the Inkatha Freedom Party (IFP), one of the political parties in the forefront of the antiapartheid movement.

With that first step into activism, her life began to transform. Shezi's main responsibility as a member of the IFP was to attend meetings and bring information about the party's activities and agenda back to her peers at school. She continued that work until graduation, and then returned home, where her interest in politics deepened. Soon she joined the Soweto Youth Congress, which was affiliated with the national South African Youth Congress (SAYCO), a leading antiapartheid youth organization. Increasingly active in the underground group Umkhonto we Sizwe, she began to covertly transport ammunition that would be used to destroy government buildings.

An informer gave Shezi's name to authorities, and in September 1988, the South African police arrived at her home to arrest her—but first they beat her. The men struck her with such brutality that her mother, who had been forced to watch, cried out to them, "Do not kill her in front of me. Show me her body when you are finished."

Shezi survived the attack and was thrown into a van with other detainees. Their destination was the central police station of Johannesburg, legendary for its brutality. The authorities called her a terrorist and demanded that she point them to antiapartheid movement leaders and reveal the location of stockpiled weapons. She claimed she did not know. She was beaten again, and officers poured a mixture of water and acid onto a sack that had been placed over her head. As the acid washed down her face, she could feel her skin burning and peeling; the acid was so strong that it caused permanent vision loss in one eye.

But it wasn't over. The horror was taken to another level when four white policemen entered the cold, dark room where Shezi had been delivered, almost naked and in chains, and they each raped her.

During those moments of horror, Shezi says that the only way she could think to survive was to detach herself from the physical reality of her experience. "A small voice inside said to me: to survive, you have to remove your soul from your body," she explained more than a decade later. "I took control of my soul. I decided to separate my soul from my body. It was as if a supernatural power had taken control of me. On the spot, I decided that nobody is going to get control of my soul. As they huffed and puffed on top of me I quietly separated my body from my soul. It was as if my soul was looking down on my body being abused."

In the months to come, Shezi found some solace in the companionship of other political prisoners, many of whom had also been brutalized. Their shared experience and common commitment to end apartheid kept her going. "During lunch, we would be taken out to the sun, and we would meet each other on the grounds and talk to each other. So when I was in jail, there were so many people that were helping me to cope with the situation."

Shezi was released from prison after her family learned where she was being held and hired an attorney to demand her freedom. When the taxi dropped her off close to home, she was almost too frail to walk, and shuffled along the road to her front door. It took her an agonizing hour to complete what should have been a five-minute stroll. "That is when I had to face the reality that I've been raped, I've been tortured. That is when the pain started."

For almost a year after she returned, Shezi rarely went outside, instead locking herself inside a dark room and sobbing for hours on end. Haunted by nightmares, plagued by anger and fear, she soon found herself directing her rage at Ayanda and Mbalizethu, whose lives had been torn apart by her arrest. "I was no longer a loving mother," she said. "I found myself shouting, screaming, throwing abusive language, and violent all the time. Anything that reminded me of the rape made me throw a tantrum. I would bang my child's head against the wall and only feel bad about it afterwards." Her daughter's simple request for a slice of bread could make Shezi irritable enough to deliver a slap.

Divisions in her family only made things worse, with some relatives refusing to associate with someone they called a terrorist. Shezi was the only politically active person in the household, and no one understood what she had gone through, or what she was going through still. Rage and violence, emotional responses that had never before been part of her life, seemed to consume her. "I was very angry, angry at myself, angry at the system. I was telling myself, 'You looked for it. Why did you get involved in politics?' "

Shezi could no longer handle sexual intimacy, and her partner ended their relationship. She found herself incapable of loving or being loved by another man. If a man so much as approached her to say hello, she feared his intentions. She fought with her parents and continued to strike her children with such ferocity that her daughter, too, became a victim of trauma. Shezi seemed to have lost any chance at a normal life. "Everything came to a standstill," she recalled. "I felt like some part of me had been taken away."

It stayed that way for many years. She says her favorite books and music offered some respite, but long after the framework of apartheid had been dismantled, she continued to be enraged that her captors were free while she was trapped by their actions. Thoughts of revenge often haunted her. "In my head, I was thinking that if I could find one of those people who did this to me, I'll kill him. I wanted the revenge so much for that pain that I went through."

At the same time, she bore a heavy burden of guilt and frequently turned the blame inward, as if she were somehow responsible. So great was her shame that she told no one all the details of her time in prison. "I just kept it in myself," she says. "I thought it was going to be my secret. I thought I'd done something that I deserved to be treated like that. . . . All along I had thought I could keep this inside myself and just retain it up until I die." But the silence threatened to consume her. "It took me ten years before I could realize that I've got this baggage inside me that is eating me up," she acknowledged.

Slowly and with great difficulty, she began reaching out to others, becoming a counselor at the Khulumani Support Group, a grass-roots organization established to support the victims of apartheid. But though she encouraged others to share their stories, at first she revealed little of her own. Her colleagues and clients knew she had been imprisoned and tortured, but she held on to the secret of the gang rape for a long time before finally telling her story in a one-on-one counseling session.

It would be another two years before she was ready to tell the world. In July 1997, Shezi stepped onto the witness stand at a hearing of the Human Rights Violation Committee of the Truth and Reconciliation Commission, established by South Africa's unity government to reveal the abuses of apartheid and promote reconciliation. There, she described in excruciating detail the torture she'd endured. It was the first time even her own mother heard the full story.

Shezi says she thought that offering her testimony of the crimes committed against her would help her continue the healing process, but she admits it did the opposite. "You rip open the closed wounds and you are left with a gaping wound," she recalls. The role of the Truth and Reconciliation Commission was to gather information, not to promote healing, she says, and it provided no mechanism for ensuring that victims who testified had adequate access to emotional support afterward. There was little in the way of reconciliation—no compensation was offered to help her rebuild her life—and truth was also scarce. The officers who had abused her maintained their innocence and were never prosecuted.

The real breakthrough in her long, slow road to healing came in the late 1990s, when she participated in a theatrical piece entitled *The Story I Am About to Tell*, reciting her testimony from the Truth and Reconciliation Commission to audiences in South Africa and Europe. She was joined onstage by Duma Kumalo, a human rights activist who had been sentenced to death following a violent protest march in Sharpeville, and Catherine Mlangeni, whose son had been killed by a policeman in charge of a government death squad. "Standing on the stage, relating what happened, dramatizing it, it made me laugh at myself," says Shezi. "Yes, at the end of the performance, I cry, I will be broken. But as time goes on, repeatedly telling my story and telling and telling—it made me able to acknowledge myself, to love myself, and to be able to tell my story without crying."

Talking, she says, is healing. "If you don't talk, share your pain with others, you are not going to reach that."

Today Shezi, now in her early fifties, is again talkative and capable of laughter. Her irritability and anger have dissipated, she says, replaced by acceptance and forgiveness. "You have to forgive the perpetrators so that you can be able to move on, and also forgive yourself. You have to say, 'Well, it happened. I have to pick myself up and move on with life.'"

This forgiveness should not be interpreted as compassion for the perpetrators, but rather the need to act compassionately toward herself. Shezi recognized that holding tight to her anger and fear was harming only her, and she didn't want to wake up every day as a victim. Forgiveness, she says, also makes it possible to be of use to others. "You have to put a distraction to the traumatic events that you have experienced so that you'll be able to move on with life. Because if you are still holding on to the past and the pain, you are not going to be able to be there for somebody else who needs your help."

Shezi will never forget what happened to her. But she has been able to gaze directly into the past and reject its dominance over her life. To shake free of someone else's evil grip demanded an internal reckoning with her personal experience of trauma, "to reconcile with your inner self," she says. "Accept that this happened, and I cannot change it. So you have to accept, live with it, and learn how to deal with it."

A TOOLKIT BORROWED FROM BUDDHISM

Buddhism began in India two and a half millennia ago, when its founder, Siddhartha Gautama, taught others the methods he had used to reach enlightenment himself. A religion that excludes the notion of an all-powerful creator, Buddhism developed into richly diverse traditions as it spread throughout Asia. Recently, it has been the subject of much inquiry from scientists, therapists, and doctors who are studying the mechanisms of attention and emotion as these relate to a variety of mental conditions.

Buddhism offers as its essential premise the belief that liberation from mental suffering, whether driven by worldly desires and the trappings of ego or by hatred and enmity, is both the goal of spiritual enlightenment and a practical possibility. This would appear to be a useful model for attaining freedom from the captured mind. But is it really possible to quiet the neurological circuitry that biases attention to salient stimuli?

I posed this question to Evan Thompson, a philosopher whose work encompasses cognitive science and philosophy of mind as well as Asian philosophy. "Buddhism would certainly claim that you can unbias attention," said Thompson. This goal can be accomplished, he continued, through mental training, including meditation or other contemplative practices, and by ethical training: "how you live in the world and how you behave."

He went on to describe how Buddhism strategically substitutes one form of capture for another: "In Tibetan Buddhism and in Indian Buddhism generally, they would use the term 'applying an antidote.' For example, an antidote to anger might be generating a mental state of wishing somebody well, or of compassion or loving-kindness. You're shifting the bias from a negative to a positive, so you're still working with a biased or referential function." There are also meditation practices, common to yogic traditions as well as Buddhism, that shift attention from external sensory stimuli and focus it inward, whether on the breath or a repeated sound or mental image. "They're all ways of disengaging attention from outer, fluctu-ating sensory stimuli and anchoring it in something that's internal, and that gains an increasing mental stability the more that you prac-tice." The more you can stabilize your attention through such in-tense concentration, the quieter, more focused, and more peaceful the mind becomes.

Thompson also describes a different type of Buddhist practice: a form of mindfulness meditation that cultivates an open, nonreactive awareness. The goal is "a balanced, poised state of mind that's ready to respond or act in a positive way but is not biased by either sensory salience or affective salience." In this type of meditation, though your eyes are open, you do not focus on an object; rather, you maintain mental stability and an alert clarity of awareness. When a distraction arises, whether an external intrusion on the senses or an inner thought, you acknowledge it and then release it. You do not let it hold your focus or carry your thoughts off in a new direction. This "open monitoring" practice, as Thompson calls it, aims to improve emotional balance, or what Buddhists call equanimity. A mind that can process stimuli without bias can, in theory, keep an even keel through favorable or unfavorable winds.

Formal training in the meditation techniques that lead to this open, nonjudgmental awareness is traditionally undertaken by Buddhist monks who have dedicated themselves to years of practice in the structured environment of a monastery. But mindfulness practice has also proven effective in surprisingly small doses. The basic techniques have been adapted, often secularized and stripped of Buddhist references, for clinical applications; for self-help workshops; in elementary school classrooms, corporate training sessions, and even military training. Some question whether lifting these techniques out of the ethical context in which they were traditionally taught may have unforeseen consequences, but there is strong evidence for their benefits in mental health.

I spoke to Jon Kabat-Zinn, professor of medicine emeritus at the University of Massachusetts Medical School, who pioneered the use of mindfulness meditation in the treatment of chronic pain, depression, and stress-related disorders. I asked him, as I had Thompson, if one can ever really break free of the mechanism of capture. "The short answer to your question is yes, I do think that it's possible. That's the essence of the work that my colleagues and I have been doing for many years. What's required is what we call an orthogonal rotation in consciousness, where all of a sudden you're looking with different kinds of lenses. While everything is the same externally and internally, including your addictive impulses, you're recruiting another dimension of your humanity that allows you to make choices that you never, ever thought were available to you. And that is the freedom: the freedom to actually see the addictive impulse and to cultivate a muscle to a sufficient degree of strength where you can tilt in a different and much more profoundly healthy direction."

How does one achieve this freedom? A typical format for Kabat-Zinn's training in mindfulness-based stress reduction, and related programs that have substantially reduced participants' risk of relapse in major depressive disorder, is an eight-week course that combines weekly group sessions with daily individual practice. Can one really learn something so radical as escape from capture in this time frame, or is it another process of substitution?

"It's a lot more complicated than that," answers Kabat-Zinn. He explains that the mindfulness practices adapted from Buddhism depend on meta-awareness rather than substitution of thought patterns, and that research on treatment for depression has also gone far beyond the familiar substitution strategies of cognitive therapy. "It's really a different paradigm, where you're bringing awareness to thought. Rather than substituting a more positive thought for a more negative thought, you're looking into the direct essence of thought and recognizing that your thoughts are actually little 'secretions' of the mind that come and go extremely quickly, and that only have power over us if we actually believe that they are true—that their content is indistinguishable from fact."

So how do you loosen the hold of thoughts that are preoccupied with the self, which is busy generating a story about your failures and how you're a fraud? The key is not to argue the facts; trying to think your way out of depressive rumination only deepens the ruts of the neural pathways where wheels have been spinning too long. Instead, as Kabat-Zinn suggests, you allow those thoughts but stand back from them and recognize "that toxic pattern of self-deprecation is really just a thought habit. It has no more actuality than anything else." Moment by moment, you call upon a grounded bodily aware-ness that sees thoughts as just thoughts and not facts. And that shift in awareness, that orthogonal rotation of consciousness, will be visible in the brain via neuroimaging techniques. "If you cultivate other ways of being in relationship with the present moment than with your thoughts, you see different networks in the cortex light up that don't have to do with a narrative self-referencing."

It's clear that the mechanism of capture may play a constructive if paradoxical role in this learned freedom from capture. As Kabat-Zinn explains, "What we're talking about is a trainable skill to access other aspects of our being that change our relationship to the things that capture our attention." It's not that your attention avoids capture by salient stimuli, but you don't lose sight of what's happening. "You know that it's being captured and you can, in some sense, at least play at or experiment with not falling into that usual mindless, habitual, automatic pattern. You exercise that muscle over and over again, then you can actually shift the default mode . . . to realize that, 'Oh, here I go, that old pattern's emerging, but if I hold it in awareness, I don't have to collapse into it.'" And in that moment, you're free to make a different choice.

Many have asked whether the liberation offered by Buddhist practice involves abandoning the rich textures of our emotional lives. After all, if capture lies at the root of great art and literature, and of many other spiritual and religious traditions, what do we stand to lose by loosening its bonds? If we learn to decrease our reactivity and reflexive judgment, do we also decrease our range of emotional responses in a way that handicaps creative potential or limits the full scope of human experience?

I asked Evan Thompson. Capture, he offered, "would imply an inability to be flexible and to disengage and reorient appropriately should a situation demand it. You could say an artist is captured by a certain vision of beauty or a mystic is captured by the ecstasy of a beatific vision of the divine. That would be capture if the artist or the mystic were unable to be flexible and engage appropriately with others and instead were single-mindedly or maniacally pursuing his or her vision. It could be a powerful artistic or mystical vision, but if it doesn't have that flexibility, then it would be a form of capture." Capture in that sense may not necessarily be experienced as negative, but it would lack "resilience, flexibility, sensitivity to others, the ability to act in ethically appropriate ways. . . . Because if you're captured you're bound up and you have a lesser range of movement, less of a repertoire of possibilities for your thinking and your acting."

In response to the same question, Kabat-Zinn reminds me first that mindfulness is not a philosophical inquiry but a practical exercise that must be experienced to be fully understood. So, from his own experience of long practice, he offers this insight: "By coming back over and over and over again to the frame—to the awareness that holds thought or affect, or that holds the experience of interconnected patterns that are usually below the level of our awareness—they become more real for us. The salience gets more nuanced and richer, rather than less. And then, of course, you have a more expanded field of possibilities for responding rather than reacting." Our thoughts and emotions become no less vivid or richly textured; if anything, we can perceive them in far more subtle detail. They only lose their power to hold us in thrall.

IS THERE FREEDOM FROM CAPTURE?

In working with her patients, Danielle Roeske, a psychotherapist who also has a background in philosophy, uses an integrative approach that does not limit her to any one therapeutic school. She believes passionately in people's ability to change, but this does not prohibit her from feeling that "freedom" is too strong a word to use when it comes to loosening the bonds of capture.

"Freedom from capture signals a permanency that makes me uneasy, because I don't think it works that way," she explains. "I don't think we work that way. No matter how much change happens, we never discard parts of who we are. To me, freedom would indicate those parts are gone. And that doesn't happen. I think release is a better way of thinking about it, because then what was dominant, whether it be an object of capture or concept of capture, is no longer the dominant force. And that is much more my sense of what happens in transformation."

When I press Roeske on the issue of freedom versus release, she explains that she doesn't see it as a bad thing.

"I think it's a more acceptable sense of how we can grow and develop as individuals," she explains. "And I think when there's a change in one's relationship to affective experience, when people go through a transformation of how they experience themselves, it's not uncommon for their sense of their history to be rewritten as well. For past events that were considered traumatic or horrible, there can be a clearing for a new understanding of those experiences so that they're less frightening. They're not forbidden anymore, and might actually be seen as avenues for growth. Because when we bury things inside, it's not like they go away. They just haunt us in different ways."

Perhaps the most difficult challenge in beginning this process of release, as Roeske prefers to call it, is identifying what is causing an internal struggle. "Simple as this might seem, it's not an easy place for one to get to," she points out. "As the power of capture takes hold of one's internal authority, it can appear as the only solution. This accounts for the baffling reality that makes us so resistant to change."

Often, people don't reach this point until the pain caused by their capture begins to overwhelm. "Gathering the courage to change is perhaps one of the more difficult parts of transformation, as it requires a deliberate leap into the unknown," she remarks. "To say that such an act requires a gesture of faith seems to put the cart before the horse. Faith is what follows after new experience has been had, while hope is the quality that comes right before it."

Taking such a risk is crucial, however, as it is the way in which we open ourselves to something meaningful enough, and therefore salient enough, to change long-held perceptions.

"It takes time to incur a true repatterning and reshaping," Roeske concedes. "While there are those few who, for reasons unknown, have profound and immediate revelations, for most individuals the path is slow to unfold and transformation is only a hindsight discovery. But it is no exaggeration to say that the world is transformed for those released from the confines of capture. While the increments might be small, with sufficient accumulation a revolution of perspective takes place."

Yet even when such a drastic shift in perspective takes place, there is always the possibility of return. That is, the long-established neural patterns remain in place even as new ones form.

"Perhaps the point here is that one is never rendered invulnerable," Roeske offers. "When people enter recovery through, for example, an AA model, when they stop drinking and when their lives are together and they're upstanding and functional and life is good—it's all the more important that they continue to work the recovery program. And that's not because there is something disingenuous about their change or transformation; it's because the nature of the addiction is that it's always there. And if there's not a continued reinforcement of a new way of being and understanding, there will be a gradual pull to go back to the foundational tried-and-true escape through drugs and alcohol."

Roeske also believes that the community aspect of Alcoholics Anonymous may offer an advantage in breaking the spell of capture: "Similar to religious traditions and Eastern practices, AA is useful in a way that psychotherapy can be limited. There's an important feeling of unity, of being part of something. I think this is important to the surrender process. And I think when there's a sense of capture, it usually eclipses any sense of unity. If the mind is captured, it is only going to do what it does under the state of capture—until there's some other force, whether it be a deity or a person or a suggestion that's allowed in, in an authentic way."

I ask Roeske if it is in that moment that another affective experience will take place that allows for change.

"That's right, but it's twofold," she responds. "For example, if alcohol has been the solution, anytime there's emotional discomfort, which is not uncommon, or extreme joy—any strong affect—then there's a real need to be able to tolerate those feelings without going to that solution. And so it means creating space where those affects can be experienced in a new way, which is painful but also liberating. Our sense of ourselves is only based on our experience of ourselves—we cannot feign a persona as an act of faith that something more integral will emerge. Such 'acting' will likely backfire if there is not at least some belief that one can become the person that he or she aspires to be."

When you've seen patients make this kind of shift, I ask her, how difficult has it been to maintain?

"There are no people, even if they really have a sense of peace in their lives, who won't get triggered," Roeske says. "There has to be a real desire for presence, so that rather than reflexively engaging, or falling into the overdeveloped neural pattern, there's actually room to choose something different, or there's room to allow the affective experience to pass through without ascribing the behaviors of the past or the beliefs of the past to it."

Roeske's earlier use of the word "surrender" caught my attention; I ask her how this word, in the context of capture, differs from "release" in her mind.

"I think surrender is what happens when there is just that fragment of willingness to allow something else in—to stop fighting the feeling and let yourself feel the impulse without giving in. It's kind of a funny thing, because it seems like it's work, but it's actually an opportunity to stop controlling things, to stop trying to control your affect or your experience, and to allow yourself the experience. Then release is the quality of experience that is able to be felt after."

A MODEST FORM OF AUTONOMY

Human experience is, by nature, chaotic.

At any given moment the brain is bombarded with an astounding variety of stimuli. These stimuli arrive steadily at the gates of the mind without form or structure. Faced with this barrage of disconnected perceptions, we need to shape them, to give them meaning—ultimately, to construct a sense of self.

In order to do this, we engage in a continual process of automatic filtering, with rapid-fire actions affecting where to aim our attention.

This process has practical applications, such as helping us to become aware of the child who has just run in front of our car, as well as much more profound implications for our lives. The same filtering process that helps us avoid an accident is also responsible for selecting which experiences come to define each of our individual realities. Where one person might experience defeat in the face of rejection, another is inspired to act with greater determination. Where David Foster Wallace felt a deep sense of fraudulence and inadequacy, another person might have experienced a profound sense of accomplishment.

Over the course of a lifetime, each of us creates a coherent account out of the jumbled, often fragmentary chaos of life—the ever-evolving narrative of our lives. Over time, certain characters and experiences emerge as central, while others prove tangential, soon to be forgotten. Without self-created storylines, the trajectory of our lives would feel like a constant scattering of random details. Joan Didion gets to the heart of it in few words: "We tell ourselves stories in order to live."

An essential question, then, is how our stories are conceived.

I believe that capture is the instinctive process by which we wrest a narrative from our disordered surroundings. And therein lies the challenge: though capture permits us to make sense of the world, it does so largely outside rational thought. These stories, then—the stories of our lives—do not always feel like the ones we intended to tell.

Is there freedom from capture? Can we throw a switch and see the entire stage, every trapdoor and spike and rafter, for what it is? In the most basic sense, the answer is no. Attention is, by its very nature, selective and self-reinforcing. Our environments—historical, economic, physical—dictate what becomes salient for each of us, forging patterns that in turn determine how we experience the world and, ultimately, who we become.

Yet a more modest form of autonomy is within our grasp. When we understand capture for what it is, a neural mechanism that both influences and is influenced by our experiences, we can become more than passive chroniclers bound to unhappy or troubling narratives. By becoming aware of the ways in which we deploy our attention, we may even develop a beneficial flexibility of mind, one that allows us simultaneously to tell varied, sometimes even contradictory stories.

We can influence this process not by accepting a static diagnosis, such as "anxiety" or "depression," but by actively changing what occupies our attention. It is possible to put ourselves in the way of a more positive influence, and its attendant stimuli, in order to overcome a prior capture. This happens, in effect, when the new form of capture becomes so important to us that the last one loses its magnetism.

For those with more serious afflictions, change often comes in the form of a conversion experience. Focusing on others (as Chris Ware was unexpectedly able to do when he became a father) or being grasped by feelings of an otherworldly love (as Simone Weil was) can mitigate the condemnatory or accusatory nature of a self-focused capture.

The power to will, however, is not enough to sustain change. The challenge is to draw strength from something other than mere self-discipline—or condemnation. Lasting change occurs when we let go of such isolating pressures and allow ourselves to feel support and connection instead of preoccupation with the self. This transformation of the self often occurs through sacrifice, service, love, belief in a cause, or membership in a community.

To avoid unnecessary emotional suffering, we must first understand the innate forces that govern our minds. Only then will we begin to see how we might use these forces to our advantage. Indeed, we can gradually reshape our minds, even transform our way of experiencing the world, though overcoming one form of capture often depends on discovering another.

Once we understand the underlying mechanism of capture, we can reduce our vulnerability to its most pernicious forms, and perhaps even discover a way to move beyond them.

NOTES

The following pages contain notes providing source references and support for the book. Included as well are several extensive essays, notably on Ludwig Wittgenstein, Carl Jung, and Paul Tillich. These thinkers are key to our understanding of capture. Wittgenstein is a powerful counterpoint to David Foster Wallace; Carl Jung must be considered within any context that includes Freud; and Paul Tillich helps to illuminate the underlying psychology of religious belief. I decided to exclude them from the main body of the text as I felt their stories were a distraction from the flow of the narrative. Still, they are vital to my thesis, and I offer their stories here, instead, as further elucidation to the work that precedes these notes.

[Notes that appear within brackets are the sources for the information contained within that note.]

CHAPTER 1: A HUMAN MYSTERY

THE TERRIBLE MASTER

Sources

Interviews by the Author
Brayboy, Martin (Nov. 2011).
Brooke, Fred (Nov. 2011).
Colmar, Dave (Nov. 2011).
Costello, Mark (Nov. and Dec. 2011).
Desai, Rajiv (Nov. 2011).
DeVries, William (Nov. 2011).

Estell, Dan (Nov. 2011).
Flygare, John (Nov. 2011).
Friedman, Matt (Nov. 2011).
Harris, Kymberly (Nov. and Dec. 2011).
Javit, Daniel (Nov. 2011).
Larson, Nate (Nov. 2011).
Maehr, Marty (Nov. 2011).
McLagan, Charlie (Nov. 2011).
Parker, Andrew (Nov. 2011).
Peterson, Dale (Nov. 2011).
Spaulding, Ralph (Oct. 2011).
Stumpf, Fred (Dec. 2011).
Wallace, Amy (Nov. and Dec. 2011).
Wallace, James and Sally (Nov. and Dec. 2011).
Washington, Corey (Nov. 2011).

David Foster Wallace Works
Wallace, David Foster. *Brief Interviews with Hideous Men*. New York: Little, Brown, 2007.
———. *The Broom of the System*. New York: Penguin, 2004.
———. *Consider the Lobster and Other Essays*. New York: Little, Brown, 2005.
———. *Fate, Time, and Language: An Essay on Free Will*. New York: Columbia University Press, 2011.
———. *Girl with Curious Hair*. New York: Norton, 1989.
———. *Infinite Jest*. New York: Little, Brown, 1996.
———. *Oblivion*. New York: Little, Brown, 2004.
———. *The Pale King*. New York: Little, Brown, 2011.
———. "The Planet Trillaphon as It Stands in Relation to the Bad Thing." *Amherst Review* 12 (1984), originally accessed at http://quomodocumque.files.wordpress.com/2008/09/wallace_amherst_review_the_planet.pdf.
———. *A Supposedly Fun Thing I'll Never Do Again*. New York: Little, Brown, 1997.
———. *This Is Water: Some Thoughts, Delivered on a Significant Occasion, About Living a Compassionate Life*. New York: Little, Brown, 2009.

David Foster Wallace Materials and Correspondence
Books, with handwritten marginalia, that belonged to David Foster Wallace, including Lewis Hyde's *The Gift: Imagination and the Erotic Life of Property*; and Theodore Isaac Rubin's *Compassion and Self-Hate: An Alternative to Despair*. Harry Ransom Book Collection, University of Texas, Austin.
David Foster Wallace Papers. Personal and Career-Related Series 1997–2008. Containers 31.6 [essays and exams]; 31.8 [correspondence, 1992–2007]; 31.9 [early schoolwork]; 31.14 [personal journal papers, 1996, undated]. Harry Ransom Center, University of Texas, Austin.
Bonnie Nadell Collection of David Foster Wallace. Correspondence to Bonnie Nadell (Sept. 20, no year; Oct. 4, 1988; May 26, 1989; Dec. 4, 2007; March 27, 2008; July 24, 2006; Oct. 10, 1995; Feb. 1, no year). Containers 1.1, 1.2. Harry Ransom Center, University of Texas, Austin.
Corey Washington personal property. Letters and audiotapes, David Foster Wallace to Corey Washington. Courtesy of Corey Washington.

Don DeLillo Papers. Series II Correspondence, 1959–2008; Subseries A Alphabetical Files, 1962–2003; Individual Correspondence of David Foster Wallace with replies from DeLillo, 1992–2003. Container 101.10. Harry Ransom Center, University of Texas, Austin.

William E. Kennick Papers. Nine letters to Professor William Kennick from David Foster Wallace (AC 1985), 1985–2002 (one undated item). Box 1, Folder 4, Amherst College Library Archives and Special Collections.

David Foster Wallace Biographies, Interviews, Articles

Bustillos, Maria. "Inside David Foster Wallace's Private Self-Help Library." *The Awl*, April 5, 2011. http://www.theawl.com/2011/04/inside-david-foster-wallaces-private-self-help-library.

Lipsky, David, and David Foster Wallace. *Although of Course You End Up Becoming Yourself: A Road Trip with David Foster Wallace.* New York: Broadway Books, 2010.

Max, D. T. *Every Love Story Is a Ghost Story: A Life of David Foster Wallace.* New York: Viking, 2012.

McCaffery, Larry. "A Conversation with David Foster Wallace." *Review of Contemporary Fiction* 13, no. 2 (Summer 1993).

Quomodocumque. "A Letter from David Foster Wallace, Maybe." *Quomodocumque*, Sept. 14, 2008. https://quomodocumque.wordpress.com/2008/09/14/a-letter-from-david-foster-wallace-maybe/.

Schmeidel, Stacey. "Brief Interview with a Five Draft Man." *Amherst* magazine, Spring 1999. Republished at https://www.amherst.edu/aboutamherst/magazine/extra/node/66410.

Wallace, David Foster. Interviewed by Charlie Rose, March 27, 1997: https://www.youtube.com/watch?v=mLPStHVi0SI, https://www.youtube.com/watch?v=wDIVX7pNwGE, https://www.youtube.com/watch?v=vAT9V2wHx3M, https://www.youtube.com/watch?v=Cjf27-uY0Ss.

Wallace, David Foster. Interviewed on German television station. ZDF, 2003: https://www.youtube.com/watch?v=FkxUY0kxH80.

Notes

4   **phenomenon in his short story:** Wallace, *Oblivion*, 142–43.
4   **"trying to appear impressive":** Ibid., 147.
5   **"fraudulence paradox":** Ibid.
5   **"Grandiosity":** Bustillos, "Inside David Foster Wallace's Private Self-Help Library."
6   **"What goes on inside":** "Good Old Neon," in Wallace, *Oblivion*, 151.
6   **2005 commencement address:** May 21, 2005, Kenyon College, Gambier, OH.

CAPTURE

Notes

6   **more than two decades:** David A. Kessler, *A Question of Intent: A Great American Battle with a Deadly Industry* (New York: PublicAffairs, 2001); David A. Kessler, *The End of Overeating: Taking Control of the Insatiable American Appetite* (Emmaus, PA: Rodale, 2009).
6   **eat until we feel sick:** I came to realize that many people were experiencing a similar problem with eating as they were encountering with smoking. They re-

ported reaching for a bag of Doritos or another slice of pizza with the same blind compulsion as smokers lighting up a cigarette.

As part of our research efforts, my colleagues and I identified three common characteristics underlying both these behaviors: First, there was a loss of control, a giving over to a powerful force that was beyond the scope of consciousness. Second, those who fell under this spell had a very difficult time stopping it. And, third, those suffering were simultaneously preoccupied by their compulsions and unable to resist them.

8   **If I could shut it out:** Daniel M. Wegner, *White Bears and Other Unwanted Thoughts: Suppression, Obsession, and the Psychology of Mental Control* (New York: Viking, 1989); Sadia Najmi and Daniel M. Wegner, "Thought Suppression and Psychopathology," in *Handbook of Approach and Avoidance Motivation*, ed. Andrew J. Elliot (New York and London: Psychology Press, Taylor and Francis Group, 2008), 447.

THE NATURE OF MENTAL DISTRESS

Sources

Interview by the Author
Wallace, James and Sally (Nov. and Dec. 2011).
Work Referenced
Wallace, David Foster. *Infinite Jest*. New York: Little, Brown, 1996.

Notes

11   **"burning windows . . . two terrors":** Wallace, *Infinite Jest*, 696.
11   **"Socrates would say":** In *Protagoras*, Plato's Socrates argues that we never do what we do not want to; we always behave in ways that we believe to be rational and good, though our beliefs often prove mistaken. Knowledge is thus "able to govern man": "Whoever learns what is good and what is bad will never be swayed by anything to act otherwise than as knowledge bids." Reason, in other words, supplies the ultimate impulse to action. Socrates rejected the notion that reason is ever too weak to rule the passions, though our rational powers may be hobbled by mistaken assumptions. (At their most powerful, the emotions may distort the dictates of reason, but they cannot overcome them.) If we do not do what is supremely rational or good, Socrates argues, it is because we are mistaken; if I act wrongly, I am simply beholden to an erroneous idea.
       [**"Whoever learns what is good":** Plato, *Protagoras*, trans. W. R. M. Lamb (Cambridge, MA: Harvard University Press, 1962), cited in David L. Schaefer, "Wisdom and Morality: Aristotle's Account of Akrasia," *Polity* 21, no. 2 (Winter 1998), footnote 17, http://www.jstor.org/stable/3234805.
       impulse to action: Susan Sauvé Meyer, review of *Akrasia in Greek Philosophy: From Socrates to Plotinus*, Christopher Bobonich and Pierre Destrée, eds., in *Notre Dame Philosophical Review*, Jan. 18, 2008, http://ndpr.nd.edu/news/23317/?id=12183.
       an erroneous idea: Schaefer, "Wisdom and Morality."]
11   **"Plato seemed to get":** Unlike Socrates, Plato did not believe that all unjust or immoral action was the result of a mistaken idea. Rather, he suggested that childhood memories—of a particular satisfaction, or of something that once as-

suaged our pain—could trigger a similar desire in our adult selves, even when the object of that desire was no longer in our best interests. What once gave us comfort or soothed our hunger came to hold an almost irrational attraction for us. Remembered pleasures (a mother's breast, a ripe fruit) can thus lead us toward gluttony or intemperance.

To convey the sheer strength of such vestigial impulses, Plato related the story of Leontius, the son of Aglaion. Walking along the North Wall of the city, Leontius saw a number of corpses lying at the feet of their executioners. He found himself disgusted by the sight, but at the same time he felt drawn to it. At first he covered his face, but soon, overcome by curiosity, he opened his eyes and ran toward the corpses, crying, "Look for yourselves, you evil wretches, take your fill of the beautiful sight!" Even for Plato, reason could not tame this potent mixture of horror and delight, or resolve the underlying conflict between dispassionate contemplation and aesthetic ecstasy.

[in our best interests: Christopher Bobonich and Pierre Destrée, eds., *Akrasia in Greek Philosophy: From Socrates to Plotinus* (Leiden: Brill, 2007).]

THE SEARCH FOR A COMMON MECHANISM

Sources

Interview by the Author
Bunney, Steve (2013).

CHAPTER 2: THE HISTORICAL AND SCIENTIFIC CONTEXT OF CAPTURE

Notes

18 **"We pass through":** St. Teresa of Avila, *The Interior Castle, The Fourth Mansion* (New York: Penguin, 2003), chapter 1.

18 **anguish or insanity:** Giuseppe Roccatagliata, *A History of Ancient Psychiatry* (New York: Greenwood, 1986).

18 **religious belief:** Bronislaw Malinowski, *Magic, Science, and Religion, and Other Essays* (Garden City, NY: Doubleday, 1948), A23.

18 **anatomy:** Bennett Simon, "Mind and Madness in Classical Antiquity," in Edwin R. Wallace and John Gach, eds., *History of Psychiatry and Medical Psychology: With an Epilogue on Psychiatry and the Mind-Body Relation* (New York: Springer, 2008), 182.

19 **invisible fluids:** Ibid., 183.

19 **Rollo May:** Rollo May, *Love and Will* (New York: Norton, 1969), 146.

19 **"eternal battle":** Ibid., 127, footnote 8.

19 **"conclusions of our reason":** David Hume, "Moral Distinctions Not Derived from Reason," in Russ Shafer-Landau, ed., *Ethical Theory: An Anthology*, 2nd ed. (New York: Wiley, 2012), 10.

19 **anatomical or structural basis:** G. E. Berrios, "The Psychopathology of Affectivity: Conceptual and Historical Aspects," *Psychological Medicine* 15, no. 1 (1985): 745–58.

20 **"The passions . . . of our health":** Sir Alexander Crichton, *An Inquiry into the Nature and Origin of Mental Derangement: Comprehending a Concise*

System of the Physiology and Pathology of the Human Mind and a History of the Passions and Their Effects, vol. 2 (London: Cadell and Davies, 1798), 98–99.

20   **directed our behavior:** Ibid., 67–68.

WILLIAM JAMES AND ATTENTION

Sources

Works Referenced

Allen, Gay Wilson. *William James: A Biography*. New York: Viking, 1967.

Barnard, William G. *Exploring Unseen Worlds: William James and the Philosophy of Mysticism*. Albany: State University of New York Press, 1997.

Bjork, Daniel W. *William James: The Center of His Vision*. Washington, DC: American Psychological Association, 1997.

Carrette, Jeremy, ed. *William James and the Varieties of Religious Experience*. New York: Routledge, 2005.

Davies, John Llewelyn, and David James Vaughan. *The Republic of Plato*. London: Macmillan, 1902.

DeArmey, Michael H., and Stephen Skousgaard, eds. *The Philosophical Psychology of William James*. Washington, DC: Center for Advanced Research in Phenomenology and University Press of America, 1986.

Feinstein, Howard M. *Becoming William James*. Ithaca, NY: Cornell University Press, 1999.

Gale, Richard M. *The Divided Self of William James*. Cambridge, UK: Cambridge University Press, 2007.

James, William. "Are We Automata?," *Mind* 4 (1879): 1–22.

———. *Essays in Radical Empiricism*. Lincoln: University of Nebraska Press, 1996.

———. *The Principles of Psychology*, vol. 1. New York: Cosimo, 2007.

———. *The Works of William James: Manuscript Lectures*. Cambridge, MA: Harvard University Press, 1988.

James, William, and Henry James. *The Letters of William James: Two Volumes Combined*. New York: Cosimo Classics, 2008.

Kamber, Richard, ed. *William James: Essays and Lectures*. New York: Pearson, 2007.

Myers, Gerald E. *William James: His Life and Thought*. New Haven, CT: Yale University Press, 1986.

Perry, Ralph Barton. *The Thought and Character of William James*. Nashville, TN: Vanderbilt University Press, 1976.

Proudfoot, Wayne, ed. *William James and a Science of Religions*. New York: Columbia University Press, 2004.

Putnam, Ruth Anna, ed. *The Cambridge Companion to William James*. New York: Cambridge University Press, 1997.

Richardson, Robert D. *William James: In the Maelstrom of American Modernism*. New York: First Mariner, 2006.

Seigfried, Charlene Haddock. *William James's Radical Reconstruction of Philosophy*. Albany: State University of New York Press, 1990.

Taylor, Eugene. *William James on Exceptional Mental States: The 1896 Lowell Lectures*. New York: Scribner, 1982.

Wallace, Edwin R., and John Gach, eds. *History of Psychiatry and Medical Psychology: With an Epilogue on Psychiatry and the Mind-Body Relation.* New York: Springer, 2008.

Notes

20 **coerce attention:** William James, *Manuscript Lectures,* 57.
21 **"continual verge of suicide" . . . intertwined:** Richardson, *William James,* 83.
21 **"steadily deteriorating":** Perry, *Thought and Character,* 119.
21 **"born again":** Richardson, *William James,* 176.
21 **battling his demons:** Ibid., 315–20.
21 **MD degree in 1869:** Wallace and Gach, *History of Psychiatry,* 15.
21 **"embraced the medical profession":** Feinstein, *Becoming William James,* 165.
21 **"little fitted by nature":** Ibid., 214.
22 **"fear of my own existence":** Carrette, *William James and the Varieties of Religious Experience,* 157.
22 **"greenish skin":** Richardson, *William James,* 117.
22 **"quivering fear":** Ibid., 118.
22 **"first act of free will":** Ibid., 120.
23 **"individual reality":** Ibid., 122.
23 **"great effusion":** William and Henry James, *The Letters of William James,* 169.
23 **"health to his bones":** Ibid., 170.
23 **"Despair lames":** Richardson, *William James,* 91.
23 **"Zeus and Fate":** In *Hippolytus,* the Greek playwright Euripides explores the role of the gods in manipulating the human mind. Suddenly consumed by an illicit desire for her stepson, Phaedra seems to lose control of her mind. In Euripides's play, the chorus approaches this case as a psychiatrist might, positing a series of possible explanations for Phaedra's mental condition:

> *Is it Pan's frenzy that possesses you*
> *Or is Hecate's madness upon you, maid?*
> *Can it be the holy Corybantes*
> *Or the mighty Mother who rules the mountains:*
> *Are you wasted in suffering thus,*
> *For a sin against Dictynna, Queen of the Hunters?*
> *Are you perhaps unhallowed, having offered*
> *no sacrifice to her from taken victims?*

All the chorus's diagnoses involve the wrath of the gods: teasing out an explanation becomes a matter of figuring out which god Phaedra has offended. In each of these possible scenarios, indignation leads a god to "possess" the mind of a mortal, depriving him or her of the powers of reason.
["**Is it Pan's . . . victims":** Euripides, *Euripides I: Alcestis, The Medea, The Heracleidae, Hippolytus,* trans. R. Lattimore, R. Warner, R. Gladstone, and D. Grene (Chicago: University of Chicago Press, 1955).]
23 **Aristotle:** Among the Greek philosophers, Aristotle articulated the most nuanced and flexible—in a word, modern—understanding of human motivation. Like Plato, Aristotle challenged the Socratic notion that, in a truly virtuous person, all motivations are necessarily consistent with reason. "It is problematic," he wrote, "how someone with correct understanding can lack self-mastery." In other words, perfectly rational people often behave impetuously

and soon come to regret their actions, and it is not always possible to ascribe their behavior to faulty reasoning. According to Aristotle, in these moments our passions "actually alter our bodily condition, and in some men even produce fits of madness." Aristotle named this phenomenon *akrasia*, or weakness of will. Here, Aristotle complicated the Socratic view of impulsive, or irrational, behavior. Akratic behavior is the result not of a faulty assumption but of "something else . . . by nature contrary to reason, which fights and resists it. For exactly as with paralytic limbs which when their owners decide to move them to the right take off in the wrong direction. . . so it is in the case of the soul."

This description may sound similar to Plato's theory of a divided soul, but unlike Plato, Aristotle sought to understand human motivation in psychological (as opposed to metaphysical) terms. Indeed, for Aristotle, ethics was above all a matter of attention. In the *Ethics*, he argues that our mode of perceiving the surrounding world dictates how we relate to and act within it. In other words, how a person attends determines how he acts. Hence the qualitative distinction between rote memorization and sustained contemplation, or the pleasures of a philosophical conversation with friends. In the first case, attention proceeds automatically; in the second, however, it becomes highly curated, even choreographed. Likewise, the courageous or just person "attends to the right aspects of his interactions with others"—emotional and intellectual substance as opposed to, say, outward appearance—and responds accordingly.

The Aristotelian understanding of just or virtuous behavior can be rephrased in terms of the kinds of attention that particular stimuli foster or preclude. A sensual pleasure, for instance, "engage[s] our attention and move[s] us to act in a way that is specific to *the sort of sensual pleasure it is*." In other words, attention is not merely a passive state of receptivity but a form of active selection: I choose not only the objects of my attention but *how* I attend to them. The virtuous person directs his attention toward the most worthy objects, in a process of sustained, deliberate, and dispassionate contemplation.

For Aristotle, the human soul is composed not of separate, warring faculties but of multiple forms of attention. Intentional objects—intellectual, affective, sensual—have distinctive ways of "affecting us and moving us to act." Virtue is a matter of attending correctly, or of exercising the form of attention most appropriate to a given object. For the truly virtuous, there is no clear distinction between the self and the manifold objects it perceives: their internal receivers have been so perfectly tuned to the signals of the world that they become indistinguishable from the people, books, images, and ideas that animate them. For Aristotle, then, virtue and even justice are matters primarily of perceiving rather than reasoning. Careful attention to suffering will inspire me to generosity, just as contemplation of the beautiful will activate my desire to create.

["**self-mastery**": Kenneth Dorter, "Weakness and Will in Plato's Republic," in Tobias Hoffmann, ed., *Weakness of Will from Plato to the Present* (Washington, DC: Catholic University of America Press, 2008), 2.

"**fits of madness**": Aristotle, *Aristotle's Nicomachean Ethics*, ed. Robert C. Bartlett and Susan D. Collins (Chicago: University of Chicago Press, 2011), 1147a, 18–24.

"**something else . . . of the soul**": Jessica Dawn Moss, *Aristotle on the Apparent Good: Perception, Phantasia, Thought, and Desire* (Oxford: Oxford University Press, 2012).

conversation with friends: "Transcendence Without God: On Atheism and Invisibility," in *Philosophers Without Gods*, ed. Louise M. Antony (Oxford: Oxford University Press, 2007), 122.

responds accordingly: Ibid.

"sensual pleasure it is": G. H. von Wright, *The Varieties of Goodness* (New York: Humanities Press, 1963); Heda Segvic and Myles Burnyeat, *From Protagoras to Aristotle: Essays in Ancient Moral Philosophy* (Princeton, NJ: Princeton University Press, 2009), 98.

"moving us to act": Segvic and Burnyeat, *From Protagoras*, 98.]

24   **consciousness:** For James, consciousness was characterized by a rushing stream of thoughts and perceptions—fleeting, idiosyncratic, often irrational and unrelated—that nonetheless cohere into the impression of a whole, a continuous life. This model of the mind would influence generations not only of scientists and psychologists but of novelists, poets, and painters, especially those associated with the turn-of-the-century aesthetic known as modernism. The art of James Joyce and Marcel Proust, Virginia Woolf and T. S. Eliot—all bears the unmistakable imprint of James's thought. Joyce's famously difficult novel *Ulysses* is one attempt to record the buzzing, fragmented blur of thought as it ricochets from one object to another, often propelled forward by loose associations rather than logical connections:

> a quarter after what an unearthly hour I suppose they're just getting up in China now combing out their pigtails for the day well soon have the nuns ringing the angelus they've nobody coming in to spoil their sleep except an odd priest or two for his night office the alarm clock next door at cockshout clattering the brains out of itself let me see if I can doze off 1 2 3 4 5 what kind of flowers are those they invented like the stars the wallpaper in Lombard street was much nicer . . .

As Molly Bloom's attention wanders from the hour of the day to an imagined scene in China and finally to her bedroom wallpaper, the principle guiding her attention is all but impossible to discern. Literary critics can (and indeed have) posited trains of associative logic to connect one thought to the next, though Joyce's text only hints at such connections, providing a richly Jamesian view of mental life. Thoughts bubble up to the surface of the mind, quite independent of any perceptible act of will, and then quickly recede back into its depths.

At the same time, James believed that the human mind obeyed certain natural laws that, like the molecular weight of oxygen or the acceleration of gravity, could be uncovered through diligent observation. James's predecessors in both Europe and America had likewise aspired to make psychology just as scientifically rigorous as Newton's physics and Galileo's astronomy. To do so, they developed a set of tools to measure psychological reactions, though the results of their tests often proved more confounding than illuminating. Human minds, it seemed, were more dissimilar than alike. These scientists, however, managed to conclude that stimuli brought about physical changes in the nervous system and in particular the brain; such changes, in turn, led to thoughts, feelings, and actions. Everything we thought, felt, or did, in other words, was the result of physiological reactions to the surrounding world.

This theory, in turn, led to a host of philosophical, and practical, problems for nineteenth-century science. If we are, in the words of Thomas Huxley, "mere machines or automata," then how can we be responsible for our actions? Are we simply actors in a neurochemical play that we did not write? The challenge for James, then, was to account for consciousness—for the awareness of ourselves as thinking creatures who can reason, make decisions, and act accordingly—while at the same time acknowledging the irrational, uncontrollable, and even random character of much of our mental life. For James, understanding the neural processes that produced consciousness did not entail demoting man to the status of mere automaton.

Something crucial was, after all, missing in a purely materialist understanding of human behavior. A materialist would be forced to conclude that if we could somehow study the nervous system of Shakespeare, taking into account his entire environment, we would "be able to show why at a certain period of his life his hand came to trace on certain sheets of paper those crabbed little black marks which we, for shortness' sake, call the manuscript of *Hamlet*. We should understand the rationale of every erasure and alteration therein, and we should understand all this without in the slightest degree acknowledging the existence of the thoughts in Shakespeare's mind." If all mankind's greatest achievements, and most singular failures, are indeed traceable to electrical impulses, then the very notion of literary greatness quickly loses its meaning. In defending the achievements of Shakespeare against the behavioralist assumptions of his day, James latched on to a concept that is crucial to understanding why we do what we do—one that has still not been fully explored.

Like any good student of Darwin, James began his investigation by framing his question in evolutionary terms: Of what use to a nervous system is a superadded consciousness? Why might consciousness have evolved in the first place? Its purpose, James concluded, was to allow us to attend selectively to one (or two or ten) of the infinite array of stimuli available to the mind at any given moment and, just as important, to ignore the rest. The key to understanding consciousness, then, was "*interest* and *selective attention*." What separates a human mind from an automaton is this very particularity, the capacity to "choose out of the manifold experiences present to it at a given time some one for particular accentuation. . . . From its simplest to its most complicated forms, it exerts this function with unremitting industry."

On what principle does the mind make such a selection? The process often seems to occur without our input or even awareness, but James insisted that it occurs nonetheless. What James called "interest," the principle guiding selective attention, could likewise be termed "investment" or "care":

> Millions of items of the outward order are present to my senses which never properly enter into my experience. Why? Because they have no *interest* for me. . . . Only those items which I *notice* shape my mind—without selective interest, experience is an utter chaos. Interest alone gives accent and emphasis, light and shade, background and foreground—intelligible perspective, in a word. It varies in every creature, but without it the consciousness of every creature would be a gray chaotic indiscriminateness, impossible for us even to conceive.

For James, then, volition was nothing but attention. The "whole drama of the voluntary life hinges on the amount of attention" accorded to a given object;

"each of us literally chooses, by his ways of attending to things, what sort of universe he shall . . . inhabit."

Still, the precise nature of interest, or the principle guiding attention, remained murky: Why do I become invested in one stimulus rather than another? If "my experience is what I agree to attend to," then one would imagine the world to be very different from the one we know, in which human attention often veers away from the most pressing matters and even behaves in ways that seem arbitrary or self-defeating. America's greatest philosopher and psychologist struggled to account for this seeming contradiction, ultimately admitting that "the last word of psychology here is ignorance."

["mere machines": Thomas Henry Huxley, *Science and Culture, and Other Essays* (New York: Appleton, 1882).

"Shakespeare's mind": William James, *The Principles of Psychology* (New York: Holt, 1890), 132.

"unremitting industry": William James, "Are We Automata?," *Mind* 4, no. 13 (1879).

"Millions . . . to conceive": James, *The Principles of Psychology*, 403.

"whole drama" . . . "inhabit": Ibid., 453.

"the last word": Ibid., 454.]

24   rushing stream of thoughts: Richardson, *William James*, 197.

25   "intelligible perspective": Myers, *William James*, 182.

25   lecture he gave in 1896: William James, *Manuscript Lectures*, 56–57.

25   Ribot: In developing his model of the *idée fixe*, Ribot relied on the theories of earlier French psychologists Philippe Pinel and Jean-Étienne Esquirol. In the late eighteenth century, Pinel defined the passions as affective processes of long duration; emotions, on the other hand, were "sudden and intense commotions of the mind" with definite beginnings and ends. For Pinel, passions (emotions that outlasted their natural duration) inevitably took shape around a fixed idea, or a core obsession. A student of Pinel's and the author of the first modern textbook on psychiatry, Jean-Étienne Esquirol, brought his mentor's theory to its logical conclusion, arguing that virtually all forms of mental anguish were the result of monomania. Ascribing a diverse array of psychiatric conditions to the *idée fixe*, Esquirol saw monomania as a form of delirium, or a "partial lesion" of the will; this lesion soon became a "magnetic core" that focused attention and directed behavior. Patients would seize upon a false principle and pursue it relentlessly, albeit without deviating from an underlying logic: despite their illness, they could deduce legitimate consequences from their preoccupations and, outside the immediate realm of their delusions, act, think, and reason normally.

Esquirol ascribed an incredibly diverse range of mental states (paranoia, excessive lust, suicide, drunkenness, pyromania, and homicide) to monomania, though he separated these *idées fixes* into various subcategories. In melancholy, for instance, sorrowful and depressing thoughts became the sole object of attention, gradually warping the relationship between mind and reality. The melancholic recedes into himself, eventually living entirely within the echo chamber of his own mind. Other forms of monomania, by contrast, encompassed obsessions and compulsions—with train timetables, say, or strategies for longevity, with family genealogy or the pursuit of wealth. In all these cases, attention became both highly concentrated and increasingly immobile; it could not be easily displaced from one object onto another.

One of Dr. Esquirol's most vexing patients, Madame T., was forty years old. At the age of sixteen, she had experienced a "slight attack of melancholy." Nine months after the birth of her first child, she became melancholic once again and even "contemplated the destruction of her child." While watching her infant daughter sleep, she experienced a strong desire to murder the baby. Profoundly disturbed by such thoughts and afraid that she might act on them, she sought the help of Dr. Esquirol: "It is better that I should die, than this dear innocent."

Just one month after their meeting, Madame T. was found in her bedroom, almost dead from the fumes of burning charcoal. The patient described herself as helpless in the face of her melancholy: "Would that I could persuade myself that I were sick; but I cannot: I am a wretched mother." Once Madame T. was hospitalized, Dr. Esquirol finally succeeded in convincing her that her fears were the result of disease. Her condition slowly improved. When she left Dr. Esquirol's care and returned home, however, her fantasies of infanticide gradually resumed.

Esquirol recognized a link between the sufferings of Madame T. and the symptoms of his other patients. It is a link that remains worth pursuing. One hundred years after Dr. Esquirol published his account of this case, the American Psychiatric Association asked its members just how far psychiatry had progressed since his time. The response: "We have indeed little cause for boasting."

As scientists continued to grapple with the emotional life, writers and humanists used very different methods to make sense of the same phenomena. Prose and poetry, however, did not suffer from the limitations that nineteenth-century science did; indeed, literature was often more successful in capturing the felt experience of the passions, of irrational and obsessive thought. Dr. Esquirol even cited Miguel de Cervantes's picaresque masterwork *Don Quixote* as a prime depiction of the *idée fixe*. The novel's title character lives under the spell of monomania, in this case an obsession with the chivalry that "prevailed over nearly the whole of Europe" in the Middle Ages.

Later generations of philosophers, poets, and novelists would explore the same questions as Cervantes: in the late eighteenth century, for instance, the British journalist and novelist William Godwin described in almost painful detail the process by which the *idée fixe* takes hold. The monomaniacal mind, Godwin conjectured, is captured by an object that lies just beyond its reach; true satisfaction proves elusive, as the force of desire becomes increasingly insatiable. Godwin's daughter, Mary Shelley, adopted and refined her father's theme in the Gothic novel *Frankenstein*. Here, the three main characters have widely disparate *idées fixes*: conquering the North Pole, seeking revenge, and bestowing life on a mechanical creation. Consumed by his quest to uncover the "physical secrets of the world," Victor Frankenstein finds himself in the thrall of an all-consuming "passion, which afterwards ruled my destiny" and transformed "bright visions of careful extensive usefulness into gloomy and narrow reflections upon [my]self." As all other objects of interest recede beyond the horizon, Frankenstein's mind becomes filled by "one thought, one conception, one purpose."

What is striking in many of these literary depictions is that the *idée fixe* evolves in response to the social norms and political climate of a given epoch. Esquirol observed that monomania "borrowed" its objects from whatever passions were in vogue at the time: superstition in the Middle Ages; amorous and gallant courage immediately following the Crusades; religious zeal during the Reformation. He even suggested that governments monitor "monomaniacal epidemiology" in order to better understand "political convulsions" and, in particular, the unbridled ambition that reigned in the wake of the French Revolution. Esquirol observed that everyday French men and women of his era increasingly saw themselves as "emperors or kings, empresses or queens."

Marie-Henri Beyle, better known as Stendhal, captured this particular form of monomania, *monomanie ambitieuse*, in *The Red and the Black*. Stendhal's Julien is obsessed with worldly success to the exclusion of all other interests: even during romantic encounters, thoughts of such success would not fade. Julien bears an uncanny resemblance to Esquirol's patients, having suffered a devastating reversal of fortune before falling into delusions of grandeur: "Their ideas are exaggerated. Their passions are very strong. They are dominated by ambition and pride."

While some *idées fixes* corresponded to very specific historical circumstances— Stendhal's *monomanie ambitieuse*, for instance, could be seen as a response to the rampant individualism of postrevolutionary France—others seemed to float above history, exercising a hold over the European mind for centuries. H.F., the protagonist of Daniel Defoe's *A Journal of the Plague Year*, is captured by an insatiable and dangerous curiosity; like Plato's corpse-seeking Leontius, he cannot overcome his obsession with death, even when a plague epidemic endangers the welfare of his family. Rather than flee the sickly streets of London, H.F. returns again and again to scenes of death and dying: "The shrieks of women and children at the windows and doors of their houses. . . . Curiosity led me to observe things more than usually, and indeed I walked a great way where I had no business." He even visits a mass grave of plague victims, in defiance of "a strict order to prevent people coming to those pits." Through a combination of cunning deception and unabashed begging, he manages to gain entry to the grave site; he is driven toward the decaying corpses by an inexorable force: "I had been pressed in my mind to go." As Defoe scholars have noted, H.F.'s morbid passion presents almost as grave a danger to his welfare as the plague itself.

As a psychiatric diagnosis, monomania was doomed to fail. At one point as many as 20 percent of patients entering French psychiatric institutions were diagnosed as monomaniacal. As Arthur Conan Doyle recognized, Esquirol's theory was infinitely elastic and therefore of limited use: "A man who had read deeply about Napoleon, or who had possibly received some hereditary family injury through the great war, might conceivably form such an 'idée fixe' and under its influence be capable of any fantastic outrage." Others were more measured in their critique of the concept. For the librarian and author Charles Nodier, monomania came in two forms: the harmless and the militant. Nodier observed that Dr. Esquirol's diagnosis lost favor because patients were often found to have more than one *idée fixe*. The very notion of multiple *idées fixes* operating simultaneously seemed paradoxical: was a fixed idea truly "fixed" if the brain could jump from one object to another? Psychiatrists soon realized that they could not identify the practical limits of passion, monomania, and madness.

Monomania was therefore easily mocked. In Nodier's 1831 story "The Bibliomaniac," a doctor describes a case of "bibliomaniacal typhus": the patient, a zealous book collector, is convinced that monsters with ghastly scissors are methodically reducing the margins of one of his most prized books by exactly one and a half inches. The patient then realizes in terror that his copy of Virgil is one third of a line short: "'A third of a line!' he repeated, shaking his fist furiously at the heavens, like an Ajax or a Capaneus." Nodier cannot resist caricaturing the psychiatric establishment, in all its pseudoscientific sanctimony. Perhaps not surprisingly, the patient falls into a deep depression: "Suffering was all that was left of life for the poor man. He merely repeated, from time to time, 'a third of a line!' as he gnawed at his hands." The diagnosis: "A plague on books and typhus."

These critiques did not, however, diminish the importance of Esquirol's fundamental observation: a force within our minds can resist conscious control, redirecting attention and behavior. For the French psychologist Pierre Janet, this force was not simply a rare form of pathology; rather, it was an incontrovertible part of everyday life. Janet was fascinated by the many forms of suffering that lay just behind the appearance of normalcy:

> Obsessive shame for one's facial features; obsessive shame of speaking; obsession about being a child; obsessive guilt toward the cat's escape; obsession about religious crimes; obsessive thoughts about forming a pact with the devil; obsession about intestinal worms. Mania for metaphysical research; mania for the perfect love; mania for predictions and oaths; mania to summon up perfectly a visual recollection.

For Janet, past experiences and memories were ultimately responsible for dictating which of these many *idées fixes* took hold in a given patient. One of his first patients, a forty-year-old woman named Justine, had an obsessive fear of cholera, which Janet was able to link to her childhood. Her mother had been a nurse, and in her youth Justine had helped her care for cholera patients. At the age of seventeen, she first saw the sickeningly blue-gray corpses of those who had succumbed to the disease.

In the stories of Justine and others like her, Janet recognized a second clue about the nature and treatment of monomania. Philippe Pinel, Janet's predecessor at the Salpêtrière Hospital in Paris, had observed that strong emotions could sometimes break the hold of an *idée fixe*, essentially snapping a patient back to normal. Building on the observations of his mentor, Dr. Janet concluded that obsessive attention need not be fixed at all but could be reshaped, deflected, or even transferred onto benign objects. A forty-one-year-old woman who had recently lost her husband, CK suffered from all sorts of hypochondriacal obsessions: headache, fatigue, generalized anxiety. All her suffering disappeared, however, when she was reunited by chance with an old teacher who suffered herself from certain obsessions and idiosyncrasies. The two began living together and were able to assuage each other's anxieties, essentially shortcircuiting their erstwhile obsessions with their own suffering through a process of empathic identification and benevolent care.

These observations led Janet to his most important conclusion about monomania: the *idée fixe* was above all a matter of attention. He reasoned that, at any moment, we are aware of only a fraction of the infinite variety of stimuli bombarding our brains; whereas a normal mind moves easily from one stimulus to another, the attention of monomaniacal patients adheres, with greater and greater tenacity, to a single stimulus. In these patients, according to Janet, certain thoughts, or cognitive "nuclei," would break off, or dissociate, from normal consciousness. These dissociated ideas could be simple thoughts or fully formed independent personalities. In Janet's theory, these dissociated fragments were not governed by the higher forms of reasoning that controlled everyday behavior, because they occurred outside the immediate field of consciousness: "The idea, like a virus, develops in a corner of the personality inaccessible to the subject, works subconsciously, and brings about all disorders of hysteria and of mental disease." The *idée fixe* was defined, in other words, by a failure of attention: an inability to divorce oneself from the immediate moment, to see one's own thoughts from the bird's-eye perspective of reason.

In 1889, Janet wrote *L'Automatisme psychologique*, in which he argues that his patients' *idées fixes* were all controlled by subconscious forces and took shape in response to memories of traumatic events from the distant past. The book was highly prescient: in many ways, it foreshadowed the central claims of psychoanalysis. Janet argued that the *idée fixe*, though hidden from consciousness, silently influences everything a person thinks, feels, and does.

[beginnings and ends: Louis C. Charland, "Reinstating the Passions: Arguments from the History of Psychopathology," in *The Oxford Handbook of Philosophy of Emotion* (Oxford: Oxford University Press, 2009), http://www.oxfordhandbooks.com/view/10.1093/oxfordhb/9780199235018.001.0001/oxfordhb-9780199235018-e-11. (I want to credit Professor Charland for his help in my understanding Dr. Ribot.)

"partial lesion": Jean-Étienne Dominique Esquirol, *Mental Maladies. A Treatise on Insanity*, trans. Ebenezer K. Hunt (Philadelphia: Lea and Blanchard, 1845), 320.

reason normally: Ibid.

subcategories: Ibid.

"dear innocent": Ibid., 375–76.

"wretched mother": Ibid., 376.

"boasting": Ibid.

Cervantes's: Ibid., 331.

William Godwin: William Godwin, *The Adventures of Caleb Williams* (Boston: Fields, Osgood, 1869), 168, http://catalog.hathitrust.org/Record/011636679, cited by Lennard J. Davis, *Obsession: A History* (Chicago: University of Chicago Press, 2008), 55.

"one thought": Davis, *Obsession*, 74–75.

"emperors or kings, empresses or queens": Ibid., 201.

Red and the Black: Kathleen Kete, "Stendhal and the Trials of Ambition in Postrevolutionary France," *French Historical Studies* 28, no. 3 (Summer 2005).

"ambition and pride": Ibid.

welfare of his family: Daniel Defoe and Edward Wedlake Brayley, *A Journal of the Plague Year* (London: Tegg, 1848); Daniel Defoe and Louis A. Landa, *A Journal of the Plague Year, Being Observations or Memorials of the Most Remarkable Occurrences, as Well Publick as Private, Which Happened in London During the Last Great Visitation in 1665* (London: Oxford University Press, 1969).

"I had been pressed": Ibid.

plague itself: Geoffrey M. Sill, *The Cure of the Passions and the Origins of the English Novel* (Cambridge, UK: Cambridge University Press, 2001), 115.

diagnosed as monomaniacal: Jan Ellen Goldstein, *Console and Classify: The French Psychiatric Profession in the Nineteenth Century* (Cambridge, UK: Cambridge University Press, 1987), 154.

critique of the concept: Sir Arthur Conan Doyle, *The Return of Sherlock Holmes* (Leipzig: Tauchnitz, 1905), 212.

harmless and the militant: Charles Nodier, "Reverie psychologique de la monomanie reflective," in *L'Amateur de livres*, ed. Jean-Luc Steinmetz (Paris: Castor Astral, 1993), 48–49, cited in Marina Van Zuylen, *Monomania: The Flight from Everyday Life in Literature and Art* (Ithaca, NY: Cornell University Press, 2005), 63.

easily mocked: Van Zuylen, *Monomania*, 64. (I want to credit Professor Zuylen for pointing out a number of examples of monomania.)

"Obsessive shame . . . recollection": Ibid., 22.

benevolent care: Pierre Janet and Fulgence Raymond, *Les obsessions et la psych-asthénie*, 2 vols. (Paris: Alcan, 1903), cited in Van Zuylen, *Monomania*.

"mental disease": Pierre Janet, *L'Automatisme psychologique* (Paris: Alcan, 1889), 436, cited in Henri F. Ellenberger, *The Discovery of the Unconscious: The History and Evolution of Dynamic Psychiatry* (New York: Basic, 1970), 149, 178. (I want to credit Dr. Ellenberger for his analysis of Dr. Janet.) Giuseppe Roccatagliata, *A History of Ancient Psychiatry* (Westport, CT: Greenwood Press, 1986).

distant past: Janet, *L'Automatisme psychologique*, 436, cited in Ellenberger, *Discovery of the Unconscious*, 360.]

25 **mental illness:** In his 1896 Lowell Lecture, James adopts Janet and Bénédict Augustin Morel's notion that all mental illnesses, from the neurotic to the psychopathic, exhibit similar characteristics. The mentally ill "show fear, anger, pity, tears and fainting"; they demonstrate excessive responses, are oversensitive, and subject to impulses, obsessive ideas, and phobias. When a given mental function goes completely awry, the resultant state is unmistakably legible to the physician and could thereby elucidate more familiar forms of psychological distress. As an example, James discussed the doubts that accompany daily activities: making sure the door is locked, the gas is turned off, or that no one is hiding under the bed. In each of these cases, a mental image gives rise to a doubt, which in turn leads to an action that relieves the initial doubt. Nothing about the cycle is inherently dangerous or pathological, though James understood that in certain patients such doubts could lead to uncontrollable anxiety or behaviors that today would be labeled obsessive-compulsive. In such a "morbid case" the initial doubt returns and returns, no matter how firmly the gas nozzle is clamped or how many times the bed has been checked. These normally reassuring actions no longer provide relief, as an all-powerful imagination short-circuits the faculty of critical reasoning.

To illustrate this process, James related a number of case studies from the European and American literature of his day:

One involved a young man who developed an obsession with counting everything at the dinner table: spoonfuls, mouthfuls, apple seeds. Eventually he came to dread the very act of eating, the bell announcing mealtime.

A sixty-year-old man compulsively researched the private life of the actresses whom he had seen perform; he eventually became so distressed by the compulsion that he stopped going to the theater. But a similar compulsion soon took the place of the first, forcing him to inquire into the private life of every pretty woman he saw. Every time he encountered a beautiful woman on the street, he was seized by anxiety; his fixed idea would dissipate only if a friend told him that the woman was, in fact, plain.

A forty-year-old male had since childhood attached prophetic significance to certain objects or gestures. A particular necklace would bring happiness; failing to touch a certain stone would invite evil. Not reading a certain line, or not forming a particular letter a certain way, would spell disaster. For twenty years, the man would travel to the railroad station every Sunday to kick a particular post with his foot three times.

In all these obsessions and compulsions, James noted a tendency to "shade into a general anxious melancholy." To describe this tendency, he used the German term *Grübelsucht*, or addiction to brooding, defined as a constant urge to inquire and question, which ultimately gave rise to an "endless web of meaningless questionings." But such an endless web might tell us something about the finite web of more meaningful questions that all of us confront on a daily basis.

26   **complete insanity:** Louis C. Charland, "Reinstating the Passions: Arguments from the History of Psychopathology," *Oxford Handbooks Online*, July 19, 2015, http://www.oxfordhandbooks.com/view/10.1093/oxfordhb/9780199235018 .001.0001/oxfordhb-9780199235018-e-11; Peter Goldie, ed., *The Oxford Handbook of Philosophy of Emotion* (Oxford: Oxford University Press, 2009), 254, http://www.oxfordhandbooks.com/view/10.1093/oxfordhb/9780199235 018.001.0001/oxfordhb-9780199235018.

26   **toward mental illness:** Théodule-Armand Ribot, *The Psychology of the Emotions* (New York: Charles Scribner's Sons, 1903), 226–27.

26   **"powers of the will":** Ibid.

27   **brain cells:** Ibid.

27   **"assault of the everyday":** Marina Van Zuylen, *Monomania: The Flight from Everyday Life in Literature and Art* (Ithaca, NY: Cornell University Press, 2005), 22.

27   **Henry Parkhurst:** James, *Manuscript Lectures*, 57.

27   **"human nature":** Taylor, *William James on Exceptional Mental States*, 148.

27   **"hot place":** Carrette, *William James and the Varieties of Religious Experience*, 193.

27   **"comparative chaos . . . divine":** Ibid., 167.

27   **"inner citadel":** Richardson, *William James*, 391.

27   **"really believing":** Perry, *Thought and Character*, 159.

28   **"explosive intensity":** Carrette, *William James and the Varieties of Religious Experience*, 170.

28   **"mystical overbeliefs":** Myers, *William James*, 460.

In such forms of spiritual experience James found what he had failed to identify in his psychological writings. By religion, James does not mean institutional religion, or the feelings of those for whom religious faith is merely a "dull habit"; instead the term refers to the experience of religion as the "very inner citadel," a bulwark against the falsifications and vicissitudes of the everyday. "The whole point lies in really believing," James writes; it is the belief in belief, or in the viability and significance of belief, that ultimately matters. For those who believe in this way—believe without compulsion—religion brings to the surface of con-

scious life otherwise hidden regions of the mind: "the irrational part . . . vital
needs and mystical overbeliefs."

[**vicissitudes:** Ralph Barton Perry, *The Thought and Character of William James*,
new pbk. ed. (Nashville, TN: Vanderbilt University Press, 1996), 259.
ultimately matters: Ibid.
"overbeliefs": Ibid., 258.]

28   **"power beyond":** Carrette, *William James and the Varieties of Religious Experience*, 503.
28   **"new will . . . did not grasp it":** Ibid., 169.
28   **"stimulus or passion":** Ibid., 172.

FREUD AND DRIVE

Notes

29   **"demand made":** Jay R. Greenberg and Stephen A. Mitchell, *Object Relations
in Psychoanalytic Theory* (Cambridge, MA: Harvard University Press, 1983),
21.
29   **counterforces:** Ibid., 21, 23.
29   **measure of energy:** Ibid., 22.
29   **"the mental and physical":** George Dowell, "Freud's Concept of Unconscious
Mind" (PhD diss., Yale University, 1968), 8.
29   **depths of the unconscious:** W. L. Northridge, *Modern Theories of the Unconscious* (London: K. Paul, Trench, Trubner, 1924), 99–105.
30   **needs and urges:** Greenberg and Mitchell, *Object Relations*, 24, citing Sigmund
Freud, *An Outline of Psycho-Analysis* (London: Hogarth, 1949), 148–49.
30   **invisible terrain:** Northridge, *Modern Theories*, 99–105.
30   **content of our drives:** The psychiatrist Carl Jung was all too familiar with the
uncanny. After struggling for years against a nightmarish tide of visions, Jung
resigned his professorship in 1913. The images and voices in his mind had left
him shattered. "All, all was needed for this earthquake and eruption of spirit
within himself," recalled one colleague.

Indeed, the visions were eerily akin to the delusions and hallucinations regularly reported by his psychotic patients. When "peculiar reactions" began to
disturb his writing, he learned "to distinguish between myself and the interruption." As soon as a "vulgar or banal" vision interrupted his train of thought, he
calmly reminded himself, "It is perfectly true that I have thought in this stupid
way at some time or other, but I don't have to think that way now. I must not
accept this stupidity as mine."

Soon the intrusions became more formidable. Instead of resisting their incursion, however, Jung gradually came to accept the images as bearers of meaning. "I did my best," Jung later recounted, "not to lose my head but to find some
way to understand these strange things. I stood helpless before an alien world:
everything in it seemed difficult and incomprehensible. I was living in a constant state of tension; often I felt as if gigantic blocks of stone were tumbling
down upon me. One thunderstorm followed another."

In October, while I was alone on a journey, I was suddenly seized by an
overpowering vision: I saw a monstrous flood covering all the northern
and low-lying lands between the North Sea and the Alps. . . . I saw the

mighty yellow waves, the floating rubble of civilization, and the drowned bodies of uncounted thousands. Then the whole sea turned to blood.

Two weeks later the vision recurred, "still more violent than before." "I wrestled with it," Jung wrote of the experience, "but it held me fast."

In a 1925 lecture, Jung described the inevitable sense of estrangement that besets the unquiet mind. "One begins to watch one's [own] mind," he explained. "One begins to observe the autonomous phenomena in which one exists as a spectator, or even as a victim. It is very much as if one [had] stepped out of the protection of his house into an antediluvian forest and was confronted by all the monsters that inhabit the latter. . . . It is as though one [had given] up one's freedom of will."

The fauna of this antediluvian forest became the *prima materia* for Carl Jung's lifelong investigation of the unconscious. "The years when I was pursuing my inner images," he once said, "were the most important in my life. In them everything essential was decided."

Once his initial zeal in copying out the content of these visions subsided, Jung could not help asking himself, "What is this I am doing? It certainly is not science: what is it?"

"This is art," a voice replied.

What of this voice? "My conclusion," said Jung, "was that *It* must be soul in the primitive sense." He described the entrance of this interlocutor into his psychic life: "It was like the feeling of an invisible presence in the room one enters." Jung soon came to imagine himself boring a hole, as if in a cave, to find the source of these mysterious images. Rather than trying to control his mind, he actively sought out these waking dreams, "believing so thoroughly in a fantasy that it [led me] into further fantasy." The more he worked, the deeper into his psyche he descended.

His notebooks from this period record visions of increasing complexity, populated by omens and symbolic imagery. Sinking along a gray rock face, Jung stood "in black dirt up to my ankles in a dark cave." Seized by fear, he nonetheless ventured into a dark cave, climbing through a narrow crack in the rock. The cave echoed with "frightful" wails and screams, though no one was in sight. At its base trickled a stream of "dark water," and beyond it gleamed a "luminous red stone." Wading through the "muddy water," Jung moved toward the stone, as if propelled by gravity. "I hold the stone in my hand," he recalled, "peering around inquiringly. I do not want to listen to the voices, they keep me away. But I want to know."

Soon a "bloody head" floats to the surface of the dark stream—"someone wounded, someone slain"—and a black scarab bobs into view. Thousands of serpents crawl along the walls of the cave, "veiling the sun." As night falls, a stream of blood surges from the river: "I withdrew from the hole, and then blood came gushing from it as from a severed artery."

What sort of insight might this symbolically laden landscape offer? Perhaps, Jung conjectured to his audience, the red crystal was a stone of wisdom, and the serpents somehow related to ancient Egypt. In private, however, Jung remained completely flummoxed by the iconography of his visions. "Heal the wounds that doubt inflicts on me, my soul," he implored the spirits that haunted him. "My spirit is a spirit of torment, it tears asunder my contemplation, it would dismantle everything and rip it apart. I am still a victim of my thinking. When can I order my thinking to be quiet, so that my thoughts, those unruly hounds, will crawl to my feet? How can I ever hope to hear your voice louder, to see your face clearer, when all my thoughts howl?"

What was the driving force behind Jung's fantasies? "From the beginning," he wrote, "I had conceived my voluntary confrontation with the unconscious as a scientific experiment which I myself was conducting and in whose outcome I was vitally interested. Today I might equally well say that it was an experiment which was being conducted on me." At any given moment the results of the experiment seemed inconclusive, as throughout his life Jung was beset by an ever-evolving cast of mystical visions: "I find myself again on the desert path. It was a desert vision, a vision of the solitary who has wandered down long roads. There lurk invisible robbers and assassins and shooters of poison darts."

Convinced that "a supreme meaning" lay behind his imaginings, Jung spent years trying to decipher his visions. The search often took the form of conversations with the Soul, which he duly recorded in his diary:

Jung: I am ready.... What is it? Speak!
Soul: Why have you received the revelation? You should not hide it.
Jung: But you are not thinking that I should publish what I have written. That would be a misfortune. And who would understand it?
Soul: Above all your calling comes first.
Jung: But what is my calling?
Soul: The new religion and its proclamation.

Three days later, the Soul chided Jung for relying on Reason to proclaim this new religion. "The way is symbolic," it intoned. Jung admitted that during these years it often seemed as if he were "living in an insane asylum of my own making." His professional practice was peopled not by the good burghers of Zurich but by "centaurs, nymphs, satyrs, gods and goddesses, as though they were patients and I was analyzing them."

But it was not only these characters of his visions that had captured Jung; he trained his focus on art as well, asking just why certain forms of aesthetic experience can so profoundly transfix the mind. "Whoever speaks in primordial images," Jung explained, "speaks with a thousand voices; he enthralls and overpowers, while at the same time he lifts the idea he is seeking to express out of the occasional and the transitory into the realm of the ever-enduring." The artist who trades in this shared imagery "transmutes our personal destiny."

Like art, religion held such power over our minds because it traded in these primordial images. Friedrich Schleiermacher's theory of religious experience was not lost on Jung, who even in childhood had sensed the importance of numinous experience in everyday family life. (Jung's uncle Johann Sigismund von Jung had married Schleiermacher's younger sister.) Jung's mother, Emilie, regularly protected his grandfather from bothersome ghosts, sitting beside him as he wrote his sermons. Indeed, Jung's maternal grandparents sometimes fell silent for hours on end, claiming to have seen ghosts. "I have always suspected," Jung once said, "that my blessed grandfather laid a very strange egg into my mixture."

Jung's mother kept a diary of all her supernatural encounters, and as a child, Jung reported "hearing things walking around the house at night." Foreboding figures in top hats, weeping women, and yawning graves plagued the young boy's mind. Strange incidents (the cracking of a heavy walnut table, the shattering of a bread knife) were seen as evidence of the occult. Jung's fascination with numinous experience deepened in adolescence, and when he began his medical training, he elected to study psychiatry. On first opening Krafft-Ebing's textbook of psychiatry to begin studying for the state examinations, Jung came to appreciate the "subjective character" of the discipline. "My heart suddenly began to pound," he recalled. "I had to stand up and draw a deep breath. My excitement was intense, for it had become clear to me, in a flash of illumination, that for me the only possible goal was psychiatry. Here alone the two currents of my interest could flow together and in a united stream dig their own bed. Here was the empirical field common to biological and spiritual facts, which I had everywhere sought and nowhere found. Here at last was the place where the collision of nature and spirit became a reality." Even more intriguing, perhaps, was the lack of a coherent theory of mind among psychiatrists: "No one really knew anything about it, and there was no psychology which regarded man as a whole and included his pathological variations in the total picture."

Jung soon apprenticed himself to two professors who specialized in the psychology of religion. His medical school dissertation focused on séances that he and his fifteen-year-old female cousin conducted at home. One frequent spirit-guest was a "small but fully grown black haired woman, of markedly Jewish type, clothed in white garments, her head wrapped in a turban." (The spirit had apparently been seduced by both Goethe and King David.) Jung soon came to believe that all forms of numinous experience had their origins in the mind; even God was a product of the unconscious. "No religion has survived or ever will without mystery, to which the devotee is most intimately bound," he said to his fellow students.

While Jung's remarks engendered considerable debate at school, they reflected a long-standing theological tradition dating back to the fifteenth century, according to which the divine inheres above all in the experience of the believer. Years later, while speaking at Yale, Jung would affirm this view: "In speaking of religion, I must make clear from the start what I mean by that term. Religion, as the Latin word denotes, is a careful and scrupulous observation of what Rudolf Otto aptly termed the *numinosum*, that is, a dynamic agency or affect not caused by an arbitrary act of will. On the contrary, it seizes and controls the human subject, who is always rather its victim than its creator. The *numinosum*—whatever its cause may be—is an experience of the subject independent of his will."

Religion was, then, no more than "a peculiar attitude of the human mind" that "transcends the ordinary categories of space, time and causality." For Jung, this sort of experience, albeit rare, had a particularly important role in treating mental illness. "Inasmuch as you attain to the numinous experiences," he wrote, "you are released from the curse of pathology. Even the very disease takes on a numinous character." An ecstatic release from the bonds of the everyday—the laws of cause and effect, the boundaries of space and time—promised to heal all sorts of psychic wounds. This is not to say that psychotherapy must be religious in character; rather, Jung found within the fabric of day-to-day life the kernel of a prerational spirituality. In "conversions, illuminations, emotional shocks, blows of fate" lay the possibility of a different understanding of the self and its relation to the surrounding world.

It remained unclear, however, just where such experiences originated. What was this "dynamic agency," beyond the scope of the will? Here, Jung struggled to make sense of the inevitable paradox of mysticism. "Whatever the nature of these numinous experiences may be," Jung observed, "they all have one thing in common: they relegate their source to a region outside of consciousness," even as they exert "a peculiar alteration [in] consciousness." The numinous is, in other words, experienced simultaneously as self and as other, rooted in some "pre-rational power" that rattles the mind but does not originate in it.

The iconography of particular religions sought to tame the inherent strangeness of all mystical experience. Only by anthropomorphizing the divine can we convey just how intimate our experience of the numinous necessarily is. The desire for such intimacy was, for Jung, "a force as real as hunger and the fear of death." In the German verb *ergriffen* (to seize or to grasp), Jung found a suitable linguistic form for this desire: "So a savior is one who seizes, the *Ergreifer* who catches people like objects and whirls them into a form which lasts as long as the whirlwind lasts, and then the thing collapses and something new must come." The numinous, then, reorients our experience of ourselves (the facts about ourselves that once seemed immutable, everlasting) even as it refuses to obey the dictates of the will.

Such experiences could, of course, prove profoundly destabilizing. "Wherever the psyche is set violently oscillating by a numinous experience," Jung warned, "there is a danger that the thread by which one hangs may be torn." In the wake of a mystical vision, one man may "tumble into an absolute affirmation" while another falls into "an equally absolute negation." For Jung, the *numinosum* was "dangerous because it lures men to extremes, so that a modest truth is regarded as the truth and a minor mistake is equated with fatal error."

Jung did not limit the numinous to religious or aesthetic experience. Jung cited the example of the angry man who fantasizes about killing his enemy. Such a fantasy would lead certain people to "assume that they are potential murderers, to believe themselves wholly wrong, children of the devil." "In short," Jung continued, "you cannot live with yourself, unless you understand yourself as a sort of givenness, a datum . . . an objective fact." When such a datum—once solid and impervious to change—changes before our eyes, our very understanding of ourselves as autonomous beings is shaken. It is as if we were rulers of a land only partially known to ourselves, in charge of an unknown number of subjects, whose very identity remains largely a mystery to us. From time to time we are nonetheless reminded that we indeed have subjects to govern and nations to protect. All one can say, then, is *I find myself as the ruler of a country which has unknown borders and unknown inhabitants, possessing qualities of which I am not entirely aware.* In such a moment we are thrown "out of [our] subjectivities . . . confronted with a situation in which [we] are a sort of prisoner": prisoner-kings of unknown lands.

Jung understood how readily the boundaries that separate one from another dissolve. When he read in Goethe about Faust's murder of Philemon and Baucis, Jung felt that he himself had helped commit the murder: "This strange idea alarmed me and I regarded it as my responsibility to atone for this crime." Years later, when the visions began to descend on Jung, Philemon figured prominently: "Philemon and other figures of my fantasies brought home to me the crucial insight that there are things in the psyche which I do not produce, but which produce themselves and have their own life. . . . Philemon represented a force which was not I. In my fantasies I held conversations with him, and he said things which I had not consciously thought."

Jung soon turned to the historical antecedents of his psychological theories and, in particular, to Gnosticism, which held that esoteric knowledge of Christ would lead to salvation. The psychoanalyst became fascinated by this "yearning for the hidden self"; the Gnostics believed that the soul was trapped in the human body, the essential "spark" disguised by matter. To Jung, the Gnostics seemed to possess a precocious understanding of human psychology—our need to transcend the facts of earthly existence without completely leaving that existence behind.

In the later decades of his life, Jung changed course and undertook a sustained study of alchemy, which he came to see as an extension of Gnosticism. Whereas Gnosticism sought to liberate the soul from the prison of matter, alchemists promised to transform base metals into gold. Both movements emphasized the interplay between the concrete, familiar world and the realm of the numinous; like the Gnostics, the alchemists sought in the stuff of the everyday a hidden sublimity. "The experiences of the alchemists were in a sense my experiences," Jung wrote, "and their world was my world." He argued that the trans-historical allure of mystical imagery, symbols, and myths was rooted in a basic fact about the human soul. Tracing the process whereby certain images took hold in his own mind, he soon found himself in an infinite regress that David Foster Wallace would have immediately recognized. Though he developed an entire lexicon to describe this process (*complexes, archetypes, shadow, anima*), Jung returned again and again to this underlying feature of human behavior. What does it mean to be an individual if our thoughts, feelings, and behaviors are not always within our control? What is the nature of this controlling force?

As a therapist, Jung tried to foster in his patients a dialogue between the conscious and unconscious minds. One of his goals was to integrate these two modes of perceiving and understanding such that the self identified equally with both. For Jung, merely subordinating the unconscious to rational control would only exacerbate pathology, insofar as the numinous was an integral part of human experience. Jungian psychoanalysis thus stripped religion of dogma, leaving intact the underlying desire for—indeed, the necessity of—transcendence. The therapist was, then, an educator of the soul; his primary task was to show patients "the way to the primordial experience which most clearly befell St. Paul . . . on the road to Damascus."

On a bright Sunday afternoon in the summer of 1916, the doorbell of Jung's house began ringing wildly. Jung and his two housekeepers looked to see who was there, but found no one at the door. Still, the doorbell rang. "Then I knew that something had to happen," Jung later recounted. "The whole house was filled as if there were a crowd present, crammed full of spirits. They were packed deep right up to the door, and the air was so thick it was scarcely possible to breathe. . . . They cried out in chorus, 'We have come back from Jerusalem where we found not what we sought.'"

["spirit within himself":** Laurens van der Post, *Jung and the Story of Our Time* (New York: Knopf, 1977), 157.

"stupidity as mine": Carl G. Jung, *Introduction to Jungian Psychology: Notes of the Seminar on Analytical Psychology Given in 1925* (Princeton, NJ: Princeton University Press, 2012), 49.

"One thunderstorm followed another": Carl G. Jung, *The Essential Jung*, ed. Anthony Storr (Princeton, NJ: Princeton University Press, 2013), 78.

"In October . . . turned to blood": Ibid., 77.

"it held me fast": Jung, *Introduction to Jungian Psychology*, 44, note 9.

"freedom of will": Ibid., 40.

"everything essential was decided": Carl G. Jung, quoted in Peter Homans, *Jung in Context: Modernity and the Making of a Psychology* (Chicago: University of Chicago Press, 1995), 162.

he descended: Jung, *Introduction to Jungian Psychology*, 45–51.

"But I want to know": Ibid., 52, note 9.

"severed artery": Ibid., 52.

"all my thoughts howl?": Carl G. Jung, *The Red Book: A Reader's Edition*, trans. Mark Kyburz, ed. Sonu Shamdasani (New York: Norton, 2012), 148.

"conducted on me": Carl G. Jung, quoted in John Chodorow, *Encountering Jung on Active Imagination* (Princeton, NJ: Princeton University Press, 2015), 26.

"poison darts": Jung, *The Red Book*, 159.

"I am ready . . . its proclamation": Ibid., 61.

"way is symbolic": Ibid., 62.

"analyzing them": Ibid., 12.

"personal destiny": Carl G. Jung, *The Spirit in Man, Art, and Literature*, trans. R. F. C. Hull (Princeton, NJ: Princeton University Press, 1966), 82.

Friedrich Schleiermacher's: See later note, chapter 7, on Rudolf Otto.

family life: Jay Sherry, "Carl Gustav Jung: Avant-Garde Conservative" (PhD diss., Freie Universität Berlin, 2008).

wrote his sermons: Aniela Jaffé, "The Psychic World of C. G. Jung," in Nandor Fodor, ed., *Freud, Jung, and Occultism* (New Hyde Park, NY: University Books, 1971), 188; see also F. X. Charet, *Spiritualism and the Foundations of C. G. Jung's Psychology* (Albany: State University of New York Press, 1993), 68; Aniela Jaffé, *From the Life and Work of C. G. Jung* (Einsiedeln, Switzerland: Daimon, 1989), 2.

"into my mixture": Carl G. Jung, Letter to Professor O. Schrenk (Nov. 18, 1953), in Gerhard Adler, ed., *Letters of C. G. Jung*, vol. 2 (London: Routledge, 2015), 132.

evidence of the occult: Charet, *Spiritualism*, 71.

"became a reality": Carl G. Jung, quoted in Anthony Stevens, *Archetype Revisited: An Updated Natural History of the Self* (London: Routledge, 2015), 23.

"in the total picture": Carl G. Jung, quoted in Thomas Stephen Szasz, *The Myth of Psychotherapy: Mental Healing as Religion, Rhetoric, and Repression* (Syracuse, NY: Syracuse University Press, 1978), 163.

"intimately bound": Carl G. Jung, *The Collected Works of Carl Jung*, vol. 1, ed. and trans. Gerhard Adler and R. F. C. Hull (Princeton, NJ: Princeton University Press, 2014), 33.

experience of the believer: Franz Pfeiffer, ed., *Works of Meister Eckhart* (Whitefish, MT: Kessinger, 2010), 8.

"independent of his will": Jung, *Essential Jung*, 239.

"time and causality": Carl G. Jung, *Psychology and Religion* (New Haven, CT: Yale University Press, 1938), 5.

"numinous character": Carl G. Jung, quoted in Ann Belford Ulanov, *Spirit in Jung* (Einsiedeln, Switzerland: Daimon, 2005), 165.

surrounding world: Jung, *Collected Works*, vol. 11, 183.

"pre-rational power": J. Harley Chapman, *Jung's Three Theories of Religious Experience* (Lewiston, NY: Edwin Mellen Press, 1988), 91.

"fear of death": Jung, *Collected Works*, vol. 7, 249.

"something new must come": Carl G. Jung, *Nietzsche's Zarathustra: Notes of the Seminar Given in 1934–1939*, vol. 1, ed. James L. Jarrett (Princeton, NJ: Princeton University Press, 1988), 1030.

"fatal error": C. G. Jung, *Memories, Dreams, Reflections*, ed. Aniela Jaffé, trans. Richard and Clara Winston (New York: Vintage Books, 2011), 151.

unknown lands: Fodor, *Freud, Jung, and Occultism*, 214.

"atone for this crime": Jung, *Nietzsche's Zarathustra*.

"consciously thought": Fodor, *Freud, Jung, and Occultism*, 223.

disguised by matter: Robert Segal, *The Allure of Gnosticism: The Gnostic Experience in Jungian Philosophy and Contemporary Culture* (Chicago: Open Court, 1999), 2–3, 29–33.

"their world was my world": Jung, *Memories, Dreams, Reflections*, 205; see also Segal, *The Allure of Gnosticism*, 26.

equally with both: Jeffrey C. Miller, *The Transcendent Function: Jung's Model of Psychological Growth Through Dialogue with the Unconscious* (Albany: State University of New York Press, 2004), 5.

"road to Damascus": Carl G. Jung, Letter to Pastor Walther Uhsadel (Aug. 18, 1936), in Adler, ed., *Letters of C. G. Jung*, vol. 1, 216–17.

"'not what we sought'": Jung, *The Red Book*, 41.]

30   **representation of desire:** Helen Block Lewis, *Freud and Modern Psychology* (New York: Plenum, 1981), 2.

30   **person sees the world:** Freud became friends with the French essayist Romain Rolland. Rolland tried to explain to Freud that *la sensation océanique* (the sense of losing oneself to a limitless expanse) lay at the heart of all religion. In his seminal 1929 work, *Civilization and Its Discontents*, Freud recast this mystical sensation in crude psychoanalytic terms as a "primary narcissistic union between mother and infant." In other words, the experience of the divine was a mere regression to a childlike state.

Freud and Rolland's correspondence reveals the tension between the worldview of a believer whose understanding of religion necessarily emphasized felt experience, and the psychoanalytic theory of his close friend and confidant, who saw all forms of religion as an expression of repressed sexual drives and infantile desires. Still, Freud respected Rolland's intellect and sought to deepen his own understanding of spiritual experience.

Freud had previously sent Rolland a copy of his 1927 book, *The Future of an Illusion*, in which the psychoanalyst speculates on the origins of religion. His friend responded with restraint: "Your analysis of religions is fair. But I would have liked to see you analyze spontaneous religious *feeling* or, more exactly religious *sensation*." For Rolland, this sensation had little to do with dogma or institutions or scripture or even morality; rather, it encompassed "all sensation of the eternal (which may very well not be eternal, but simply without perceptible limits, and in that way oceanic)." Freud had missed the boat: he understood religion as a product of social forces rather than a deeply felt, personal experience. The oceanic feeling, Rolland added, is "subjective in character" and, as such, it should hold a particular interest for the theorist of the mind.

Rolland readily admitted that he was quite familiar with *la sensation océa-nique*. "Throughout my whole life I have never lacked it; and I have always found it a source of vital renewal." Indeed, Rolland saw no contradiction be-tween religiosity and the life of the mind; though he subscribed to no particular sect or institutionalized credo, he described himself as "profoundly religious," comparing his very particular sense of the divine to "an underground bed of water which I feel surfacing under the bark." Moreover, this oceanic feeling had nothing to do with an afterlife: "Personally," Rolland wrote, "I aspire to eternal rest; survival has no attraction for me. But the sensation that I feel is thrust upon me as a fact. It is a *contact*."

Freud waited some eighteen months before responding to Rolland's letter. "Your remarks about a feeling you describe as 'oceanic' have left me no peace," he finally admitted to his friend. Indeed, he would use Rolland's description as a starting point in *Civilization and Its Discontents*, where he began to recast, if not completely abandon, his earlier interpretation of religion. Despite his newfound interest in mystical experience, Freud still insisted that psychosex-ual conflict was "the true source of religious sentiments," which merely gave outward form to the repressive self-loathing of the superego.

Freud compared Rolland's *sensation océanique* to "the consolation offered by an original and somewhat eccentric dramatist to his hero who is facing a self-inflicted death." In the moment before the fatal blow, the hero is reassured: "We cannot fall out of this world." As the hour of death nears, there rises "an indis-soluble bond," a sense "of being one with the external world as a whole." Unable to find any trace of the oceanic feeling in himself, Freud relegated mystical union to the realm of coping mechanisms, which disguise the bitter finality of our fate, the truth of our ultimate solitude. Yet Freud, who regularly plumbed the depths of his own psyche to make sense of human suffering, could not ex-plain why he could not experience this feeling.

In fact, the failure is not at all surprising: Freud was so convinced by his own theory of human development that he would necessarily experience religion as yet another extension of his worldview. For him, religion was merely another item on the long list of human foibles that stemmed from unconscious conflict. The unremitting suspicion, and the equally strong certainty, with which Freud analyzed all manner of human thoughts and behaviors made it impossible for him to experience the mind as anything but Freudian.

["mother and infant": William B. Parsons, *The Enigma of the Oceanic Feeling* (Oxford: Oxford University Press, 1998), 501.

"in that way oceanic": Romain Rolland, letter to Sigmund Freud (1927), quoted in D. J. Fisher, *Romain Rolland and the Politics of Intellectual Engagement* (Berkeley: University of California Press, 1988), 9.

"vital renewal": Ibid.

"It is a contact": Ibid., 9–10; see also Parsons, *Enigma of Oceanic Feeling*, 173.

"have left me no peace": Freud, letter to Romain Rolland (1929), quoted in Par-sons, 174.

repressive self-loathing of the superego: Freud, quoted in Michael Palmer, *Freud and Jung on Religion* (London: Routledge, 2003), 37.

"the external world as a whole": Freud, *Civilization and Its Discontents*, trans. and ed. James Strachey (New York: Norton, 1961), 12–15; see also Albert L. Blackwell, *The Sacred in Music* (Louisville, KY: Knox, 1999), 97.]

31 **"legendary...shaken":** Sigmund Freud, *The Interpretation of Dreams* (London: Allen and Unwin, 1954), 544.

31 **in the interim:** Jerome Carl Wakefield, "Do Unconscious Mental States Exist? Freud, Searle, and the Conceptual Foundations of Cognitive Science" (PhD diss., University of California, Berkeley, 2001).

31 **compelling whole:** Thomas C. Caramagno, *The Flight of the Mind: Virginia Woolf's Art and Manic-Depressive Illness* (Berkeley: University of California Press, 1992), 29, citing Steven E. Goldberg, *Two Patterns of Rationality in Freud's Writings* (Tuscaloosa: University of Alabama Press, 1988).

31 **"When we invent...fiction":** David Ballin Klein, *The Unconscious—Invention or Discovery? A Historico-Critical Inquiry* (Santa Monica, CA: Goodyear, 1977), 2, citing Edward Bradford Titchener, *A Textbook of Psychology* (1910; repr. Delmar, NY: Scholars' Facsimiles and Reprints, 1980), 2.

32 **"We do not possess":** D. Rapaport, "On the Psycho-Analytic Theory of Affects," *International Journal of Psychoanalysis* 34, no. 3 (1953): 177.

32 **"largely unfocused":** Peter H. Knapp, "Book Essay: Some Contemporary Contributions to the Study of Emotions," *Journal of the American Psychoanalytic Association* 35, no. 1 (Feb. 1987).

32 **everyday mental life:** Antti Revonsuo, *Consciousness: The Science of Subjectivity* (New York: Psychology Press, 2010), 60.

32 **conscious mental activity:** Ibid., 61.

32 **scientific findings:** Sonu Shamdasani, "Epilogue: The 'Optional' Unconscious," in Angus Nicholls and Martin Liebscher, eds., *Thinking the Unconscious: Nineteenth-Century German Thought* (Cambridge, UK: Cambridge University Press, 2010), 294.

34 **virtually undetectable:** S. Tomkins, "What Are Affects?," in Eve Kosofsky Sedgwick and Adam Frank, eds., *Shame and Its Sisters: A Silvan Tomkins Reader* (Durham, NC: Duke University Press, 1995), 33, 54.

34 **models of psychopathology:** Arthur Freeman, *Comprehensive Handbook of Cognitive Therapy* (New York: Plenum, 1989).

34 **errors in processing information:** Daniel David, Steven J. Lynn, and Albert Ellis, *Rational and Irrational Beliefs: Research, Theory, and Clinical Practice* (New York: Oxford University Press, 2010); Albert Ellis, *Growth Through Reason: Verbatim Cases in Rational-Emotive Therapy* (Palo Alto, CA: Science and Behavior Books, 1971); Albert Ellis, *Humanistic Psychotherapy: The Rational-Emotive Approach* (New York: Julian, 1973).

34 **lay behind them:** Aaron T. Beck, *Depression: Causes and Treatment* (Philadelphia: University of Pennsylvania Press, 1972); Aaron T. Beck, *Cognitive Therapy and the Emotional Disorders* (New York: International Universities Press, 1976); Aaron T. Beck, *Cognitive Therapy of Depression* (New York: Guilford, 1979); Aaron T. Beck, *Cognitive Therapy of Substance Abuse* (New York: Guilford, 1993); Aaron T. Beck, *Schizophrenia: Cognitive Theory, Research, and Therapy* (New York: Guilford, 2009); Aaron T. Beck, Gary Emery, and Ruth L. Greenberg, *Anxiety Disorders and Phobias: A Cognitive Perspective*, 15th anniversary ed. (Cambridge, MA: Basic, 2005).

The Science Underlying Capture

Sources

Works Referenced
See scientific articles in the notes.

Notes

38 **tuning based on experience:** M. Fahle and T. Poggio, *Perceptual Learning* (Cambridge, MA: MIT Press, 2002); Y. Sasaki, J. E. Nanez, and T. Watanabe, "Advances in Visual Perceptual Learning and Plasticity," *Nature Reviews Neuroscience* 11, no. 1 (2010): 53–60; J. A. Cromer, J. E. Roy, and E. K. Miller, "Representation of Multiple, Independent Categories in the Primate Prefrontal Cortex," *Neuron* 66, no. 5 (2010): 796–807.

38 **responding more vigorously:** R. Ptak, "The Frontoparietal Attention Network of the Human Brain: Action, Saliency, and a Priority Map of the Environment," *Neuroscientist* 18, no. 5 (2012): 502–15; L. Itti and C. Koch, "Computational Modeling of Visual Attention," *Nature Reviews Neuroscience* 2, no. 3 (2001): 194–203.

39 **what matters more or less:** P. Roelfsema, A. van Ooyen, and T. Watanabe, "Perceptual Learning Rules Based on Reinforcers and Attention," *Trends in Cognitive Sciences* 14, no. 2 (2010): 64–71; B. A. Anderson, P. A. Laurent, and S. Yantis, "Value-Driven Attentional Capture," *Proceedings of the National Academy of Sciences* 108, no. 25 (2011): 10367–71.

39 **out of place:** C. S. Gilbert and W. Li, "Top-Down Influences on Visual Processing," *Nature Reviews Neuroscience* 14, no. 5 (2013): 350–63; C. Summerfield and T. Egner, "Expectation (and Attention) in Visual Cognition," *Trends in Cognitive Sciences* 13, no. 9 (2009): 403–9.

39 **what is threatening:** S. J. Bishop, "Neural Mechanisms Underlying Selective Attention to Threat," *Annals of the New York Academy of Sciences* 1129 (2008): 141–52.

39 **awareness:** G. Rees, "Neural Correlates of Consciousness," *Annals of the New York Academy of Sciences* 1296 (2013): 4–10; N. Tsuchiya and J. van Boxtel, "Introduction to Research Topic: Attention and Consciousness in Different Senses," *Frontiers in Psychology* 4 (2013): 249; A. Treisman and G. Gelade, "A Feature-Integration Theory of Attention," *Cognitive Psychology* 12 (1980): 97–136; L. Itti and C. Koch, "Computational Modeling of Visual Attention," *Nature Reviews Neuroscience* 2, no. 3 (2001): 194–203; M. Corbetta and G. L. Shulman, "Control of Goal-Directed and Stimulus-Driven Attention in the Brain," *Nature Reviews Neuroscience* 3, no. 3 (2002): 201–15. While attention and awareness often go hand in hand, there are lots of studies that demonstrate the two are dissociable. Some studies have demonstrated awareness without attention and attention without awareness (e.g., Hsieh et al., *Psychological Science*, 2011; Lamme, *Trends in Cognitive Sciences*, 2003; Wyart and Tallon-Baudry, *Journal of Neuroscience*, 2008; Watanabe et al., *Science*, Nov. 2011).

39 **our eyes are open:** J. S. Werner and L. M. Chalupa, eds., *The New Visual Neurosciences* (Cambridge, MA: MIT Press, 2013). This statement is supported also by many studies that have elicited responses from visual cortex neurons in anesthetized animals.

39   **use the mechanism of selective attention:** R. Desimone and J. Duncan, "Neural Mechanisms of Selective Visual Attention," *Annual Review of Neuroscience* 18 (1995): 193–222; See also A. C. Nobre and S. Kastner, eds., *The Oxford Handbook of Attention* (Oxford: Oxford University Press, 2014).

39   **Separate neural networks:** D. Sridharan, D. J. Levitin, and V. Menon, "A Critical Role for the Right Fronto-Insular Cortex in Switching Between Central-Executive and Default-Mode Networks," *Proceedings of the National Academy of Sciences* 105, no. 34 (2008): 12569–74, published ahead of print, Aug. 22, 2008, doi:10.1073/pnas.0800005105. If by "internal stimuli," we are referring to reactivated memories or imagery of past stimuli, then the same system controls responses to both external and internal stimuli. If by "internal stimuli," we are referring to thoughts and feelings about other people in particular, then this is hypothesized to be controlled by the medial frontal cortex and the Default Mode Network. This DMN is separate from the attention systems discussed.

39   **attention is controlled:** M. Corbetta and G. L. Shulman, "Control of Goal-Directed and Stimulus-Driven Attention in the Brain," *Nature Reviews Neuroscience* 3, no. 3 (2002): 201–15. These two systems interact with each other; indeed, the separation between the two systems is not clear in some instances (see Malacuso and Doricchi, *Frontiers in Human Neuroscience*, 2013; E. Awh et al., *Trends in Cognitive Sciences*, 2012).

39   **play a role in decision making:** J. Duncan, "The Multiple-Demand (MD) System of the Primate Brain: Mental Programs for Intelligent Behaviour," *Trends in Cognitive Sciences* 14, no. 4 (2010): 172–79; D. T. Stuss and R. T. Knight, eds., *Principles of Frontal Lobe Function* (New York: Oxford University Press, 2013).

39   **a stimulus is detected:** M. Corbetta and G. L. Shulman, "Control of Goal-Directed and Stimulus-Driven Attention in the Brain," *Nature Reviews Neuroscience* 3, no. 3 (2002): 201–15; L. Itti, C. Koch, and E. Niebur, "A Model of Saliency-Based Visual Attention for Rapid Scene Analysis," *IEEE Transactions on Pattern Analysis and Machine Intelligence* 20, no. 11 (1998): 1254–59.

40   **learned or experienced:** A. Baddeley, "Working Memory: Looking Back and Looking Forward," *Nature Reviews Neuroscience* 4, no. 10 (2013): 829–39.

40   **One part of working memory:** M. Petrides, "Lateral Prefrontal Cortex: Architectonic and Functional Organization," *Philosophical Transactions of the Royal Society of London B: Biological Sciences* 360, no. 1456 (2005): 781–95.

40   **vital for learning:** N. Osaka, R. Logie, and M. D'Esposito, eds., *The Cognitive Neuroscience of Working Memory* (Oxford: Oxford University Press, 2007); A. D. Wagner, "Working Memory Contributions to Human Learning and Remembering," *Neuron* 22, no. 1 (1999): 19–22.

40   **close interaction:** A. Gazzaley and A. C. Nobre, "Top-Down Modulation: Bridging Selective Attention and Working Memory," *Trends in Cognitive Sciences* 16, no. 2 (2012): 129–35; D. Soto, J. Hodsoll, P. Rotshtein, and G. W. Humphreys, "Automatic Guidance of Attention from Working Memory," *Trends in Cognitive Sciences* 12, no. 9 (2008): 342–48; E. Awh, E. K. Vogel, and S.-H. Oh, "Interactions Between Attention and Working Memory," *Neuroscience* 139, no. 1 (2006): 201–8.

40   **more likely to remain in working memory:** A. Gazzaley and A. C. Nobre, "Top-Down Modulation: Bridging Selective Attention and Working Memory," *Trends in Cognitive Sciences* 16, no. 2 (2012): 129–35.

40   **Conversely:** These two concepts might be one and the same. One view is that selective attention is the mechanism by which items remain or become activated in working memory. (See Awh et al., *Neuroscience*, 2006; Postle, *Neuroscience*, 2006.)

40   **Both objects remain:** Only objects, concepts, etc., that are behaviorally relevant at the moment are held in working memory.

41   **conduit to lasting recall:** A. D. Wagner, "Working Memory Contributions to Human Learning and Remembering," *Neuron* 22, no. 1 (1999): 19–22; P. E. Wais and A. Gazzaley, "Distractibility During Retrieval of Long-Term Memory: Domain-General Interference, Neural Networks and Increased Susceptibility in Normal Aging," *Frontiers in Psychology* 5, no. 280 (2014): 1–12.

41   **processed for storage:** H. Eichenbaum, *The Cognitive Neuroscience of Memory: An Introduction*, 2nd ed. (Oxford: Oxford University Press, 2011). Working memory is the mechanism that brings long-term memories back into active use.

41   **stimuli that seize our attention:** Salient stimuli can capture attention and cause arousal. M. Corbetta and G. L. Shulman, "Control of Goal-Directed and Stimulus-Driven Attention in the Brain," *Nature Reviews Neuroscience* 3, no. 3 (2002): 201–15; P. Vuilleumier, "How Brains Beware: Neural Mechanisms of Emotional Attention," *Trends in Cognitive Sciences* 9, no. 12 (2005): 585–94. Affectively salient stimuli are detected more easily in threshold conditions, facilitate detection of stimuli in their spatial region, and can capture attention even when it is detrimental to one's current task. S. L. Nielsen and I. G. Sarason, "Emotion, Personality, and Selective Attention," *Journal of Personality and Social Psychology* 41, no. 5 (1981): 945–60; J. L. Armony and R. J. Dolan, "Modulation of Spatial Attention by Fear-Conditioned Stimuli: An Event-Related fMRI Study," *Neuropsychologia* 40, no. 7 (2002): 817; A. Richards and I. Blanchette, "Independent Manipulation of Emotion in an Emotional Stroop Task Using Classical Conditioning," *Emotion* 4, no. 3 (2004): 275–81. Affectively salient stimuli can produce increased heart rate and skin conductance, and increased subjective ratings of emotional arousal. H. D. Critchley, P. Rotshtein, Y. Nagai, J. O'Doherty, C. J. Mathias, and R. J. Dolan, "Activity in the Human Brain Predicting Differential Heart Rate Responses to Emotional Facial Expressions," *NeuroImage* 24, no. 3 (2005): 751–62; R. D. Lane, P. M. Chua, and R. J. Dolan, "Common Effects of Emotional Valence, Arousal, and Attention on Neural Activation During Visual Processing of Pictures," *Neuropsychologia* 37, no. 9 (1999): 989–97.

41   **stand out:** Salience refers to the fact that a stimulus stands out when compared with other stimuli in the environment. L. Itti and C. Koch, "A Saliency-Based Search Mechanism for Overt and Covert Shifts of Visual Attention," *Vision Research* 40, no. 10 (2000): 1489–506; R. M. Todd, W. A. Cunningham, A. K. Anderson, and E. Thompson, "Affect-Biased Attention as Emotion Regulation," *Trends in Cognitive Sciences* 16, no. 7 (2012): 365–72.

41   **seemingly neutral stimuli:** Learning (the association of two stimuli) can make neutral stimuli salient. An extensive body of work on fear conditioning—the pairing of an unconditioned stimulus (a stimulus that is in itself aversive to the organism) with a conditioned stimulus (one that signals the presence of the unconditioned stimulus) such that the conditioned stimulus comes to elicit the same response as the unconditioned stimulus—has demonstrated sensory cortex plasticity. C. Steinberg, C. Dobel, H. T. Schupp, J. Kissler, L. Elling, C. Pantev,

and M. Junghofer, "Rapid and Highly Resolving: Affective Evaluation of Olfactorily Conditioned Faces," *Journal of Cognitive Neuroscience* 24, no. 1 (2012): 17–27; R. J. Dolan, H. J. Heinze, R. Hurlemann, and H. Hinrichs, "Magnetoencephalography (MEG) Determined Temporal Modulation of Visual and Auditory Sensory Processing in the Context of Classical Conditioning to Faces," *NeuroImage* 32, no. 2 (2006): 778–89; D. A. Pizzagalli, L. L. Greischar, and F. J. Davidson, "Spatio-Temporal Dynamics of Brain Mechanisms in Aversive Classical Conditioning: High-Density Event-Related Potential and Brain Electrical Tomography Analyses," *Neuropsychologia* 41, no. 2 (2003): 184–94; W. Skrandies and A. Jedynak, "Associative Learning in Humans—Conditioning of Sensory-Evoked Brain Activity," *Behavioural Brain Research* 107, no. 1 (2000): 1–8; V. Miskovic and A. Keil, "Acquired Fears Reflected in Cortical Sensory Processing: A Review of Electrophysiological Studies of Human Classical Conditioning," *Psychophysiology* 49, no. 9 (2012): 1230–41.

42 **neutral stimulus acts as a cue:** Cues that predict the arrival of salient stimuli can themselves become salient stimuli. This is the process of classical conditioning. E.g., S. Maren, "Neurobiology of Pavlovian Fear Conditioning," *Annual Review of Neuroscience* 24 (2001): 897–931; A. Dickinson, *Contemporary Animal Learning Theory* (Cambridge, UK: Cambridge University Press, 1980); N. J. Mackintosh, *Conditioning and Associative Learning* (Oxford: Oxford University Press, 1983); C. R. Gallistel, *The Organization of Learning* (Cambridge, MA: MIT Press, 1990); L. A. Real, "Animal Choice Behavior and the Evolution of Cognitive Architecture," *Science* 253, no. 5023 (1991): 980; G. M. Edelman, "Linking Brain to Behavior: Value and Selection in Neural Populations," in *Proceedings of the Course on Neuropsychology: The Neuronal Basis of Cognitive Function*, Fidia Research Foundation (New York: Thieme Medical Publishers, 1992), 55–66; O. Sporns, N. Almássy, and G. Edelman, "Plasticity in Value Systems and Its Role in Adaptive Behavior," *Adaptive Behavior* 8, no. 2 (2000): 129–48; V. Miskovic and A. Keil, "Acquired Fears Reflected in Cortical Sensory Processing: A Review of Electrophysiological Studies of Human Classical Conditioning," *Psychophysiology* 49, no. 9 (2012): 1230–41.

42 **becomes salient itself:** D. Bindra, "How Adaptive Behavior Is Produced: A Perceptual-Motivation Alternative to Response Reinforcement," *Behavioral and Brain Sciences* 1 (1978): 41–91; K. Berridge, "Reward Learning: Reinforcement, Incentives, and Expectations," in *Psychology of Learning and Motivation*, D. L. Medin, ed. (New York: Academic Press, 2001), 223–78; A. Öhman, A. Hamm, and K. Hugdahl, "Cognition and the Autonomic Nervous System: Orienting, Anticipation, and Conditioning," in J. T. Cacioppo, L. G. Tassinary, and G. G. Berntson, eds., *Handbook of Psychophysiology* (New York: Cambridge University Press, 2000); M. L. Leathers and C. R. Olson, "In Monkeys Making Value-Based Decisions, LIP Neurons Encode Cue Salience and Not Action Value," *Science* 338, no. 6103 (2012): 132–35.

E. H. Castellanos et al., "Obese Adults Have Visual Attention Bias for Food Cue Images: Evidence for Altered Reward System Function," *International Journal of Obesity* 33, no. 9 (2009): 1063–73. Smokers find smoking-related stimuli salient. B. Bradley, M. Field, K. Mogg, and J. De Houwer, "Attentional and Evaluative Biases for Smoking Cues in Nicotine Dependence: Component Processes of Biases in Visual Orienting," *Behavioural Pharmacology* 15, no. 1 (2004): 29–36; R. N. Ehrman, S. J. Robbins, and M. Bromwell, "Comparing Attentional Bias to

Smoking Cues in Current Smokers, Former Smokers, and Non-Smokers Using a Dot-Probe Task," *Drug and Alcohol Dependence* 67, no. 2 (2002): 185–91. There is an extensive literature showing attentional bias in drug and alcohol addiction, e.g., M. Field and W. M. Cox, "Attentional Bias in Addictive Behaviors: A Review of Its Development, Causes, and Consequences," *Drug and Alcohol Dependence* 97, no. 1–2 (2008): 1–20; M. A. Miller and M. T. Fillmore, "The Effect of Image Complexity on Attentional Bias Towards Alcohol-Related Images in Adult Drinkers," *Addiction* 105, no. 5 (2010): 883–90.

In some situations, cues can be used as incentives to reinforce new behavior (in the same way as rewarding stimuli). V. Lovic, B. T. Saunders, L. M. Yager, and T. E. Robinson, "Rats Prone to Attribute Incentive Salience to Reward Cues Are Also Prone to Impulsive Action," *Behavioural Brain Research* 223, no. 2 (2011): 255–61; R. N. Cardinal, J. A. Parkinson, J. Hall, and B. J. Everitt, "Emotion and Motivation: The Role of the Amygdala, Ventral Striatum, and Prefrontal Cortex," *Neuroscience and Biobehavioral Reviews* 26, no. 3 (2002): 321–52; K. Berridge, "Reward Learning: Reinforcement, Incentives, and Expectations," in *Psychology of Learning and Motivation*, ed. D. L. Medin (New York: Academic Press, 2001), 223–78.

42   **involuntary component:** When salient stimuli capture attention and affect how we feel, the process can feel as if it is automatic or beyond our control. Salient stimuli can induce an orienting response that is fast and automatic. L. Carretié, J. A. Hinojosa, M. Martín-Loeches, F. Mercado, and M. Tapia, "Automatic Attention to Emotional Stimuli: Neural Correlates," *Human Brain Mapping* 22, no. 4 (2004): 290–99; P. J. Lang, M. Davis, and A. Öhman, "Fear and Anxiety: Animal Models and Human Cognitive Psychophysiology," *Journal of Affective Disorders* 61, no. 3 (2000): 137–59; T. Brosch, D. Sander, G. Pourtois, and K. R. Scherer, "Beyond Fear: Rapid Spatial Orienting Toward Positive Emotional Stimuli," *Psychological Science* 19, no. 4 (2008): 362–70. See also L. Pessoa and R. Adolphs, "Emotion Processing and the Amygdala: From a 'Low Road' to 'Many Roads' of Evaluating Biological Significance," *Nature Reviews Neuroscience* 11, no. 11 (2010): 773–83.

It is more difficult to disengage attention from salient stimuli. E.g., E. Fox, R. Russo, and K. Dutton, "Attentional Bias for Threat: Evidence for Delayed Disengagement from Emotional Faces," *Cognition and Emotion* 16, no. 3 (2002): 355–79; D. Derryberry and M. A. Reed, "Temperament and Attention: Orienting Toward and Away from Positive and Negative Signals," *Journal of Personality and Social Psychology* 66, no. 6 (1994): 1128.

42   **areas of the brain:** The areas of the brain that encode for salience are connected via the insula, hypothalamus, and brainstem. These brain areas help to regulate hormones and to produce autonomic increasing heart rate and skin temperature. The locus coeruleus in the pons of the brainstem is connected to several regions involved in affective salience, such as the orbitofrontal cortex, anterior cingulate cortex, and amygdala. G. Aston-Jones, J. Rajkowski, W. Lu, Y. Zhu, J. D. Cohen, and R. J. Morecraft, "Prominent Projections from the Orbital Prefrontal Cortex to the Locus Coeruleus in Monkey," *Society for Neuroscience Abstracts* 28 (2002): 86–89; J. Rajkowski, W. Lu, Y. Zhu, J. Cohen, and G. Aston-Jones, "Prominent Projections from the Anterior Cingulate Cortex to the Locus Coeruleus in Rhesus Monkey," *Society for Neuroscience Abstracts* 26 (2000): 838–15; G. Aston-Jones and J. D. Cohen, "An Integrative Theory of Locus Coeruleus–

Norepinephrine Function: Adaptive Gain and Optimal Performance," *Annual Review of Neuroscience* 28 (2005): 403–50; C. W. Berridge and B. D. Waterhouse, "The Locus Coeruleus–Noradrenergic System: Modulation of Behavioral State and State-Dependent Cognitive Processes," *Brain Research Reviews* 42, no. 1 (2003): 33–84. The insula is connected reciprocally with the amygdala and anterior cingulate cortex. E. Mufson, M. Mesulam, and D. Pandya, "Insular Interconnections with the Amygdala in the Rhesus Monkey," *Neuroscience* 6, no. 7 (1981): 1231–48; J. R. Augustine, "The Insular Lobe in Primates Including Humans," *Neurological Research* 7, no. 1 (1985): 2–10; J. R. Augustine, "Circuitry and Functional Aspects of the Insular Lobe in Primates Including Humans," *Brain Research Reviews* 22, no. 3 (1996): 229–44. The hypothalamus has strong connections with the amygdala. W. J. H. Nauta, "Fibre Degeneration Following Lesions of the Amygdaloid Complex in the Monkey," *Journal of Anatomy* 95, no. 4 (1961): 515–31; J. E. Krettek and J. L. Price, "Amygdaloid Projections to Subcortical Structures Within the Basal Forebrain and Brainstem in the Rat and Cat," *Journal of Comparative Neurology* 178, no. 2 (1978): 225–54; D. G. Amaral et al., "Some Observations on Hypothalamo-Amygdaloid Connections in the Monkey," *Brain Research* 252 (1982): 13–27; T. Ono et al., "Topographic Organization of Projections from the Amygdala to the Hypothalamus of the Rat," *Neuroscience Research* 2, no. 4 (1985): 221–39. The brainstem and hypothalamus are part of the autonomic system regulation, which is responsible for reflexive actions, arousal, and orienting e.g., G. Gabella, *Structure of the Autonomic Nervous System* (London: Chapman and Hall, 1976); G. A. G. Mitchell, *Anatomy of the Autonomic Nervous System* (Edinburgh: Livingstone, 1953); A. Öhman, A. Hamm, and K. Hugdahl, "Cognition and the Autonomic Nervous System: Orienting, Anticipation, and Conditioning," in *Handbook of Psychophysiology*, J. T. Cacioppo, J. G. Tassinary, and G. G. Berntson, eds. (New York: Cambridge University Press, 2000), 48.

42 **heart rate:** Stimuli can evoke an affective response. Affectively salient stimuli are those that result in emotional responses, including physiological changes and changes in subjective feeling states. R. M. Todd, W. A. Cunningham, A. K. Anderson, and E. Thompson, "Affect-Biased Attention as Emotion Regulation," *Trends in Cognitive Sciences* 16, no. 7 (2012): 365–72. Viewing positive and negative arousing stimuli leads to activation of the sympathetic nervous system as evidenced by increased heart rate, skin conductance, and pupil dilation. M. M. Bradley, L. Miccoli, M. A. Escrig, and P. J. Lang, "The Pupil as a Measure of Emotional Arousal and Autonomic Activation," *Psychophysiology* 45, no. 4 (2008): 602–7; P. J. Lang, M. K. Greenwald, M. M. Bradley, and A. O. Hamm, "Looking at Pictures: Affective, Facial, Visceral, and Behavioral Reactions," *Psychophysiology* 30, no. 3 (1993): 261–73.

42 **brightness, color:** Salience can refer to the physical characteristics of an object—luminescence, color, shape, motion, novelty, or other marked differences from other objects. It can also refer to the attitudes, perspectives, goals, needs, or desires of the perceiver. S. E. Taylor and S. T. Fiske, "Salience, Attention, and Attribution: Top of the Head Phenomena," *Advances in Experimental Social Psychology* 11 (1978): 249–88.

Bottom-up salience is related to the inherent, low-level features of a stimulus and captures our attention by standing out from its surroundings. A. M. Treisman and G. Gelade, "A Feature-Integration Theory of Attention," *Cognitive*

Psychology 12, no. 1 (1980): 97–136; L. Itti, C. Koch, and E. Niebur, "A Model of Saliency-Based Visual Attention for Rapid Scene Analysis," *IEEE Transactions on Pattern Analysis and Machine Intelligence* 20, no. 11 (1998): 1254–59. In "pop-out" visual search tasks, a unique stimulus (due to, e.g., color, size, shape, or direction) can be found regardless of the number of distractors. R. Desimone and J. Duncan, "Neural Mechanisms of Selective Visual Attention," *Annual Review of Neuroscience* 18, no. 1 (1995): 193–222.

42   **printer on a stove:** M. L. H. Vo and J. M. Henderson, "Does Gravity Matter? Effects of Semantic and Syntactic Inconsistencies on the Allocation of Attention During Scene Perception," *Journal of Vision* 9, no. 8 (2009): article 418.

42   **unexpected:** D. E. Berlyne, "Novelty, Complexity, and Hedonic Value," *Perception and Psychophysics* 8, no. 5 (1970): 279–86; E. N. Sokolov, "Higher Nervous Functions: The Orienting Reflex," *Annual Review of Physiology* 25, no. 1 (1963): 545–80; E. Courchesne, S. A. Hillyard, and R. Galambos, "Stimulus Novelty, Task Relevance and the Visual Evoked Potential in Man," *Electroencephalography and Clinical Neurophysiology* 39, no. 2 (1975): 131–43; C. Escera, K. Alho, I. Winkler, and R. Näätänen, "Neural Mechanisms of Involuntary Attention to Acoustic Novelty and Change," *Journal of Cognitive Neuroscience* 10, no. 5 (1998): 590–604.

42   **desires:** The perceiver who has a certain goal, such as looking for her phone or watching the door, focuses attention on features relevant to that goal. At the level of the brain, frontoparietal networks modulate visual cortex activation to increase activation in regions of sensory cortex coding for goal-relevant features and decrease activation in regions coding for competing stimuli. M. Corbetta and G. L. Shulman, "Control of Goal-Directed and Stimulus-Driven Attention in the Brain," *Nature Reviews Neuroscience* 3, no. 3 (2002): 201–15; D. M. Beck and S. Kastner, "Top-Down and Bottom-Up Mechanisms in Biasing Competition in the Human Brain," *Vision Research* 49, no. 10 (2005): 1154–65.

Salient stimuli also include stimuli that can change how we feel. These stimuli are emotionally arousing and receive enhanced perceptual processing. J. Markovic, A. K. Anderson, and R. M. Todd, "Tuning to the Significant: Neural and Genetic Processes Underlying Affective Enhancement of Visual Perception and Memory," *Behavioural Brain Research* 259 (2014): 229–41; A. K. Anderson, "Affective Influences on the Attentional Dynamics Supporting Awareness," *Journal of Experimental Psychology: General* 134, no. 2 (2005): 258; E. A. Phelps, "Emotion and Cognition: Insights from Studies of the Human Amygdala," *Annual Review of Psychology* 57 (2006): 27–53; G. Pourtois, A. Schettino, and P. Vuilleumier, "Brain Mechanisms for Emotional Influences on Perception and Attention: What Is Magic and What Is Not," *Biological Psychology* 92, no. 3 (2013): 492–512.

42   **attitudes:** Salient stimuli can also refer to "emotional" or "affective" stimuli. Such stimuli have "valence," either positive or negative. A. K. Anderson, "Affective Influences on the Attentional Dynamics Supporting Awareness," *Journal of Experimental Psychology: General* 134, no. 2 (2005): 258–81. Affectively salient stimuli result in increased activation in sensory cortices and enhanced attentional processing. A. K. Anderson and E. A. Phelps, "Lesions of the Human Amygdala Impair Enhanced Perception of Emotionally Salient Events," *Nature* 411, no. 6835 (2001): 305–9; M. M. Bradley, D. Sabatinelli, P. J. Lang, J. R. Fitzsimmons, W. King, and P. Desai, "Activation of the Visual Cortex in Moti-

vated Attention," *Behavioral Neuroscience* 117, no. 2 (2003): 369–80; P. J. Lang, M. M. Bradley, J. R. Fitzsimmons, B. N. Cuthbert, J. D. Scott, B. Moulder, et al., "Emotional Arousal and Activation of the Visual Cortex: An fMRI Analysis," *Psychophysiology* 35, no. 2 (1998): 199–210; R. D. Lane, P. M. Chua, and R. J. Dolan, "Common Effects of Emotional Valence, Arousal, and Attention on Neural Activation During Visual Processing of Pictures," *Neuropsychologia* 37, no. 9 (1999): 989–97; D. Sabatinelli, M. M. Bradley, J. R. Fitzsimmons, and P. J. Lang, "Parallel Amygdala and Inferotemporal Activation Reflect Emotional Intensity and Fear Relevance," *NeuroImage* 24, no. 4 (2005): 1265–70; D. Grandjean, D. Sander, G. Pourtois, S. Schwartz, M. L. Seghier, K. R. Scherer, et al., "The Voices of Wrath: Brain Responses to Angry Prosody in Meaningless Speech," *Nature Neuroscience* 8, no. 2 (2005): 145–46; T. Ethofer, J. Bretscher, M. Gschwind, B. Kreifelts, D. Wildgruber, and P. Vuilleumier, "Emotional Voice Areas: Anatomic Location, Functional Properties, and Structural Connections Revealed by Combined fMRI/DTI," *Cerebral Cortex* 22, no. 1 (2012): 191–200; T. Ethofer, D. Van De Ville, K. Scherer, and P. Vuilleumier, "Decoding of Emotional Information in Voice-Sensitive Cortices," *Current Biology* 19, no. 12 (2009): 1028–33.

42   **opportunity:** Rewards and punishments are salient stimuli. Studies on monkeys have found that the relative reward value of a stimulus predicted primary visual cortex activity. The timing and magnitude of the effect on visual cortex neurons was similar to that of selective attention. L. Stanisor, C. van der Togt, C. M. A. Pennartz, and P. R. Roelfsema, "A Unified Selection Signal for Attention and Reward in Primary Visual Cortex," *Proceedings of the National Academy of Sciences* 110, no. 22 (2013): 9136–41; J. T. Serences and S. Saproo, "Population Response Profiles in Early Visual Cortex Are Biased in Favor of More Valuable Stimuli," *Journal of Neurophysics* 104, no. 1 (2010): 76–87. Food-deprived individuals show increased selective attention to food-related stimuli, and overweight and obese individuals show an attentional bias to food. K. Mogg, B. P. Bradley, H. Hyare, and S. Lee, "Selective Attention to Food-Related Stimuli in Hunger: Are Attentional Biases Specific to Emotional and Psychopathological States, or Are They Also Found in Normal Drive States?," *Behaviour Research and Therapy* 36, no. 2 (1998): 227–37; E. H. Castellanos, E. Charboneau, M. S. Dietrich, S. Park, B. P. Bradley, K. Mogg, et al., "Obese Adults Have Visual Attention Bias for Food Cue Images: Evidence for Altered Reward System Function," *International Journal of Obesity* 33, no. 9 (2009): 1063–73.

42   **major life events:** Stimuli associated with major life events can become salient. For example, individuals with PTSD show attentional bias and increased amygdala sensitivity to stimuli related to their trauma. Researchers have used a Stroop test (in which a participant is instructed to name the color of a word and ignore the word's semantic content) to look at selective processing of certain stimuli. Reaction times in a Stroop task are slower for words that a participant finds salient because attention is directed away from word color. Individuals with PTSD following a motor vehicle accident, for instance, show a selective Stroop interference for words associated with the accident. J. G. Beck, J. B. Freeman, J. C. Shipherd, J. L. Hamblen, and J. M. Lackner, "Specificity of Stroop Interference in Patients with Pain and PTSD," *Journal of Abnormal Psychology* 110, no. 4 (2001): 536–43. Selective Stroop interference has also been found in Vietnam

veterans and rape victims with PTSD. R. J. McNally, S. P. Kaspi, B. C. Riemann, and S. B. Zeitlin, "Selective Processing of Threat Cues in Posttraumatic Stress Disorder," *Journal of Abnormal Psychology* 99, no. 4 (1990): 398–402; E. B. Foa, B. O. Rothbaum, D. S. Riggs, and T. B. Murdock, "Treatment of Posttraumatic Stress Disorder in Rape Victims: A Comparison Between Cognitive-Behavioral Procedures and Counseling," *Journal of Consulting and Clinical Psychology* 59, no. 5 (1991): 715–23.

42   **emotional state:** Salient stimuli can influence mood, and mood can influence the salience of stimuli. M. S. Clark, S. Milberg, and J. Ross, "Arousal Cues Arousal-Related Material in Memory: Implications for Understanding Effects of Mood on Memory," *Journal of Verbal Learning and Verbal Behavior* 22, no. 6 (1983): 633–49. Depressed mood, studied in individuals with subclinical depression, is associated with defocused attention, a broader attentional frame with greater sensitivity to irrelevant stimuli. U. Von Hecker and T. Meiser, "Defocused Attention in Depressed Mood: Evidence from Source Monitoring," *Emotion* 5, no. 4 (2005): 456–63. Stress-induced negative mood is associated with attentional shifting away from negative stimuli. E. H. Koster, R. De Raedt, E. Goeleven, E. Franck, and G. Crombez, "Mood-Congruent Attentional Bias in Dysphoria: Maintained Attention to and Impaired Disengagement from Negative Information," *Emotion* 5, no. 4 (2005): 446. There is evidence that positive mood broadens attention. B. L. Fredrickson, "What Good Are Positive Emotions?," *Review of General Psychology* 2, no. 3 (1998): 300–19; H. A. Wadlinger and D. M. Isaacowitz, "Positive Mood Broadens Visual Attention to Positive Stimuli," *Motivation and Emotion* 30, no. 1 (2006): 87–99.

There is evidence that salient stimuli not only compete more effectively for attentional resources but also activate arousal networks. Exposure to salient stimuli results in the release of stress hormones and the activation of the sympathetic nervous system. L. Cahill and J. L. McGaugh, "Mechanisms of Emotional Arousal and Lasting Declarative Memory," *Trends in Neurosciences* 21, no. 7 (1998): 294–99; H. D. Critchley, "Neural Mechanisms of Autonomic, Affective, and Cognitive Integration," *Journal of Comparative Neurology* 493, no. 1 (2005): 154–66.

This arousal network involves the locus coeruleus (LC)—the norepinephrine network, which modulates visual cortex activation, enhancing perception of salient stimuli. Neurons in the locus coeruleus facilitate responses to motivationally relevant stimuli and suppress responses to irrelevant stimuli. S. J. Sara, "The Locus Coeruleus and Noradrenergic Modulation of Cognition," *Nature Reviews Neuroscience* 10, no. 3 (2009): 211–23; S. J. Sara and S. Bouret, "Orienting and Reorienting: The Locus Coeruleus Mediates Cognition Through Arousal," *Neuron* 76, no. 1 (2012): 130–41; G. Aston-Jones and J. D. Cohen, "An Integrative Theory of Locus Coeruleus–Norepinephrine Function: Adaptive Gain and Optimal Performance," *Annual Review of Neuroscience* 28 (2005): 403–50. The LC is thought to enhance attention to and memory of emotional stimuli through modulation of visual cortex activity and norepinephrine release in the amygdala. J. Markovic, A. K. Anderson, and R. M. Todd, "Tuning to the Significant: Neural and Genetic Processes Underlying Affective Enhancement of Visual Perception and Memory," *Behavioural Brain Research* 259 (2014): 229–41; F. J. Chen and S. J. Sara, "Locus Coeruleus Activation by Foot Shock or Electrical Stimulation Inhibits Amygdala Neurons," *Neuroscience* 144, no. 2 (2007): 472–

81; C. W. Berridge and B. D. Waterhouse, "The Locus Coeruleus–Noradrenergic System: Modulation of Behavioral State and State-Dependent Cognitive Processes," *Brain Research Reviews* 42, no. 1 (2003): 33–84.

42 **personal experiences:** Another network is active when the individual is not focused on the external world. This network is called the default mode network (DMN) and involves the medial prefrontal cortex, posterior cingulate cortex, and medial temporal areas. J. R. Andrews-Hanna, "The Brain's Default Network and Its Adaptive Role in Internal Mentation," *Neuroscientist* 18, no. 3 (2012): 251–70; D. Ongür, A. Ferry, and J. L. Price, "Architectonic Subdivision of the Human Orbital and Medial Prefrontal Cortex," *Journal of Comparative Neurology* 460, no. 3 (2003): 425–49; R. L. Buckner, J. R. Andrews-Hanna, and D. L. Schacter, "The Brain's Default Network: Anatomy, Function and Relevance to Disease," *Annals of the New York Academy of Sciences* 1124 (2008): 1–38; P. Fransson and G. Marrelec, "The Precuneus/Posterior Cingulate Plays a Pivotal Role in the Default Mode Network: Evidence from Partial Correlation Analysis," *NeuroImage* 42, no. 3 (2008): 1178–84; A. Pfefferbaum, S. Chanraud, A.-L. Pitel, E. Müller-Oehring, A. Shankaranarayanan, D. Alsop, et al., "Cerebral Blood Flow in Posterior Cortical Nodes of the Default Mode Network Decreases with Task Engagement but Remains Higher Than in Most Brain Regions," *Cerebral Cortex* 21, no. 1 (2011): 233–44. Shulman and colleagues (1997) discovered that there was a set of regions that were less active during goal-directed tasks and more active during the passive, or "resting," periods between task sessions. G. L. Shulman, J. A. Fiez, M. Corbetta, R. L. Buckner, F. M. Miezen, M. E. Raichle, et al., "Common Blood Flow Changes Across Visual Tasks, II: Decreases in Cerebral Cortex," *Journal of Cognitive Neuroscience* 9, no. 5 (1997): 648–63. Later studies confirmed this finding. J. R. Binder, J. A. Frost, T. A. Hammeke, P. S. Bellgowan, S. M. Rao, and R. W. Cox, "Conceptual Processing During the Conscious Resting State: A Functional MRI Study," *Journal of Cognitive Neuroscience* 11, no. 1 (1999): 80–95; P. Mazoyer, L. Zago, E. Mellet, S. Bricogne, O. Etard, F. Houdé, et al., "Cortical Networks for Working Memory and Executive Function Sustain the Conscious Resting State in Man," *Brain Research Bulletin* 54, no. 3 (2001): 287–98. The DMN's increase in activity during "resting" periods is thought to reflect the intrinsic or default functioning of these regions, which becomes suppressed during goal-directed tasks. M. E. Raichle, A. M. MacLeod, et al., "A Default Mode of Brain Function," *Proceedings of the National Academy of Sciences* 98, no. 2 (2001): 676–82; M. E. Raichle and A. Z. Snyder, "A Default Mode of Brain Function: A Brief History of an Evolving Idea," *NeuroImage* 37, no. 4 (2007): 1083–90.

It is thought that the default mode network is involved in self-referential processing. Many task-irrelevant thoughts occurring during DMN activation are self-referential thoughts about the near future and past. E. Klinger and W. Cox, "Dimensions of Thought Flow in Everyday Life," *Imagination, Cognition, and Personality* 7 (1987): 105–28; E. Klinger, "Daydreaming and Fantasizing: Thought Flow and Motivation," in K. D. Markman, W. M. P. Klein, and J. A. Suhr, eds., *Handbook of Imagination and Mental Simulation* (New York: Psychology Press, 2009), 225–40; P. Delamillieure, G. Doucet, P. Mazoyer, M.-R. Turbelin, N. Delcroix, E. Mellet, et al., "The Resting State Questionnaire: An Introspective Questionnaire for Evaluation of Inner Experience During the Conscious Resting State," *Brain Research Bulletin* 81, no. 6 (2010):

565–73; J. R. Binder, J. A. Frost, T. A. Hammeke, P. S. Bellgowan, S. M. Rao, and R. W. Cox, "Conceptual Processing During the Conscious Resting State: A Functional MRI Study," *Journal of Cognitive Neuroscience* 11, no. 1 (1999): 80–95; P. Mazoyer, L. Zago, E. Mellet, S. Bricogne, O. Etard, F. Houdé, et al., "Cortical Networks for Working Memory and Executive Function Sustain the Conscious Resting State in Man," *Brain Research Bulletin* 54 (2001): 287–98; J. R. Andrews-Hanna, J. S. Reidler, C. Huang, and R. Buckner, "Evidence for the Default Network's Role in Spontaneous Cognition," *Journal of Neurophysiology* 104, no. 1 (2010): 322–35. DMN activity is correlated with poorer memory encoding unless the task involves episodic memory, in which case DMN activity is beneficial. S. M. Daselaar, S. E. Prince, N. A. Dennis, S. M. Hayes, H. Kim, and R. Cabeza, "Posterior Midline and Ventral Parietal Activity Is Associated with Retrieval Success and Encoding Failure," *Frontiers in Human Neuroscience* 3 (2009): 13; H. Kim, S. M. Daselaar, and R. Cabeza, "Overlapping Brain Activity Between Episodic Memory Encoding and Retrieval: Roles of the Task-Positive and Task-Negative Networks," *NeuroImage* 49, no. 1 (2010): 1045–54; P. Vannini, J. O'Brien, K. O'Keefe, M. Pihlajamäki, P. Laviolette, and R. A. Sperling, "What Goes Down Must Come Up: Role of the Posteromedial Cortices in Encoding and Retrieval," *Cerebral Cortex* 21, no. 1 (2011): 22–34. Some regions of the DMN, including left-lateralized regions in the medial prefrontal cortex (MPFC), medial and lateral temporal cortices, and the posterior cingulate cortex (PCC), become active in response to episodic memory and other kinds of autobiographical information. E. Svoboda, M. C. McKinnon, and B. Levine, "The Functional Neuroanatomy of Autobiographical Memory: A Meta-Analysis," *Neuropsychologia* 44, no. 12 (2006): 2189–208; S. C. Johnson, L. C. Baxter, L. S. Wilder, J. G. Pipe, J. E. Heiserman, and G. P. Prigatano, "Neural Correlates of Self-Reflection," *Brain* 125, no. 8 (2002): 1808–14; R. N. Spreng, R. A. Mar, and A. S. Kim, "The Common Neural Basis of Autobiographical Memory, Prospection, Navigation, Theory of Mind, and the Default Mode: A Quantitative Meta-Analysis," *Journal of Cognitive Neuroscience* 21, no. 3 (2009): 489–510; D. L. Schacter, D. R. Addis, and R. Buckner, "Episodic Simulation of Future Events: Concepts, Data, and Applications," *Annals of the New York Academy of Sciences* 1124 (2008): 39–60; K. B. McDermott, K. K. Szpunar, and S. E. Christ, "Laboratory-Based and Autobiographical Retrieval Tasks Differ Substantially in Their Neural Substrates," *Neuropsychologia* 47, no. 11 (2009): 2290–98; J. R. Andrews-Hanna, J. S. Reidler, J. Sepulcre, R. Poulin, and R. L. Buckner, "Functional-Anatomic Fractionation of the Brain's Default Network," *Neuron* 65, no. 4 (2010): 550–62.

The salience of a stimulus can be influenced by how closely the stimulus relates to oneself. That judgment involves primarily the medial prefrontal cortex. Several studies have found that self-referential stimuli capture attention more easily than other stimuli. N. Moray, "Attention in Dichotic Listening: Affective Cues and the Influence of Instructions," *Quarterly Journal of Experimental Psychology* 11, no. 1 (1959): 56–60; K. L. Shapiro, J. Caldwell, and R. E. Sorensen, "Personal Names and the Attentional Blink: A Visual 'Cocktail Party' Effect," *Journal of Experimental Psychology: Human Perception and Performance* 23, no. 2 (1997): 504–14; A. Mack and I. Rock, *Inattentional Blindness* (Cambridge, MA: MIT Press, 1998); F. Tong and K. Nakayama, "Robust Representations for Faces: Evidence from Visual Search," *Journal of Experimental Psychology: Human Per-*

ception and Performance 25, no. 4 (1999): 1016–35. Though see C. Devue and S. Brédart, "Attention to Self-Referential Stimuli: Can I Ignore My Own Face?," *Acta Psychologica* 128, no. 2 (2008): 290–97. Self-reference effects occur especially in cases of low attentional load and when self-referential stimuli occur rarely in the experiment. C. R. Harris and H. Pashler, "Attention and the Processing of Emotional Words and Names Not So Special After All," *Psychological Science* 15, no. 3 (2004): 171–78. Areas in the MPFC are active in response to self-relevant information, including autobiographical information, personality traits, values, emotional states, and physical attributes. W. M. Kelley, C. N. Macrae, C. L. Wyland, S. Caglar, S. Inati, and T. F. Heatherton, "Finding the Self? An Event-Related fMRI Study," *Journal of Cognitive Neuroscience* 14, no. 5 (2002): 785–94; K. N. Ochsner, J. S. Beer, E. R. Robertson, J. C. Cooper, J. D. Gabrieli, J. F. Kihsltrom, and M. D'Esposito, "The Neural Correlates of Direct and Reflected Self-Knowledge," *NeuroImage* 28, no. 4 (2005): 797–814; K. N. Ochsner, K. Knierim, D. H. Ludlow, J. Hanelin, T. Ramachandran, G. Glover, and S. C. Mackey, "Reflecting upon Feelings: An fMRI Study of Neural Systems Supporting the Attribution of Emotion to Self and Other," *Journal of Cognitive Neuroscience* 16, no. 10 (2004): 1746–72; P. Fossati, S. J. Hevenor, S. Graham, C. Grady, M. L. Keightley, F. Craik, and H. Mayberg, "In Search of the Emotional Self: An fMRI Study Using Positive and Negative Emotional Words," *American Journal of Psychiatry* 160, no. 11 (2003): 1938–45; G. Northoff, A. Heinzel, M. de Greck, F. Bermpohl, H. Dobrowolny, and J. Panksepp, "Self-Referential Processing in Our Brain—A Meta-Analysis of Imaging Studies on the Self," *NeuroImage* 31, no. 1 (2006): 440–57; D. A. Gusnard, E. Akbudak, G. L. Shulman, and M. E. Raichle, "Medial Prefrontal Cortex and Self-Referential Mental Activity: Relation to a Default Mode of Brain Function," *Proceedings of the National Academy of Sciences* 98, no. 7 (2001): 4259–64.

42   **Sometimes:** Stimuli become salient based on past learning, experience, and memories. J. S. Bruner and C. C. Goodman, "Value and Need as Organizing Factors in Perception," *Journal of Abnormal and Social Psychology* 42, no. 1 (1947): 33; R. M. Todd, W. A. Cunningham, A. K. Anderson, and E. Thompson, "Affect-Biased Attention as Emotion Regulation," *Trends in Cognitive Sciences* 16, no. 7 (2012): 365–72; M. Mather and L. L. Carstensen, "Aging and Attentional Biases for Emotional Faces," *Psychological Science* 14, no. 5 (2003): 409–15; V. Miskovic and A. Keil, "Acquired Fears Reflected in Cortical Sensory Processing: A Review of Electrophysiological Studies of Human Classical Conditioning," *Psychophysiology* 49, no. 9 (2012): 1230–41; R. J. Dolan, "The Human Amygdala and Orbital Prefrontal Cortex in Behavioural Regulation," *Philosophical Transactions of the Royal Society B: Biological Sciences* 362, no. 1481 (2007): 787–99.

Traumatic events affect which stimuli we find salient. E.g., M. Vythilingam, K. S. Blair, D. McCaffrey, M. Scaramozza, M. Jones, M. Nakic, et al., "Biased Emotional Attention in Post-Traumatic Stress Disorder: A Help as Well as a Hindrance?," *Psychological Medicine London* 37, no. 10 (2007): 1445; J. G. Beck, J. B. Freeman, J. C. Shipherd, J. L. Hamblen, and J. M. Lackner, "Specificity of Stroop Interference in Patients with Pain and PTSD," *Journal of Abnormal Psychology* 110, no. 4 (2011): 536.

42   **opposite ways:** There are individual differences in what we find salient. For instance, highly anxious individuals have more difficulty directing attention away

from threatening stimuli. Y. Bar-Haim, D. Lamy, L. Pergamin, M. J. Bakermans-Kranenburg, and M. H. van IJzendoorn, "Threat-Related Attentional Bias in Anxious and Nonanxious Individuals: A Meta-Analytic Study," *Psychological Bulletin* 133, no. 1 (2007): 1. Individuals who have undergone a trauma show attentional bias toward trauma-related stimuli. E.g., S. D. Pollak and S. A. Tolley-Schell, "Selective Attention to Facial Emotion in Physically Abused Children," *Journal of Abnormal Psychology* 112, no. 3 (2003): 323; J. G. Beck, J. B. Freeman, J. C. Shipherd, J. L. Hamblen, and J. M. Lackner, "Specificity of Stroop Interference in Patients with Pain and PTSD," *Journal of Abnormal Psychology* 110, no. 4 (2011): 536.

42   **thinking:** Thinking is how we come to represent the world. Mental representation involves the "encoding" of acquired information. Our experiences can give rise to multiple simultaneous encodings. But what can be encoded? Words and symbols can be encoded, as can sounds, objects, colors, values, spatial structures, and movements. The word "water" can be encoded along with an image or concept of the clear liquid, or with the molecular structure of two hydrogen atoms linked to one oxygen, or with the concept of quenching thirst.

What did Helen Keller come to encode when her tutor spelled the word "water" in her palm while Helen's other hand was under the well pump's spigot? First, the tutor's account: "This morning while she was washing, she wanted to know the name for 'water.' I spelled w-a-t-e-r and thought no more about it until after breakfast. Afterwards, we went out to the pump house, and I made Helen hold her mug under the spout while I pumped. As the cold water gushed forth filling the mug, I spelled 'w-a-t-e-r' in Helen's free hand. The word coming so close upon the sensation of the cold water rushing over her hand seemed to startle her. She dropped the mug and stood as one transfixed. A new light came into her face. She spelled 'water' several times. . . . All the way back she was highly excited and learned the name of every object she touched, so that in a few hours she had added thirty words to her vocabulary. . . . [The next morning] Helen got up like a radiant fairy. She had flitted from object to object, asking the name of everything. . . . Everything must have a name now."

Helen recounts the same episode: "As the cool stream gushed over my hand, she spelled into the other the word water, first slowly then rapidly. I stood still, my whole attention fixed upon the motion of her fingers. Suddenly I felt a misty consciousness as of something forgotten—a thrill of returning thought, and somehow the mystery of language was revealed to me. I knew that water meant the wonderful cool something that was flowing over my hand. That living word awakened my soul, gave it light, hope, joy set it free . . . As we returned to the house every object which I touched seemed to quiver with life. That was because I saw everything with the strange new sight that had come to me."

Let's take another example—the tree in Tim O'Brien's "How to Tell a True War Story": "In the mountains that day, I watched Lemon turn sideways. He laughed and said something to Rat Kiley. Then he took a peculiar half step, moving from shade into bright sunlight, and the booby-trapped 105 round blew him into a tree. The parts were just hanging there, so Dave Jensen and I were ordered to shinny up and peel him off. I remember the white bone of an arm. I remember pieces of skin and something wet and yellow that must've been the intestines. The gore was horrible and stays with me. But what wakes me up twenty years later is Dave Jensen singing 'Lemon Tree' as we threw down the parts."

The encodings associated with both "water" and "lemon tree" involve feelings as well as information. Every mental representation can encode both informational and emotional aspects. Some information is encoded without feelings. Other information is encoded along with positive or negative emotions. When feelings are attached to the information, the encoding becomes "valenced."

The important point is that once information and feelings are encoded, they become linked. Subsequent experiences that encode the information along with the feelings strengthen the association.

["Everything must have a name now": Daniel Shanahan, *Language, Feeling, and the Brain: The Evocative Vector* (New Brunswick, NJ: Transaction, 2007), 95, citing Ernst Cassirer, *An Essay on Man: An Introduction to a Philosophy of Human Culture* (New Haven, CT: Yale University Press, 1944). (I want to credit Professor Shanahan for bringing this example to my attention.)

"that had come to me": Shanahan, *Language, Feeling, and the Brain*, 95–96, citing Susanne Katherina Knauth Langer, *Philosophy in a New Key: A Study in the Symbolism of Reason, Rite, and Art* (Cambridge, MA: Harvard University Press, 1942), 62–63.

"In the mountains . . . threw down the parts": Tim O'Brien, *The Things They Carried: A Work of Fiction*, 1st Mariner Books ed. (Boston: Houghton Mifflin Harcourt, 2009).]

42   Valence: The amygdala and orbital frontal cortex are critically important in representing the valence of stimuli (whether positive or negative). The orbitofrontal cortex (OFC) codes for the reward value of a stimulus. Neurons in the OFC respond more to rewarding than neutral stimuli in various modalities, including taste, somatosensory, auditory, olfaction, and vision. J. D. Wallis, "Orbitofrontal Cortex and Its Contribution to Decision-Making," *Annual Review of Neuroscience* 30 (2007): 31–56; J. O'Doherty, E. T. Rolls, S. Francis, R. Bowtell, and F. McGlone, "Representation of Pleasant and Aversive Taste in the Human Brain," *Journal of Neurophysiology* 85, no. 3 (2001): 1315–21; D. M. Small, M. D. Gregory, Y. E. Mak, D. Gitelman, M. M. Mesulam, and T. Parrish, "Dissociation of Neural Representation of Intensity and Affective Valuation in Human Gustation," *Neuron* 39, no. 4 (2003): 701–11; J. A. Gottfried, R. Deichmann, J. S. Winston, and R. J. Dolan, "Functional Heterogeneity in Human Olfactory Cortex: An Event-Related Functional Magnetic Resonance Imaging Study," *Journal of Neuroscience* 22, no. 24 (2002): 10819–28; E. T. Rolls, M. L. Kringelbach, and I. E. De Araujo, "Different Representations of Pleasant and Unpleasant Odours in the Human Brain," *European Journal of Neuroscience* 18, no. 3 (2003): 695–703. The amygdala is sensitive to the valence of stimuli and can respond selectively based on the goals of the perceiver. W. A. Cunningham et al., "Affective Flexibility: Evaluative Processing Goals Shape Amygdala Activity," *Psychological Science* 19, no. 2 (2008): 152–60; H. Kim, L. Somerville, T. Johnstone, S. Polis, A. Alexander, L. Shin, and P. Whalen, "Contextual Modulation of Amygdala Responsivity to Surprised Faces," *Journal of Cognitive Neuroscience* 16, no. 10 (2004): 1730–45. Anderson and colleagues (2003) tried to distinguish brain regions that code for intensity from those that code for valence by matching the intensity of odors while manipulating valence. They found that the OFC codes for stimulus valence, with medial OFC responding to positive

odors and lateral OFC responding to negative odors. On the other hand, amygdala activations were correlated with participants' rating of intensity and were not related to valence. A. K. Anderson, K. Christoff, I. Stappen, D. Panitz, D. G. Ghahremani, G. Glover, J. D. Gabrieli, and N. Sobel, "Dissociated Neural Representations of Intensity and Valence in Human Olfaction," *Nature Neuroscience* 6, no. 2 (2003): 196–202.

43 **dangerous stimuli and sexually arousing stimuli:** Studies of affect-biased attention include images consistently rated at the extremes of valence and arousal. P. J. Lang, M. M. Bradley, and B. N. Cuthbert, "International Affective Picture System (IAPS): Affective Ratings of Pictures and Instruction Manual," Technical Report A-8 (Gainesville: University of Florida, 2008).

43 **stimuli with the greatest salience:** Attention is influenced by salient stimuli and vice versa. Affectively salient stimuli capture attention more easily when stimuli are at the threshold of awareness or when there are many stimuli in competition. S. L. Nielsen and I. G. Sarason, "Emotion, Personality, and Selective Attention," *Journal of Personality and Social Psychology* 41, no. 5 (1981): 945–60; J. J. Soares and A. Öhman, "Backward Masking and Skin Conductance Responses After Conditioning to Nonfeared but Fear-Relevant Stimuli in Fearful Subjects," *Psychophysiology* 30, no. 5 (1993): 460–66; A. Öhman, A. Flykt, and F. Esteves, "Emotion Drives Attention: Detecting the Snake in the Grass," *Journal of Experimental Psychology: General* 130, no. 3 (2001): 466; J. L. Armony and R. J. Dolan, "Modulation of Spatial Attention by Fear-Conditioned Stimuli: An Event-Related fMRI Study," *Neuropsychologia* 40, no. 7 (2002): 817–26; A. K. Anderson, "Affective Influences on the Attentional Dynamics Supporting Awareness," *Journal of Experimental Psychology: General* 134, no. 2 (2005): 258–81.

Attention can also influence the processing of salient stimuli. D. M. Beck and S. Kastner, "Top-Down and Bottom-Up Mechanisms in Biasing Competition in the Human Brain," *Vision Research* 49, no. 10 (2009): 1154–65; A. Holmes, M. Kiss, and M. Eimer, "Attention Modulates the Processing of Emotional Expression Triggered by Foveal Faces," *Neuroscience Letters* 394, no. 1 (2006): 48–52; L. Pessoa, S. Kastner, and L. G. Ungerleider, "Attentional Control of the Processing of Neutral and Emotional Stimuli," *Cognitive Brain Research* 15, no. 1 (2002): 31–45; K. N. Ochsner and J. J. Gross, "The Cognitive Control of Emotion," *Trends in Cognitive Sciences* 9, no. 5 (2005): 242–49. Studies using distraction have found that engaging in a distracting task reduces subjective pain and activity in regions of the brain involved in pain processing. K. N. Ochsner and J. J. Gross, "The Cognitive Control of Emotion," *Trends in Cognitive Sciences* 9, no. 5 (2005): 242–49; S. J. Bantick, R. G. Wise, A. Ploghaus, S. Clare, S. M. Smith, and I. Tracey, "Imaging How Attention Modulates Pain in Humans Using Functional MRI," *Brain* 125, no. 2 (2002): 310–19; M. Valet, T. Sprenger, H. Boecker, F. Willoch, E. Rummeny, B. Conrad, et al., "Distraction Modulates Connectivity of the Cingulo-Frontal Cortex and the Midbrain During Pain: An fMRI Analysis," *Pain* 109, no. 3 (2004): 399–408.

We are all wired to focus attention on the most salient stimuli. Visual system processing is biased toward certain stimuli (e.g., novel stimuli, emotionally salient stimuli) from the level of the receptive field and across brain systems. R. Desimone, "Visual Attention Mediated by Biased Competition in Extrastriate Visual Cortex," *Philosophical Transactions of the Royal Society of London,*

Series B: Biological Sciences 353, no. 1373 (1998): 1245–55; D. M. Beck and S. Kastner, "Top-Down and Bottom-Up Mechanisms in Biasing Competition in the Human Brain," Vision Research 49, no. 10 (2009): 1154–65; M. Mather and M. R. Sutherland, "Arousal-Biased Competition in Perception and Memory," Perspectives on Psychological Science 6, no. 2 (2011): 114–33; L. Pessoa, S. Kastner, and L. G. Ungerleider, "Attentional Control of the Processing of Neutral and Emotional Stimuli," Cognitive Brain Research 15, no. 1 (2002): 31–45; M. M. Müller, S. K. Andersen, and A. Keil, "Time Course of Competition for Visual Processing Resources Between Emotional Pictures and Foreground Task," Cerebral Cortex 18, no. 8 (2008): 1892–99.

43   we pay less and less attention: We are more easily distracted by salient stimuli. F. Dolcos and G. McCarthy, "Brain Systems Mediating Cognitive Interference by Emotional Distraction," Journal of Neuroscience 26, no. 7 (2006): 2072–79; F. Dolcos, P. Diaz-Granados, L. Wang, and G. McCarthy, "Opposing Influences of Emotional and Non-Emotional Distracters upon Sustained Prefrontal Cortex Activity During a Delayed-Response Working Memory Task," Neuropsychologia 46, no. 1 (2008): 326–35; H. Yamasaki, K. S. LaBar, and G. McCarthy, "Dissociable Prefrontal Brain Systems for Attention and Emotion," Proceedings of the National Academy of Sciences 99, no. 17 (2002): 11447–51; H. C. Ellis and P. W. Ashbrook, "Resource Allocation Model of the Effects of Depressed Mood States on Memory," in K. Fiedler and J. Forgas, eds., Affect, Cognition, and Social Behavior (Toronto: Hogrefe, 1988), 25–43.

43   vividly: Salient stimuli are remembered more vividly. Several studies have demonstrated better recall of emotional stimuli than neutral ones. E. A. Kensinger and S. Corkin, "Memory Enhancement for Emotional Words: Are Emotional Words More Vividly Remembered Than Neutral Words?," Memory and Cognition 31, no. 8 (2003): 1169–80; F. Dolcos, K. S. LaBar, and R. Cabeza, "Interaction Between the Amygdala and the Medial Temporal Lobe Memory System Predicts Better Memory for Emotional Events," Neuron 42, no. 5 (2004): 855–63.

According to McGaugh, Cahill, and colleagues' modulation hypothesis, better recall for emotional stimuli is the result of arousal modulating both initial memory encoding and long-term memory consolidation processes. J. L. McGaugh, C. K. McIntyre, and A. E. Power, "Amygdala Modulation of Memory Consolidation: Interaction with Other Brain Systems," Neurobiology, Learning, and Memory 78, no. 3 (2002): 539–52; L. Cahill and J. L. McGaugh, "Mechanisms of Emotional Arousal and Lasting Declarative Memory," Trends in Neurosciences 21, no. 7 (1998): 294–99. Todd and colleagues (2012) found that the emotional salience of a stimulus (as well as its perceptual vividness) contributes to the vividness of the image in memory. R. M. Todd, D. Talmi, T. W. Schmitz, J. Susskind, and A. K. Anderson, "Psychophysical and Neural Evidence for Emotion-Enhanced Perceptual Vividness," Journal of Neuroscience 32, no. 33 (2012): 11201–12.

43   processing emotional stimuli: These areas of the brain include a salience, or reward, network. W. W. Seeley, V. Menon, A. F. Schatzberg, J. Keller, G. H. Glover, H. Kenna, et al., "Dissociable Intrinsic Connectivity Networks for Salience Processing and Executive Control," Journal of Neuroscience 27, no. 9 (2007): 2349–56; V. Menon and L. Q. Uddin, "Saliency, Switching, Attention and Control: A Network Model of Insula Function," Brain Structure and Function 214, no. 5–6 (2010): 655–67; R. M. Todd and A. K. Anderson, "Salience,

State, and Expression: The Influence of Specific Aspects of Emotion on Attention and Perception," in *The Oxford Handbook of Cognitive Neuroscience*, vol. 2: *The Cutting Edges* (Oxford: Oxford University Press, 2013), 11.

The salience network activates in response to events such as pain and pleasure, social feedback, and the expression of emotion on someone else's face. K. Wiech, C. S. Lin, K. H. Brodersen, U. Bingel, M. Ploner, and I. Tracey, "Anterior Insula Integrates Information About Salience into Perceptual Decisions About Pain," *Journal of Neuroscience* 30, no. 48 (2010): 16324–31; A. D. Craig, "How Do You Feel? Interoception: The Sense of the Physiological Condition of the Body," *Nature Reviews Neuroscience* 3, no. 8 (2002): 655–66; N. I. Eisenberger, M. D. Lieberman, and K. D. Williams, "Does Rejection Hurt? An fMRI Study of Social Exclusion," *Science* 302, no. 5643 (2003): 290–92; R. Adolphs, D. Tranel, and A. R. Damasio, "The Human Amygdala in Social Judgment," *Nature* 393, no. 6684 (1998): 470–74.

The evaluation of the salience of stimuli takes place in areas of the brain different from where the stimuli are primarily processed. Stimuli are processed in the sensory cortex (e.g., the visual cortex in the occipital lobe). On the other hand, evaluation of an affectively salient stimulus occurs in the affective salience network, including the amygdala, orbitofrontal cortex, and locus coeruleus. L. Pessoa and R. Adolphs, "Emotion Processing and the Amygdala: From a 'Low Road' to 'Many Roads' of Evaluating Biological Significance," *Nature Reviews Neuroscience* 11, no. 11 (2010): 773–83; R. M. Todd, D. Talmi, T. W. Schmitz, J. Susskind, and A. K. Anderson, "Psychophysical and Neural Evidence for Emotion-Enhanced Perceptual Vividness," *Journal of Neuroscience* 32, no. 33 (2012): 11201–12; J. Markovic, A. K. Anderson, and R. M. Todd, "Tuning to the Significant: Neural and Genetic Processes Underlying Affective Enhancement of Visual Perception and Memory," *Behavioural Brain Research* 259 (2014): 229–41; G. Aston-Jones and J. D. Cohen, "An Integrative Theory of Locus Coeruleus–Norepinephrine Function: Adaptive Gain and Optimal Performance," *Annual Review of Neuroscience* 28 (2005): 403–50.

Top-down modulation of attention occurs in frontoparietal regions, whereas bottom-up attention to visually salient stimuli occurs in the temporoparietal and ventral frontal cortex. M. Corbetta and G. L. Shulman, "Control of Goal-Directed and Stimulus-Driven Attention in the Brain," *Nature Reviews Neuroscience* 3, no. 3 (2002): 201–15; M. Corbetta, G. Patel, and G. L. Shulman, "The Reorienting System of the Human Brain: From Environment to Theory of Mind," *Neuron* 58, no. 3 (2008): 306–24; W. W. Seeley, V. Menon, A. F. Schatzberg, J. Keller, G. H. Glover, H. Kenna, et al., "Dissociable Intrinsic Connectivity Networks for Salience Processing and Executive Control," *Journal of Neuroscience* 27, no. 9 (2007): 2349–56.

43   **instigating many of our emotions:** M. Davis and P. J. Whalen, "The Amygdala: Vigilance and Emotion," *Molecular Psychiatry* 6, no. 1 (2001): 13–34; H. Garavan, J. C. Pendergrass, T. J. Ross, E. A. Stein, and R. C. Risinger, "Amygdala Response to Both Positively and Negatively Valenced Stimuli," *Neuroreport* 12, no. 12 (2001): 2779–83; B. S. Kapp, P. J. Whalen, W. F. Supple, and J. P. Pascoe, "Amygdaloid Contributions to Conditioned Arousal and Sensory Information Processing," in J. P. Aggleton, ed., *The Amygdala: Neurobiological Aspects of Emotion, Memory, and Mental Dysfunction* (New York: Wiley-Liss, 1992), 229–54.

43   "relevance detection": The amygdala is sensitive to the affective properties of stimuli. R. Adolphs, D. Tranel, and A. R. Damasio, "The Human Amygdala in Social Judgment," *Nature* 393, no. 6684 (1998): 470–74; T. Canli, Z. Zhao, J. Brewer, J. D. E. Gabrieli, and L. Cahill, "Event-Related Activation in the Human Amygdala Associates with Later Memory for Individual Emotional Experience," *Journal of Neuroscience* 20, no. 19 (2000): 1–5; J. E. LeDoux, "Emotion Circuits in the Brain," *Annual Review of Neuroscience* 23 (2000): 155–84; D. Sabatinelli et al., "Parallel Amygdala and Inferotemporal Activation Reflect Emotional Intensity and Fear Relevance," *NeuroImage* 24, no. 4 (2005): 1265–70; J. S. Morris et al., "A Neuromodulatory Role for the Human Amygdala in Processing Emotional Facial Expressions," *Brain* 121, no. 1 (1998): 47–57; L. Pessoa et al., "Neural Processing of Emotional Faces Requires Attention," *Proceedings of the National Academy of Sciences* 99, no. 17 (2002): 11458–63.

Pessoa and Adolphs (2010) argue that the amygdala is a "convergence zone" for object information received from higher-order visual cortices via the basolateral nucleus. L. Pessoa and R. Adolphs, "Emotion Processing and the Amygdala: From a 'Low Road' to 'Many Roads' of Evaluating Biological Significance," *Nature Reviews Neuroscience* 11, no. 11 (2010): 773–83.

43   refocusing attention: The amygdala can modulate visual cortex activation, enhancing activity for emotionally salient stimuli. P. Vuilleumier, "How Brains Beware: Neural Mechanisms of Emotional Attention," *Trends in Cognitive Sciences* 9, no. 12 (2005): 585–94; E. A. Phelps and J. E. LeDoux, "Contributions of the Amygdala to Emotion Processing: From Animal Models to Human Behavior," *Neuron* 48, no. 2 (2005): 175–87.

Patients with amygdala damage and intact visual cortex lack visual cortex enhancement for affectively salient stimuli. P. Vuilleumier, M. P. Richardson, J. L. Armony, J. Driver, and R. J. Dolan, "Distant Influences of Amygdala Lesion on Visual Cortical Activation During Emotional Face Processing," *Nature Neuroscience* 7, no. 11 (2004): 1271–78; A. K. Anderson, "Affective Influences on the Attentional Dynamics Supporting Awareness," *Journal of Experimental Psychology: General* 134, no. 2 (2005): 258–81.

43   memory: The amygdala influences memory of emotional stimuli. E. A. Phelps, "Human Emotion and Memory: Interactions of the Amygdala and Hippocampal Complex," *Current Opinion in Neurobiology* 14, no. 2 (2004): 198–202; M. P. Richardson, B. A. Strange, and R. J. Dolan, "Encoding of Emotional Memories Depends on Amygdala and Hippocampus and Their Interactions," *Nature Neuroscience* 7, no. 3 (2004): 278–85; S. B. Hamann, T. D. Ely, S. T. Grafton, and C. D. Kilts, "Amygdala Activity Related to Enhanced Memory for Pleasant and Aversive Stimuli," *Nature Neuroscience* 2, no. 3 (1999): 289–93.

We acquire and process information—"implicitly learn"—about complex information and stimuli in our environment without being aware of such processing. The amygdala can process stimuli automatically, allowing for implicit fear conditioning. A. Öhman, "Automaticity and the Amygdala: Nonconscious Responses to Emotional Faces," *Current Directions in Psychological Science* 11, no. 2 (2002): 62–66; J. S. Morris, A. Öhman, and R. J. Dolan, "Conscious and Unconscious Emotional Learning in the Human Amygdala," *Nature* 393, no. 6684 (1998): 467–70; J. S. Morris, A. Öhman, and R. J. Dolan, "A Subcortical Pathway to the Right Amygdala Mediating 'Unseen' Fear," *Proceedings of the National Academy of Sciences* 96, no. 4 (1999): 1680–85; P. J. Whalen, S. L. Rauch, N. L.

Etcoff, S. C. McInerney, M. B. Lee, and M. A. Jenike, "Masked Presentations of Emotional Facial Expressions Modulate Amygdala Activity Without Explicit Knowledge," *Journal of Neuroscience* 18, no. 1 (1998): 411–18.

The information that we implicitly learn is also stored in "implicit memory" without our awareness of such processing. H. L. Roediger, "Implicit Memory: Retention Without Remembering," *American Psychologist* 45, no. 9 (1990): 1043; L. R. Squire, B. Knowlton, and G. Musen, "The Structure and Organization of Memory," *Annual Review of Psychology* 44, no. 1 (1993): 453–95; E. Tulving and D. L. Schacter, "Priming and Human Memory Systems," *Science* 247, no. 4940 (1990): 301–26. We can acquire new patterns of behavior through implicit learning. D. L. Schacter, C. Y. P. Chiu, and K. N. Ochsner, "Implicit Memory: A Selective Review," *Annual Review of Neuroscience* 16, no. 1 (1993): 159–82; D. L. Schacter, "Implicit Memory: History and Current Status," *Journal of Experimental Psychology: Learning, Memory, and Cognition* 13, no. 3 (1987): 501–18. We can implicitly learn new patterns of behavior and new emotional responses. J. A. Bargh and T. L. Chartrand, "The Unbearable Automaticity of Being," *American Psychologist* 54, no. 7 (1999): 462; F. Esteves, U. Dimberg, and A. Öhman, "Automatically Elicited Fear: Conditioned Skin Conductance Responses to Masked Facial Expressions," *Cognition and Emotion* 8, no. 5 (1994): 393–413; A. Öhman and J. J. Soares, "'Unconscious Anxiety': Phobic Responses to Masked Stimuli," *Journal of Abnormal Psychology* 103, no. 2 (1994): 231.

43   **prioritizes our conscious awareness:** The information that we "implicitly" learn guides the deployment of attention. Fear-conditioned stimuli can guide attention automatically. J. L. Armony and R. J. Dolan, "Modulation of Spatial Attention by Fear-Conditioned Stimuli: An Event-Related fMRI Study," *Neuropsychologia* 40, no. 7 (2002): 817–26; K. Mogg and B. P. Bradley, "Orienting of Attention to Threatening Facial Expressions Presented Under Conditions of Restricted Awareness," *Cognition and Emotion* 13, no. 6 (1999): 713–40. Also, implicitly learned patterns in the environment can guide visual search through a phenomenon called "contextual cueing." M. M. Chun and Y. Jiang, "Contextual Cueing: Implicit Learning and Memory of Visual Context Guides Spatial Attention," *Cognitive Psychology* 36, no. 1 (1998): 28–71; M. M. Chun and J. Yuhong, "Top-Down Attentional Guidance Based on Implicit Learning of Visual Covariation," *Psychological Science* 10, no. 4 (1999): 360–65. Although, see M. A. Kunar, S. Flusberg, T. S. Horowitz, and J. M. Wolfe, "Does Contextual Cuing Guide the Deployment of Attention?," *Journal of Experimental Psychology: Human Perception and Performance* 33, no. 4 (2007): 816.

43   **amygdala with other networks:** The orbitofrontal cortex (OFC) has bidirectional connections with the amygdala. C. Cavada et al., "The Anatomical Connections of the Macaque Monkey Orbitofrontal Cortex: A Review," *Cerebral Cortex* 10, no. 3 (2000): 220–42. It also receives input from higher-order sensory cortices. L. L. Baylis, E. T. Rolls, and G. C. Baylis, "Afferent Connections of the Caudolateral Orbitofrontal Cortex Taste Area of the Primate," *Neuroscience* 64, no. 3 (1995): 801–12; S. T. Carmichael, M. C. Clugnet, and J. L. Price, "Central Olfactory Connections in the Macaque Monkey," *Journal of Comparative Neurology* 346, no. 3 (1994): 403–34; S. T. Carmichael and J. L. Price, "Limbic Connections of the Orbital and Medial Prefrontal Cortex in Macaque Monkeys," *Journal of Comparative Neurology* 363 (1995a): 615–41; S. T. Carmichael and

J. L. Price, "Sensory and Premotor Connections of the Orbital and Medial Prefrontal Cortex of Macaque Monkeys," *Journal of Comparative Neurology* 363, no. 4 (1995b): 642–64; D. Ongur and J. L. Price, "The Organization of Networks Within the Orbital and Medial Prefrontal Cortex of Rats, Monkeys, and Humans," *Cerebral Cortex* 10, no. 3 (2000): 206–19; E. T. Rolls, S. Yaxley, and Z. J. Sienkiewicz, "Gustatory Responses of Single Neurons in the Caudolateral Orbitofrontal Cortex of the Macaque Monkey," *Journal of Neurophysiology* 64, no. 4 (1990): 1055–66.

43   **learning:** Activation of the amygdala and OFC is connected with feelings (pleasantness, disgust, fear) associated with the salient stimuli. Activation of the affective attention system is part of the process of emotion regulation. R. M. Todd, W. A. Cunningham, A. K. Anderson, and E. Thompson, "Affect-Biased Attention as Emotion Regulation," *Trends in Cognitive Sciences* 16, no. 7 (2012): 365–72. The arousal created by norepinephrine and other stress-induced neurotransmitters is connected with emotional arousal. This results in physiological changes such as increased heart rate and galvanic skin response and in pupil dilation. E.g., M. M. Bradley, L. Miccoli, M. A. Escrig, and P. J. Lang, "The Pupil as a Measure of Emotional Arousal and Autonomic Activation," *Psychophysiology* 45, no. 4 (2008): 602–7.

These parts of the brain are important not only because they encode the salience of stimuli but also because they are involved in pairing salient stimuli with other stimuli that themselves become "cues." The amygdala is involved in conditioning. E.g., R. G. Phillips and J. E. LeDoux, "Differential Contribution of Amygdala and Hippocampus to Cued and Contextual Fear Conditioning," *Behavioral Neuroscience* 106, no. 2 (1992): 274; M. T. Rogan, U. V. Stäubli, and J. E. LeDoux, "Fear Conditioning Induces Associative Long-Term Potentiation in the Amygdala," *Nature* 390, no. 6660 (1997): 604–7.

Neurons in the OFC respond to cues signaling the appearance of primary rewards (intrinsically rewarding stimuli). J. D. Wallis, "Orbitofrontal Cortex and Its Contribution to Decision-Making," *Annual Review of Neuroscience* 30 (2007): 31–56; J. D. Wallis and E. K. Miller, "Neuronal Activity in Primate Dorsolateral and Orbital Prefrontal Cortex During Performance of a Reward Preference Task," *European Journal of Neuroscience* 18, no. 7 (2003): 2069–81; S. J. Thorpe, E. T. Rolls, and S. Maddison, "The Orbitofrontal Cortex: Neuronal Activity in the Behaving Monkey," *Experimental Brain Research* 49, no. 1 (1983): 93–115; G. Schoenbaum and M. Roesch, "Orbitofrontal Cortex, Associative Learning, and Expectancies," *Neuron* 47, no. 5 (2005): 633–36. The OFC is necessary for learning new cue-reward associations, as seen in animal studies. M. Gallagher, R. W. McMahan, and G. Schoenbaum, "Orbitofrontal Cortex and Representation of Incentive Value in Associative Learning," *Journal of Neuroscience* 19, no. 15 (1999): 6610–14; A. Pears, J. A. Parkinson, L. Hopewell, B. J. Everitt, and A. C. Roberts, "Lesions of the Orbitofrontal but Not Medial Prefrontal Cortex Disrupt Conditioned Reinforcement in Primates," *Journal of Neuroscience* 23, no. 35 (2003): 11189–201. Animal studies have shown that locus coeruleus neurons respond to reinforcing stimuli and changes in reward contingencies. S. J. Sara, "The Locus Coeruleus and Noradrenergic Modulation of Cognition," *Nature Reviews Neuroscience* 10, no. 3 (2009): 211–23.

43   **motivation:** Nonhuman animal experiments have found that neurons in the monkey OFC code for expected reward value, with differing activation based

on the amount of juice to be received. J. D. Wallis and E. K. Miller, "Neuronal Activity in Primate Dorsolateral and Orbital Prefrontal Cortex During Performance of a Reward Preference Task," *European Journal of Neuroscience* 18, no. 7 (2003): 2069–81; M. R. Roesch and C. R. Olson, "Neuronal Activity Related to Reward Value and Motivation in Primate Frontal Cortex," *Science* 304 (2004): 307–10; E. T. Rolls, "The Functions of the Orbitofrontal Cortex," *Brain and Cognition* 55, no. 1 (2004): 11–29. The OFC is connected with other regions in a way that allows it to determine salience. It receives inputs from all sensory modalities and is well connected with limbic structures, including the amygdala and the cingulate gyrus. J. D. Wallis, "Orbitofrontal Cortex and Its Contribution to Decision-Making," *Annual Review of Neuroscience* 30 (2007): 31–56. A study using monetary reward and punishment found that OFC activation was correlated with the magnitude of reward or punishment. J. O'Doherty, M. L. Kringelbach, E. T. Rolls, J. Hornak, and C. Andrews, "Abstract Reward and Punishment Representations in the Human Orbitofrontal Cortex," *Nature Neuroscience* 4, no. 1 (2001): 95–102. See also C. Padoa-Schioppa and J. A. Assad, "Neurons in the Orbitofrontal Cortex Encode Economic Value," *Nature* 441, no. 7090 (2006): 223–26.

43   **decision making:** The OFC seems to integrate various features of a reward to determine its value to the organism, helping us make decisions based on the emotional significance of a stimulus. A. Bechara, H. Damasio, and A. R. Damasio, "Emotion, Decision Making, and the Orbitofrontal Cortex," *Cerebral Cortex* 10, no. 3 (2000): 295–307; J. D. Wallis, "Orbitofrontal Cortex and Its Contribution to Decision-Making," *Annual Review of Neuroscience* 30 (2007): 31–56. The OFC is thought to link a stimulus with pleasurable experience. M. L. Kringelbach, "The Human Orbitofrontal Cortex: Linking Reward to Hedonic Experience," *Nature Reviews Neuroscience* 6, no. 9 (2005): 691–702.

Cues associated with salient stimuli can themselves become salient stimuli. Wallis (2007) gives the excellent example of money, which we find salient, rewarding, and emotionally arousing, though it lacks intrinsically rewarding properties. J. D. Wallis, "Orbitofrontal Cortex and Its Contribution to Decision-Making," *Annual Review of Neuroscience* 30 (2007): 31–56. Through classical conditioning, a formerly neutral stimulus can produce an enhanced visual cortex response, for example, in the same manner as an inherently affectively salient stimulus. M. Stolarova, A. Keil, and S. Moratti, "Modulation of the C1 Visual Event–Related Component by Conditioned Stimuli: Evidence for Sensory Plasticity in Early Affective Perception," *Cerebral Cortex* 16, no. 6 (2006): 876–87. OFC and amygdala neurons respond to stimuli that indicate the presence of a reward. G. Schoenbaum, A. A. Chiba, and M. Gallagher, "Orbitofrontal Cortex and Basolateral Amygdala Encode Expected Outcomes During Learning," *Nature Neuroscience* 1 (1998): 155–59; S. J. Thorpe, E. T. Rolls, and S. Maddison, "The Orbitofrontal Cortex: Neuronal Activity in the Behaving Monkey," *Experimental Brain Research* 49, no. 1 (1983): 93–115; J. D. Wallis and E. K. Miller, "Neuronal Activity in Primate Dorsolateral and Orbital Prefrontal Cortex During Performance of a Reward Preference Task," *European Journal of Neuroscience* 18, no. 7 (2003): 2069–81; M. R. Roesch and C. R. Olson, "Neuronal Activity Related to Reward Value and Motivation in Primate Frontal Cortex," *Science* 304, no. 5668 (2004): 307–10. Furthermore, lesions of the amygdala or OFC inhibit an animal's ability to learn new behaviors with a secondary reinforcer, indicating that both the amygdala and OFC are important for learning reward contingencies and new affective salience information. A. Pears, J. A. Parkinson, L. Hopewell, B. J. Everitt, and A. C. Roberts, "Lesions of the Orbitofrontal but Not Medial Prefrontal Cortex Disrupt Conditioned Reinforcement in Primates," *Journal of Neuroscience* 23, no. 35 (2003): 11189–201; G. Schoenbaum and M. Roesch, "Orbitofrontal Cortex, Associative Learning, and Expectancies," *Neuron* 47, no. 5 (2005): 633.

Cues guide the deployment of attentional resources. M. Corbetta and G. L. Shulman, "Control of Goal-Directed and Stimulus-Driven Attention in the Brain," *Nature Reviews Neuroscience* 3, no. 3 (2002): 201–15. Cued objects are detected more easily. C. W. Eriksen and J. E. Hoffman, "The Extent of Processing of Noise Elements During Selective Encoding from Visual Displays," *Perception and Psychophysics* 14, no. 1 (1973): 155–60; M. I. Posner, C. R. Snyder, and B. J. Davidson, "Attention and the Detection of Signals," *Journal of Experimental Psychology: General* 109, no. 2 (1980): 160. Cues that signal the presence of a salient stimulus modulate attention. J. L. Armony and R. J. Dolan, "Modulation of Spatial Attention by Fear-Conditioned Stimuli: An Event-Related fMRI Study," *Neuropsychologia* 40, no. 7 (2002): 817–26; E. Koster, G. Crombez, S. Van Damme, B. Verschuere, and J. De Houwer, "Signals for Threat Modulate Attentional Capture and Holding: Fear-Conditioning and Extinction During the Exogenous Cueing Task," *Cognition and Emotion* 19, no. 5 (2005): 771–80; S. D. Smith, S. B. Most, L. A. Newsome, and D. H. Zald, "An Emotion-Induced Attentional Blink Elicited by Aversively Conditioned Stimuli," *Emotion* 6, no. 3 (2006): 523.

There are individual genetic differences in some of these systems. For example, genetic deletion variants involving the norepinephrine receptor result in a loss of receptor desensitization, leading to higher levels of norepinephrine in certain individuals and greater arousal and reactivity to salient stimuli. K. M. Small, K. M. Brown, S. L. Forbes, and S. B. Liggett, "Polymorphic Deletion of Three Intracellular Acidic Residues of the Alpha 2B-Adrenergic Receptor Decreases G Protein-Coupled Receptor Kinase-Mediated Phosphorylation and Desensitization," *Journal of Biological Chemistry* 276, no. 7 (2001): 4917–22. Individuals with the *ADRA2b* deletion variant show even greater enhancement in memory for emotional stimuli. In a free recall task following viewing neutral and emotionally salient images, all participants better remembered the emotional images, but deletion carriers had a further enhancement for the emotional images. D. J. De Quervain, I. T. Kolassa, V. Ertl, P. L. Onyut, F. Neuner, T. Elbert, et al., "A Deletion Variant of the Alpha2b-Adrenoceptor Is Related to Emotional Memory in Europeans and Africans," *Nature Neuroscience* 10, no. 9 (2007): 1137–39. In perception, deletion carriers show greater amygdala activity for negative arousing stimuli. B. Rasch, K. Spalek, S. Buholzer, R. Luechinger, P. Boesiger, A. Papassotiropoulos, et al., "Aversive Stimuli Lead to Differential Amygdala Activation and Connectivity Patterns Depending on Catechol-O-Methyltransferase Val158Met Genotype," *NeuroImage* 52, no. 4 (2010): 1712–19.

A genetic variation of the noradrenergic system is related to differential amygdala activation during encoding of emotional memories. B. Rasch, K. Spalek, S. Buholzer, R. Luechinger, P. Boesiger, A. Papassotiropoulos, and D. J.-F. de Quervain, "A Genetic Variation of the Noradrenergic System Is Related to Differential Amygdala Activation During Encoding of Emotional

Memories," *Proceedings of the National Academy of Sciences* 106, no. 45 (2009): 19191–96, published ahead of print Oct. 13, 2009, doi:10.1073/pnas.0907425106. Those with this variation also seem to possess a greater range of amygdala activation. In a study by Cousijn and colleagues, participants first viewed either a violent movie (stress condition) or a neutral movie (nonstress condition) and then completed a second task involving viewing dynamic fearful and happy faces. Nondeletion carriers showed amygdala enhancement for emotional faces only in the nonstress condition, whereas deletion carriers showed an enhancement for both conditions. The authors suggest that deletion carriers thus possess a greater range of amygdala activation while nondeletion carriers in the study are reaching the upper limit of their amygdala activation. H. Cousijn, M. Rijpkema, S. Qin, H. J. van Marle, B. Franke, E. J. Hermans, et al., "Acute Stress Modulates Genotype Effects on Amygdala Processing in Humans," *Proceedings of the National Academy of Sciences* 107, no. 21 (2010): 9867–72.

43　　**emotional response:** The most potent stimuli are those that can change how we feel. Emotionally salient stimuli, which are salient by virtue of increasing arousal, are perceived more vividly. R. M. Todd, D. Talmi, T. W. Schmitz, J. Susskind, and A. K. Anderson, "Psychophysical and Neural Evidence for Emotion-Enhanced Perceptual Vividness," *Journal of Neuroscience* 32, no. 33 (2012): 11201–12. There is evidence that we remember emotionally salient stimuli more easily and more vividly. R. M. Todd, D. Talmi, T. W. Schmitz, J. Susskind, and A. K. Anderson, "Psychophysical and Neural Evidence for Emotion-Enhanced Perceptual Vividness," *Journal of Neuroscience* 32, no. 33 (2012): 11201–12; L. Cahill and J. L. McGaugh, "A Novel Demonstration of Enhanced Memory Associated with Emotional Arousal," *Consciousness and Cognition* 4, no. 4 (1995): 410–21; K. N. Ochsner, "Are Affective Events Richly Recollected or Simply Familiar? The Experience and Process of Recognizing Feelings Past," *Journal of Experimental Psychology: General* 129, no. 2 (2000): 242–61; E. A. Kensinger and S. Corkin, "Two Routes to Emotional Memory: Distinct Neural Processes for Valence and Arousal," *Proceedings of the National Academy of Sciences* 101, no. 9 (2004): 3310–15; R. Brown and J. Kulik, "Flashbulb Memories," *Cognition* 5, no. 1 (1977): 73–99.

43　　**motivation and action:** Salient stimuli trigger the neural circuitry of the brain involved in motivation. M. M. Bradley, M. Codispoti, B. N. Cuthbert, and P. J. Lang, "Emotion and Motivation I: Defensive and Appetitive Reactions in Picture Processing," *Emotion* 1, no. 3 (2001): 276–98; P. J. Lang, M. M. Bradley, and B. N. Cuthbert, "Motivated Attention: Affect, Activation, and Action," in P. J. Lang, R. F. Simons, and M. Balaban, eds., *Attention and Orienting: Sensory and Motivational Processes* (New York: Psychology Press, 1997): 97–135; P. J. Lang and M. M. Bradley, "Emotion and the Motivational Brain," *Biological Psychology* 84, no. 3 (2010): 437–50. Salient stimuli activate the autonomic nervous system. M. M. Bradley, L. Miccoli, M. A. Escrig, and P. J. Lang, "The Pupil as a Measure of Emotional Arousal and Autonomic Activation," *Psychophysiology* 45, no. 4 (2008): 602–7. Salient stimuli also activate the limbic system, including the amygdala, the region central to motivation. A. K. Anderson and E. A. Phelps, "Lesions of the Human Amygdala Impair Enhanced Perception of Emotionally Salient Events," *Nature* 411, no. 6835 (2001): 305–9; D. H. Zald, "The Human Amygdala and the Emotional Evaluation of Sensory Stimuli," *Brain Research Reviews* 41, no. 1 (2003): 88–123; M. Davis, "The Role of the

Amygdala in Conditioned Fear," *Annual Review of Neuroscience* 15 (1992): 353–75; D. Sabatinelli, M. M. Bradley, J. R. Fitzsimmons, and P. J. Lang, "Parallel Amygdala and Inferotemporal Activation Reflect Emotional Intensity and Fear Relevance," *NeuroImage* 24, no. 4 (2005): 1265–70.

That motivational circuitry prepares us to act or not act depending on the nature of the salient stimulus, and organizes the endocrine and autonomic nervous systems' response to the salient stimuli. This circuitry includes the amygdala, which receives input from the thalamus and hippocampus and projects to the striatum, which is responsible for approach and avoidance behavior. J. F. LeDoux, "Information Flow from Sensation to Emotion Plasticity in the Neural Computation of Stimulus Values," in M. Gabriel and J. Moore, eds., *Learning and Computational Neuroscience: Foundations of Adaptive Networks* (Cambridge, MA: Bradford Books/MIT Press, 1990), 3–52; P. J. Lang and M. M. Bradley, "Emotion and the Motivational Brain," *Biological Psychology* 84, no. 3 (2010): 437–50. A large body of research has shown that projections from the thalamus to the amygdala are responsible for initiating defense behavior in response to threatening stimuli. R. J. Dolan and J. S. Morris, "The Functional Anatomy of Innate and Acquired Fear: Perspectives from Neuroimaging," in R. D. Lane and L. Nadel, eds., *Cognitive Neuroscience of Emotion* (New York: Oxford University Press, 2000), 225–41; P. Vuilleumier, J. L. Armony, J. Driver, and R. J. Dolan, "Effects of Attention and Emotion on Face Processing in the Human Brain: An Event-Related fMRI Study," *Neuron* 30, no. 3 (2001): 829–41; P. Vuilleumier and S. Schwartz, "Beware and Be Aware: Capture of Spatial Attention by Fear-Related Stimuli in Neglect," *NeuroReport* 12, no. 6 (2001): 1119–22; P. J. Whalen, "Fear, Vigilance, and Ambiguity: Initial Neuroimaging Studies of the Human Amygdala," *Current Directions in Psychological Science* 7, no. 6 (1988): 177–88; P. J. Whalen, L. M. Shin, S. C. McInerney, H. Fischer, C. I. Wright, and S. L. Rauch, "A Functional MRI Study of Human Amygdala Responses to Facial Expressions of Fear Versus Anger," *Emotion* 1, no. 1 (2001): 70. The amygdala also projects to the lateral hypothalamus, which is responsible for engaging autonomic and motor systems. T. W. Robbins and B. J. Everitt, "Neurobehavioural Mechanisms of Reward and Motivation," *Current Opinion in Neurobiology* 6, no. 2 (1996): 228–36; P. J. Lang, "The Emotion Probe: Studies of Motivation and Attention," *American Psychologist* 50, no. 5 (1995): 372. Salient stimuli activate the hypothalamus-pituitary-adrenal (HPA) axis, the endocrine system that controls stress responses. S. S. Dickerson and M. E. Kemeny, "Acute Stressors and Cortisol Responses: A Theoretical Integration and Synthesis of Laboratory Research," *Psychological Bulletin* 130, no. 3 (2004): 355. The HPA axis is modulated by activity in motivational circuits. S. Kern, T. R. Oakes, C. K. Stone, E. M. McAuliff, C. Kirschbaum, and R. J. Davidson, "Glucose Metabolic Changes in the Prefrontal Cortex Are Associated with HPA Axis Response to a Psychosocial Stressor," *Psychoneuroendocrinology* 33, no. 4 (2008): 517–29; K. Dedovic, A. Duchesne, J. Andrews, V. Engert, and J. C. Pruessner, "The Brain and the Stress Axis: The Neural Correlates of Cortisol Regulation in Response to Stress," *NeuroImage* 47, no. 3 (2009): 864–71.

The motivational circuitry readies us to take or not take action. It can activate the "fight or flight" response. P. J. Lang and M. Davis, "Emotion, Motivation, and the Brain: Reflex Foundations in Animal and Human Research," *Progress in Brain Research* 156 (2006): 3–29; W. B. Cannon, *The Wisdom of the*

Body (New York: Norton, 1932); J. M. Cedarbaum and G. K. Aghajanian, "Afferent Projections to the Rat Locus Coeruleus as Determined by a Retrograde Tracing Technique," *Journal of Comparative Neurology* 178, no. 1 (1978): 1–15; P. D. Skosnik, R. T. Chatterton Jr., T. Swisher, and S. Park, "Modulation of Attentional Inhibition by Norepinephrine and Cortisol After Psychological Stress," *International Journal of Psychophysiology* 36, no. 1 (2000): 59–68; S. J. Sara, "The Locus Coeruleus and Noradrenergic Modulation of Cognition," *Nature Reviews Neuroscience* 10, no. 3 (2009): 211–23. Stress hormones released in response to salient stimuli by the endocrine system facilitate motor neurons and prepare the body for action. S. R. White and R. S. Neuman, "Pharmacological Antagonism of Facilitatory but Not Inhibitory Effects of Serotonin and Norepinephrine on Excitability of Spinal Motoneurons," *Neuropharmacology* 22, no. 4 (1983): 489–94; C. Tsigos and G. P. Chrousos, "Hypothalamic–Pituitary–Adrenal Axis, Neuroendocrine Factors and Stress," *Journal of Psychosomatic Research* 53, no. 4 (2002): 865–71.

43   **neural pattern:** Neurons can respond selectively to salient stimuli. Single-cell physiology studies have demonstrated that there is competition for stimulus representation in the visual cortex. When a pair of stimuli both lie within a neuron's receptive field (RF), the neuron's response is not the sum of its responses to each stimulus, but rather is a weighted average of those two responses. J. H. Reynolds, L. Chelazzi, and R. Desimone, "Competitive Mechanisms Subserve Attention in Macaque Areas V2 and V4," *Journal of Neuroscience* 19, no. 5 (1999): 1736–53. Such a suppressive interaction indicates that stimuli compete for neural representation. D. M. Beck and S. Kastner, "Top-Down and Bottom-Up Mechanisms in Biasing Competition in the Human Brain," *Vision Research* 49, no. 10 (2009): 1154–65. In single-cell studies on monkeys, when a monkey paid attention to one of two stimuli in a neuron's RF, responses to the stimuli in V2, V4, and MT were similar to responses to the attended stimulus when it is presented alone. S. J. Luck, L. Chelazzi, S. A. Hillyard, and R. Desimone, "Neural Mechanisms of Spatial Selective Attention in Areas V1, V2, and V4 of Macaque Visual Cortex," *Journal of Neurophysiology* 77, no. 1 (1997): 24–42; G. H. Recanzone, R. H. Wurtz, and U. Schwartz, "Responses of MT and MST Neurons to One and Two Moving Objects in the Receptive Field," *Journal of Neurophysiology* 78, no. 6 (1997): 2904–15; J. H. Reynolds and R. Desimone, "Interacting Roles of Attention and Visual Salience in V4," *Neuron* 37, no. 6 (2003): 853–63.

The neurons that fire in response to a salient stimulus respond to features of the stimulus. A. M. Treisman and G. Gelade, "A Feature-Integration Theory of Attention," *Cognitive Psychology* 12, no. 1 (1980): 97–136; S. M. Zeki, "Functional Specialisation in the Visual Cortex of the Rhesus Monkey," *Nature* 274, no. 5670 (1978): 423–28; R. Desimone and J. Duncan, "Neural Mechanisms of Selective Visual Attention," *Annual Review of Neuroscience* 18, no. 1 (1995): 193–22. Cells in the visual cortex respond to features of stimuli such as shape, color, and motion. S. Kastner and L. G. Ungerleider, "Mechanisms of Visual Attention in the Human Cortex," *Annual Review of Neuroscience* 23, no. 1 (2000): 315–41; A. Pasupathy and C. E. Connor, "Responses to Contour Features in Macaque Area V4," *Journal of Neurophysiology* 82, no. 5 (1999): 2490–502.

The neurons that fire in response to a salient stimulus respond to a pattern, a collection of features associated with the stimulus. A. Treisman, "Feature Binding, Attention, and Object Perception," *Philosophical Transactions of the Royal Society of London, Series B: Biological Sciences* 353, no. 1373 (1998): 1295–306; K. Tanaka, "Neuronal Mechanisms of Object Recognition," *Science* 262, no. 5134 (1993): 685–88. Theories on object perception differ, but all try to explain how we recognize objects from presented visual patterns. E.g., K. Grill-Spector and R. Malach, "The Human Visual Cortex," *Annual Review of Neuroscience* 27 (2004): 649–77; D. Marr, "Early Processing of Visual Information," *Philosophical Transactions of the Royal Society of London, Series B: Biological Sciences* 275, no. 942 (1976): 483–519; D. G. Lowe, "Object Recognition from Local Scale-Invariant Features," in *Computer Vision, Proceedings of the Seventh IEEE International Conference on Computer Vision* 2 (1999): 1150–57; M. A. Peterson and G. Rhodes, eds., *Perception of Faces, Objects and Scenes: Analytic and Holistic Processes* (New York: Oxford University Press, 2003).

Pattern recognition is hierarchical. Patterns are constituents of a higher level of patterns: lines to letters to words. E. A. DeYoe and D. C. Van Essen, "Concurrent Processing Streams in Monkey Visual Cortex," *Trends in Neurosciences* 11, no. 5 (1988): 219–26; S. Kastner and L. G. Ungerleider, "Mechanisms of Visual Attention in the Human Cortex," *Annual Review of Neuroscience* 23, no. 1 (2000): 315–41. Visual cortical processing begins with simple and complex cells. Neurons higher in the hierarchy have larger receptive fields and respond to increasingly complex features of stimuli. D. H. Hubel and T. N. Wiesel, "Receptive Fields and Functional Architecture of Monkey Striate Cortex," *Journal of Physiology* 195, no. 1 (1968): 215–43; J. H. Maunsell and W. T. Newsome, "Visual Processing in Monkey Extrastriate Cortex," *Annual Review of Neuroscience* 10, no. 1 (1987): 363–401; D. J. Felleman and D. C. Van Essen, "Distributed Hierarchical Processing in the Primate Cerebral Cortex," *Cerebral Cortex* 1, no. 1 (1991): 1–47. See also S. Hochstein and M. Ahissar, "View from the Top: Hierarchies and Reverse Hierarchies in the Visual System," *Neuron* 36, no. 5 (2002): 791–804.

43 **strengthened:** Neurons that respond to a salient stimulus will increase their response to patterns associated with the stimulus. The affective salience system biases activity in sensory cortices such that a response to salient stimuli is facilitated while response to irrelevant stimuli is suppressed. C. W. Berridge and B. D. Waterhouse, "The Locus Coeruleus–Noradrenergic System: Modulation of Behavioral State and State-Dependent Cognitive Processes," *Brain Research Reviews* 42, no. 1 (2003): 33–84; S. J. Sara, "The Locus Coeruleus and Noradrenergic Modulation of Cognition," *Nature Reviews Neuroscience* 10 (2009): 211–23. One mechanism behind this may be increased extracellular norepinephrine, which decreases the spontaneous firing rate of sensory neurons but maintains response to sensory stimuli. B. D. Waterhouse and D. J. Woodward, "Interaction of Norepinephrine with Cerebro-Cortical Activity Evoked by Stimulation of Somatosensory Afferent Pathways in the Rat," *Experimental Neurology* 67, no. 1 (1980): 11–34; D. J. Woodward, H. C. Moises, B. D. Waterhouse, H. H. Yeh, and J. E. Cheun, "Modulatory Actions of Norepinephrine on Neural Circuits," *Advances in Experimental Medicine and Biology* 287 (1991): 193–208.

Patterns that are associated with salient stimuli will result in increased firing of neurons that respond to those stimuli. G. L. West, A. A. Anderson, S. Ferber, and J. Pratt, "Electrophysiological Evidence for Biased Competition in V1 for

Fear Expressions," *Journal of Cognitive Neuroscience* 23, no. 11 (2011): 3410–18; K. Rauss, S. Schwartz, and G. Pourtois, "Top-Down Effects on Early Visual Processing in Humans: A Predictive Coding Framework," *Neuroscience & Biobehavioral Reviews* 35, no. 5 (2011): 1237–253; D. M. Beck and S. Kastner, "Top-Down and Bottom-Up Mechanisms in Biasing Competition in the Human Brain," *Vision Research* 49, no. 10 (2009): 1154–65; N. Kanwisher and E. Wojciulik, "Visual Attention: Insights from Brain Imaging," *Nature Reviews Neuroscience* 1, no. 2 (2000): 91–100; S. Kastner and L. G. Ungerleider, "Mechanisms of Visual Attention in the Human Cortex," *Annual Review of Neuroscience* 23, no. 1 (2000): 315–41.

43   **our attention:** We attend selectively both internally and externally to patterns that have been associated with salient stimuli. M. Corbetta and G. L. Shulman, "Control of Goal-Directed and Stimulus-Driven Attention in the Brain," *Nature Reviews Neuroscience* 3, no. 3 (2002): 201–15. We are able to direct attention to certain stimuli in line with a particular goal. For instance, if I am looking for a green car, my visual processing will be biased toward green objects in the environment. We can also attend selectively to patterns that are in and of themselves salient, such as a color that contrasts strongly with the objects around it. If I am in a green field, my visual system biases processing toward the red flowers on the ground.

Patterns associated with features of salient stimuli are selectively attended to. P. Vuilleumier, "How Brains Beware: Neural Mechanisms of Emotional Attention," *Trends in Cognitive Sciences* 9, no. 12 (2005): 585–94; J. Driver, "A Selective Review of Selective Attention Research from the Past Century," *British Journal of Psychology* 92 (2001): 53–78; S. Kastner and L. G. Ungerleider, "Mechanisms of Visual Attention in the Human Cortex," *Annual Review of Neuroscience* 23 (2000): 315–41. Salient stimuli engage the bottom-up attentional system centered on the right temporoparietal and ventral frontal cortex. M. Corbetta and G. L. Shulman, "Control of Goal-Directed and Stimulus-Driven Attention in the Brain," *Nature Reviews Neuroscience* 3, no. 3 (2002): 201–15. Emotionally salient stimuli engage the "anterior affective system," including the orbitofrontal cortex, amygdala, and locus coeruleus. J. Markovic, A. K. Anderson, and R. M. Todd, "Tuning to the Significant: Neural and Genetic Processes Underlying Affective Enhancement of Visual Perception and Memory," *Behavioural Brain Research* 259 (2014): 229–41; G. Pourtois, A. Schettino, and P. Vuilleumier, "Brain Mechanisms for Emotional Influences on Perception and Attention: What Is Magic and What Is Not," *Biological Psychology* 92, no. 3 (2013): 492–512.

Pattern recognition associated with salient stimuli can direct selective attention automatically and outside awareness. Bias toward emotionally salient stimuli occurs even when participants are instructed to ignore them. L. Nummenmaa, J. Hyona, and M. G. Calvo, "Eye Movement Assessment of Selective Attentional Capture by Emotional Pictures," *Emotion* 6, no. 2 (2006): 257–68. Processing in V1 is biased toward affectively salient stimuli as early as fifty milliseconds after stimulus onset. G. L. West, A. A. Anderson, S. Ferber, and J. Pratt, "Electrophysiological Evidence for Biased Competition in V1 for Fear Expressions," *Journal of Cognitive Neuroscience* 23, no. 11 (2011): 3410–18. The attentional blink effect, in which participants fail to detect a second target stimulus if it is presented within about five hundred milliseconds of the first target in a rapid stream of stimuli, is diminished for emotionally salient words. A. K. Anderson, "Affective Influences on the Attentional Dynamics Supporting Awareness," *Journal of Experimental Psychology: General* 134, no. 2 (2005): 258; A. Keil and N. Ihssen, "Identification Facilitation for Emotionally Arousing Verbs During the Attentional Blink," *Emotion* 4, no. 1 (2004): 23–35.

We attend selectively to novel or unexpected patterns if there are no other stimuli that are more successful in competing for our attention. Stimuli within a single neuron's receptive field compete for processing. J. Duncan, "Cooperating Brain Systems in Selective Perception and Action," in T. Inui and J. L. McClelland, eds., *Attention and Performance XVI* (Cambridge, MA: MIT Press, 1996), 549–78; S. Kastner and L. G. Ungerleider, "The Neural Basis of Biased Competition in Human Visual Cortex," *Neuropsychologia* 39, no. 12 (2001): 1263–76; D. M. Beck and S. Kastner, "Top-Down and Bottom-Up Mechanisms in Biasing Competition in the Human Brain," *Vision Research* 49, no. 10 (2009): 1154–65. Novel and unexpected stimuli are favored in biased competition. L. Itti and P. Baldi, "Bayesian Surprise Attracts Human Attention," *Vision Research* 49, no. 10 (2009): 1295–96; C. Ranganath and G. Rainer, "Neural Mechanisms for Detecting and Remembering Novel Events," *Nature Reviews Neuroscience* 4, no. 3 (2003): 193–202.

Patterns associated with stimuli that can change how we feel (better or worse) can be selectively attended to. Goal-directed attention can bias competition in favor of relevant stimuli. A. Gazzaley, J. W. Cooney, K. McEvoy, R. T. Knight, and M. D'Esposito, "Top-Down Enhancement and Suppression of the Magnitude and Speed of Neural Activity," *Journal of Cognitive Neuroscience* 17, no. 3 (2005): 507–17. Perception is biased toward emotionally relevant stimuli. M. Mather and M. R. Sutherland, "Arousal-Biased Competition in Perception and Memory," *Perspectives on Psychological Science* 6, no. 2 (2011): 114–33; G. L. West, A. A. Anderson, S. Ferber, and J. Pratt, "Electrophysiological Evidence for Biased Competition in V1 for Fear Expressions," *Journal of Cognitive Neuroscience* 23, no. 11 (2011): 3410–18; D. Rudrauf, O. David, J. P. Lachaux, C. K. Kovach, J. Martinerie, B. Renault, et al., "Rapid Interactions Between the Ventral Visual Stream and Emotion-Related Structures Rely on a Two-Pathway Architecture," *Journal of Neuroscience* 28, no. 11 (2008): 2793–803.

44   **reinforced:** Patterns associated with salient stimuli become connected with the feelings (pleasantness, disgust, fear) associated with the salient stimuli. Stimuli acquire emotional meaning by being associated with pleasant or aversive events. J. E. LeDoux, "Brain Mechanisms of Emotion and Emotional Learning," *Current Opinion in Neurobiology* 2, no. 2 (1992): 191–97; S. Maren, "Long-Term Potentiation in the Amygdala: A Mechanism for Emotional Learning and Memory," *Trends in Neurosciences* 22, no. 12 (1999): 561–67; M. Gallagher and P. C. Holland, "The Amygdala Complex: Multiple Roles in Associative Learning and Attention," *Proceedings of the National Academy of Sciences* 91, no. 25 (1994): 11771–76; S. J. Sara, "The Locus Coeruleus and Noradrenergic Modulation of Cognition," *Nature Reviews Neuroscience* 10, no. 3 (2009): 211–23; R. M. Todd, W. A. Cunningham, A. K. Anderson, and E. Thompson, "Affect-Biased Attention as Emotion Regulation," *Trends in Cognitive Sciences* 16, no. 7 (2012): 365–72.

Our worldview is the product of past experiences with salient stimuli. Attentional filters change over time and show individual differences. R. M. Todd, W. A. Cunningham, A. K. Anderson, and E. Thompson, "Affect-Biased Attention as Emotion Regulation," *Trends in Cognitive Sciences* 16, no. 7 (2012): 365–72; R. M. Todd et al., "The Changing Face of Emotion: Age-Related Patterns of Amygdala Activation to Salient Faces," *Social Cognitive and Affective Neuroscience* 6, no. 1 (2011): 12–23. Attentional filters are affected by emotionally arousing experiences. W. A. Cunningham et al., "Aspects of Neuroticism and the Amygdala: Chronic Tuning from Motivational Styles," *Neuropsychologia* 48, no. 12 (2010): 3399–404.

What we attend to is affected by trauma. For instance, in a study by Pollak and Tolley-Schell (2003), physically abused children showed an attentional bias for angry faces. S. D. Pollak and S. A. Tolley-Schell, "Selective Attention to Facial Emotion in Physically Abused Children," *Journal of Abnormal Psychology* 112, no. 3 (2003): 323; S. D. Pollak and P. Sinha, "Effects of Early Experience on Children's Recognition of Facial Displays of Emotion," *Developmental Psychology* 38, no. 5 (2002): 784.

Attentional biases are trainable. P. T. Hertel and A. Matthews, "Cognitive Bias Modification: Past Perspectives, Current Findings, and Future Applications," *Perspectives on Psychological Science* 6, no. 6 (2011): 521–36. Using a technique called attentional bias modification (ABM), researchers have found that they can diminish participants' attentional bias toward negative stimuli. ABM is a cueing task in which targets appear more frequently in the location of neutral rather than negatively valenced stimuli. It has been shown to reduce anxiety scores in clinical and nonclinical populations (though research is still in its infancy). Y. Hakamata, S. Lissek, Y. Bar-Haim, J. C. Britton, N. A. Fox, E. Leibenluft et al., "Attention Bias Modification Treatment: A Meta-Analysis Toward the Establishment of Novel Treatment for Anxiety," *Biological Psychiatry* 68, no. 11 (2010): 982–90.

These representations in turn determine the way we see our enviroment and interpret stimuli. Attentional filters come to be applied reflexively, thereby affecting which stimuli we attend to and how, as well as impacting our emotional experience. R. M. Todd, W. A. Cunningham, A. K. Anderson, and E. Thompson, "Affect-Biased Attention as Emotion Regulation," *Trends in Cognitive Sciences* 16, no. 7 (2012): 365–72; J. S. Bruner and C. C. Goodman, "Value and Need as Organizing Factors in Perception," *Journal of Abnormal Psychology* 42 (1947): 33–44; R. H. Fazio, "On the Automatic Activation of Associated Evaluations: An Overview," *Cognition and Emotion* 15, no. 2 (2001): 115–41; W. A. Cunningham and P. D. Zelazo, "Attitudes and Evaluations: A Social Cognitive Neuroscience Perspective," *Trends in Cognitive Sciences* 11, no. 3 (2007): 97–104.

44 **feed-forward loop:** The feelings that result from a potent salient stimulus can themselves become salient. This process can in turn reinforce salience encoding and selective attention and make the neural circuits feed back on themselves. M. D. Lewis, "Bridging Emotion Theory and Neurobiology Through Dynamic Systems Modeling," *Behavioral and Brain Sciences* 28, no. 2 (2005): 169–94; E. A. Phelps, S. Ling, and M. Carrasco, "Emotion Facilitates Perception and Potentiates the Perceptual Benefits of Attention," *Psychological Science* 17, no. 4 (2006): 292–99; R. J. Dolan, "Emotion, Cognition, and Behavior," *Science* 298, no. 5596 (2002): 1191–94; R. M. Todd and A. K. Anderson, "Salience, State, and Expression: The Influence of Specific Aspects of Emotion on Attention and Perception," in *The Oxford Handbook of Cognitive Neuroscience*, vol. 2: *The Cutting Edges* (Oxford: Oxford University Press, 2013), 2, 11; M. D. Lewis and

R. M. Todd, "The Self-Regulating Brain: Cortical-Subcortical Feedback and the Development of Intelligent Action," *Cognitive Development* 22, no. 4 (2007): 406–30.

The salience of a stimulus depends on the individual's representation of the world, which in turn is determined by prior exposure and sensitivity to salient stimuli. Lewis (2005) points out the bidirectional relationship between emotion and cognitive processes and explains the relationship using a dynamical systems model. M. D. Lewis, "Bridging Emotion Theory and Neurobiology Through Dynamic Systems Modeling," *Behavioral and Brain Sciences* 28, no. 2 (2005): 169–94. Teasdale (1993) looks at the relationship between depressed mood and cognition, noting that depressive thoughts can be both antecedent and consequent to depressed mood. A depressed mood makes accessible certain thought patterns from memory and past experience (which vary by individual) and leads to a negative way of interpreting current experience. J. D. Teasdale, "Emotion and Two Kinds of Meaning: Cognitive Therapy and Applied Cognitive Science," *Behaviour Research and Therapy* 31, no. 4 (1993): 339–54.

CHAPTER 3: WHAT CAPTURES?

A Continuum from the Ordinary to Mental Illness

Sources

Bollas, Christopher. *The Shadow of the Object: Psychoanalysis of the Unthought Known*. London: Free Association Books, 1987.

Flores, Philip J. *Addiction as an Attachment Disorder*. Lanham, MD: Jason Aronson, 2004.

Hart, Susan. *The Impact of Attachment: Developmental Neuroaffective Psychology*. New York: Norton, 2011.

Holmes, Jeremy. *John Bowlby and Attachment Theory*. London: Routledge, 1993.

Marrone, Mario. *Attachment and Interaction*. London: Jessica Kingsley Publishers, 1998.

Van der Horst, Frank C. P. *John Bowlby—From Psychoanalysis to Ethnology: Unraveling the Roots of Attachment Theory*. Oxford: Blackwell, 2011.

Notes

45   **attachment:** John Bowlby spent his entire career studying attachments, the unique emotional bond that can form between human beings. Bowlby was raised by a nanny who left the family when he was four. It was, in his mother's opinion, dangerous to cater to children's incessant demands for attention and affection. At seven, Bowlby was sent to boarding school. Years later, he wrote of his early schooling: "Unhappiness in a child accumulates because he sees no end to the dark tunnel. The thirteen weeks of a term might just as well be thirteen years."

After receiving his medical degree, Bowlby decided to train as a psychoanalyst. He soon found himself caught between warring factions within child psychoanalysis; though he ultimately chose Melanie Klein as his supervisor, Bowlby could not bring himself to accept Klein's view that behavior was motivated primarily by drives. Rather, Bowlby held fast to the idea that environment, the child's experience of the surrounding world, shaped his or her inner life.

Bowlby was finally won over by Klein's concept of the "object," a person, a thing, or an idea endowed by the mind with symbolic meaning. Bowlby soon renamed such objects "attachment figures." He would dedicate much of his professional life to the interpersonal experiences that underlay such attachment figures, which were an integral part of human experience through all the phases of life: "The propensity to make strong emotional bonds to particular individuals is a basic component of human nature." Early childhood attachments were particularly important, as they influenced psychological development and mental health later in life.

Bowlby was convinced that the biology of attachment had its roots in our evolutionary history. Attachments increased the likelihood that offspring would survive and thrive; they promoted the establishment of strong emotional relationships and communal ties. Still, many of Bowlby's questions remained unanswered: What is the neurological basis for the formation of attachments? Why do they sometimes fail?

REJECTION

Sources

Works Referenced

Robinson, Jenefer. *Deeper Than Reason*. Oxford: Clarendon, 2005.
Wharton, Edith. *The Reef*. New York: Scribner, 1912.

Notes

46   **"Unexpected obstacle"**: Wharton, *The Reef*, 17. (I want to credit Professor Robinson for drawing my attention to Edith Wharton's *The Reef*.)
46   **"from Charing Cross to Dover"**: Ibid.
47   **"her unexpected face"**: Ibid., 18.

A BRUTISH FATHER

Sources

Works Referenced

Costa-Lima, Luiz. *The Limits of Voice: Montaigne, Schlegel, Kafka*. Translated by Paulo Henriques Britto. Stanford, CA: Stanford University Press, 1996.
Crumb, Robert, and David Mairowitz. *Kafka*. Seattle: Fantagraphics, 2007.
Friedlander, Saul. *Franz Kafka: The Poet of Shame and Guilt*. New Haven, CT: Yale University Press, 2013.
Gilman, Sander. *Franz Kafka: The Jewish Patient*. New York: Routledge, 1995.
Hayman, Ronald. *A Biography of Kafka*. London: Phoenix, 1981.
Janouch, Gustav. *Conversations with Kafka*. Translated by Goronwy Rees. New York: New Directions, 1971.
Kafka, Franz. *Dearest Father*. Translated by Hannah and Richard Stokes. London: Oneworld Classics, 2008.
———. *Diaries, 1910–1923*. Edited by Max Brod. New York: Shocken, 1976.
———. *Letters to Milena*. Translated by Phillip Boehm. New York: Shocken, 1990.
Murray, Nicholas. *Kafka: A Biography*. New Haven, CT: Yale University Press, 2004.
Neider, Charles. *Kafka: His Mind and Art*. London: Routledge and Kegan Paul, 1949.

Stach, Reiner. *Kafka: The Decisive Years.* Translated by Shelley Frisch. Orlando, FL: Harcourt Books, 2005.

———. *Kafka: The Years of Insight.* Translated by Shelley Frisch. Princeton, NJ: Princeton University Press, 2013.

Tiefenbrun, Ruth. *Moment of Torment: An Interpretation of Franz Kafka's Short Stories.* Carbondale: Southern Illinois University Press, 1973.

Notes

48   "God's law": Kafka, *Dearest Father*, 28.
48   "*meshugge*": Ibid., 25.
48   "could not lament": Ibid., 63.
49   "resembled a worm": Ibid., 62.
49   "the beast": Ibid., 28.
49   "partly to be annoying": Ibid., 22.
49   "I meant so little": Ibid., 23.
50   "'food on the table'": Ibid., 26.
50   "the noose is dangling": Ibid., 40.
51   "your opinion was for once the same": Ibid., 26.
51   "skinny, frail": Ibid., 24.
51   "all my past failures": Ibid.

DRINK

Sources

Works Referenced

Knapp, Caroline. *Appetites: Why Women Want.* New York: Perseus, 2003.

———. *Drinking: A Love Story.* New York: Counterpoint, 1996.

———. *The Merry Recluse: A Life in Essays.* Berkeley, CA: Counterpoint, 2004.

Thurber, Jon. "Caroline Knapp, 42; Wrote of Alcohol Struggle." *Los Angeles Times*, June 22, 2002, http://articles.latimes.com/2002/jun/22/local/me-knapp22.

Notes

51   "power of deflection": Knapp, *Drinking*, 5.
51   "I loved the sounds . . . so strong they're crippling": Ibid., 6.
52   "originated within the family": Ibid., 61.
52   "Dear Dad": Knapp, *Merry Recluse*, 125.
53   "the chase is harder to stop . . . spiritual carrot": Knapp, *Drinking*, 61.

PHYSICAL PAIN

Sources

Works Referenced

Berman, Jeffrey, and Patricia Hatch Wallace. *Cutting and the Pedagogy of Self-Disclosure.* Amherst: University of Massachusetts Press, 2007.

Carscadden, Judith S. *On the Cutting Edge: A Guide for Working with People Who Self Injure.* London: Judith S. Carscadden, 1993.

Favazza, Armando R. *Bodies Under Siege: Self-Mutilation and Body Modification in Culture and Psychiatry.* 2nd ed. Baltimore, MD: Johns Hopkins University Press, 1996.

Inckle, Kay. *Flesh Wounds? New Ways of Understanding Self-Injury.* Ross-on-Wye, UK: PCCS, 2010.

Kettlewell, Caroline. *Skin Game: A Cutter's Memoir.* New York: St. Martin's Griffin, 1999.

Levenkron, Steven. *Cutting: Understanding and Overcoming Self-Mutilation.* New York: Norton, 1998.

Plante, Lori G. *Bleeding to Ease the Pain: Cutting, Self-Injury, and the Adolescent Search for Self.* Lanham, MD: Rowman and Littlefield, 2007.

Notes

53   **"no idea, really, why"**: Kettlewell, *Skin Game*, 69.

53   **"Every day I . . . with a want like that?"**: Ibid., 55.

54   **"penny ante"**: Ibid., 59.

54   **"How many troubles"**: Ibid., 60.

54   **"razor's edge kiss"**: Ibid., 27.

55   **"cut for the cut itself"**: Ibid., 13.

55   **"elegant pain"**: Ibid., 27.

CHILDHOOD TRAUMA

Sources

Interview by the Author
"Nora" (2012, 2013, and 2014).

BLIND LOVE

Sources

Interview by the Author
"Jackie" (2014 and 2015).

OBSCENE FASCINATION

Sources

Works Referenced

"Berendzen Pleads Guilty to Obscene Calls." *Nightline*, show no. 2348, aired May 23, 1990, ABC.

Berendzen, Richard, and Laura Palmer. *Come Here: A Man Overcomes the Tragic Aftermath of Childhood Sexual Abuse.* New York: Villard, 1993.

Hewitt, Bill, and Gary Clifford. "A University President Tumbles from Grace Following Charges That He Made Obscene Phone Calls." *People*, May 14, 1990. http://www.people.com/people/archive/article/0,,20117639,00.html.

Schnarch, David M. *Constructing the Sexual Crucible: An Integration of Sexual and Marital Therapy.* New York: Norton, 1991.

Notes

65   **"head spinning"**: Berendzen and Palmer, *Come Here*, 82.

65   **"When I made a call"**: Ibid., 85.

65   **"We share everything"**: Hewitt and Clifford, "A University President."

65   sex slave: Ibid.

66   "wheel in his basement": Ibid.
66   "foreign body imprinted": "Berendzen Pleads Guilty."
66   "sexual trance": Schnarch, *Constructing the Sexual Crucible*, 77.
66   "further into the flames": Berendzen and Palmer, *Come Here*, 84.

GAMBLING

Sources

Works Referenced

Amoia, Alba Della Fazia. *Feodor Dostoevsky*. New York: Continuum, 1993.

Catteau, Jacques. *Dostoevsky and the Process of Literary Creation*. Translated by Audrey Littlewood. Cambridge, UK: Cambridge University Press, 1989.

Dostoevsky, Anna. *Dostoevsky Portrayed by His Wife: The Diary and Reminiscences of Mme. Dostoevsky*. Translated by S. S. Koteliansky. New York: E. P. Dutton, 1926.

———. *Dostoevsky: Reminiscences*. Translated by Beatrice Stillman. New York: Liveright, 1975.

Dostoevsky, Fyodor. *Fyodor Dostoevsky: Complete Letters 1860–1867*. Translated by David Allan. Berkeley: University of California Press, 1989.

———. *The Gambler*. Translated by Victor Terras. Chicago: University of Chicago Press, 1972.

———. *The Novels of Fyodor Dostoevsky*, vol. 9. Translated by Constance Garnett. New York: Macmillan, 1928.

———. *Stavrogin's Confession and the Plan of the Life of a Great Sinner*. Translated by S. S. Koteliansky and Virginia Woolf. Richmond: Hogarth, 1922.

Frank, Joseph. *Dostoevsky: A Writer in His Time*. Princeton, NJ: Princeton University Press, 2012.

———. *Dostoevsky: The Stir of Liberation: 1860–1865*. Princeton, NJ: Princeton University Press, 1986.

Freeborn, Richard. *Dostoevsky*. London: Haus, 2003.

Kjetsaa, Geir. *Fyodor Dostoevsky: A Writer's Life*. New York: Viking, 1987.

Lantz, Kenneth. *The Dostoevsky Encyclopedia*. Westport, CT: Greenwood, 2004.

Schüll, Natasha Dow. *Addiction by Design: Machine Gambling in Las Vegas*. Oxfordshire, UK: Princeton University Press, 2012.

———. "Machine Life: An Ethnography of Gambling and Compulsion in Las Vegas." PhD diss., University of California, Berkeley, 2003.

———. "Natasha Dow Schüll, Ph.D." http://scripts.mit.edu/~schull/nds/.

Wasiolek, Edward. *The Gambler*. Translated by Victor Terras. Chicago: University of Chicago Press, 1972.

Notes

67   "abominable passion": Kjetsaa, *Fyodor Dostoevsky*, 112.
67   "beautiful wife . . . some relief": Anna Dostoevsky, *Dostoevsky Portrayed by His Wife*, 62.
67   "powerless to resist": Kjetsaa, *Fyodor Dostoevsky*, 244.
67   wedding ring: Fyodor Dostoevsky, *The Gambler*, xxiv.
67   "devilish gaming": Kjetsaa, *Fyodor Dostoevsky*, 244.
67   "pawned my watch": *Fyodor Dostoevsky: Complete Letters 1860–1867*, 63.
68   opiated lull: Fyodor Dostoevsky, *The Novels of Fyodor Dostoevsky*, vol. 9, 91.
68   "utterly exhausted": Dostoevsky, *The Gambler*, 158.

69   "raking in the bank notes": Ibid.
69   on autopilot: Schüll, *Addiction by Design*, 19.
69   background and socioeconomic status: Ibid., 23.
69   "bodies disappearing": Ibid., 174.

THE BODY

Sources

Works Referenced

Devlin, Albert J., and Nancy M. Tischler, eds. *Selected Letters of Tennessee Williams: Vol. 1, 1920–1945*. Sewanee, TN: University of the South, 2000.

———. *Selected Letters of Tennessee Williams: Vol. 2, 1945–1957*. Sewanee, TN: University of the South, 2000.

Gussow, Mel. "Tennessee Williams Is Dead at 71." *New York Times*, Feb. 26, 1983, http://www.nytimes.com/books/00/12/31/specials/williams-obit.html.

New York Times. "Drugs Linked to Death of Tennessee Williams." Aug. 14, 1983, https://www.nytimes.com/books/00/12/31/specials/williams-drugs.html.

Thornton, Margaret Bradham, ed. *Tennessee Williams Notebooks*. New Haven, CT: Yale University Press, 2006.

Tischler, Nancy M. *Tennessee Williams: Rebellious Puritan*. New York: Citadel, 1961.

Weales, Gerald. *Tennessee Williams*. St. Paul: University of Minnesota Press, 1965.

Williams, Dakin. *His Brother's Keeper: The Life and Murder of Tennessee Williams*. Collinsville, IL: Dakin's Corner, 1983.

Williams, Tennessee. *Collected Stories*. New York: Ballantine, 1985.

———. *Memoirs*. New York: New Directions, 1972.

Notes

70   "Accent of a Coming Foot": Tennessee Williams, *Memoirs*, 38.
70   "I walked faster and faster": Ibid.
70   "cardiac neurosis": Thornton, *Tennessee Williams Notebooks*, 629.
70   "What a week is *behind* me": Ibid., 27.
70   "heart neurosis": Ibid., 125.
71   "fear is so much worse": Ibid.
71   "a painful jolt": Ibid., 625.
71   palpitations, jolts: Ibid., 657.
71   "A lady . . . live to be forty!": Tennessee Williams, *Memoirs*, 42.
71   *New York Times* obituary: Gussow, "Tennessee Williams Is Dead."
72   "trying to ingest": *New York Times*, "Drugs Linked."
72   "cause of death": Ibid.

A WORK OF ART

Sources

Works Referenced

Davis, Douglas M. "Miss DeFeo's Awesome Painting Is Like Living Thing Under Decay." *National Observer*, July 14, 1969.

DeFeo, Jay. Interview conducted by Paul J. Karlstrom, June 3, 1975. Smithsonian Archives of American Art. http://www.aaa.si.edu/collections/interviews/oral-history-interview-jay-defeo-13246.

Green, Jane, and Leah Levy, eds. *Jay DeFeo and the Rose*. Berkeley: University of California Press, 2003.

Jay DeFeo Trust. Letter to her mother. http://jaydefeo.org.

Miller, Dana. *Jay DeFeo: A Retrospective*. New York: Yale University Press, 2012.

Nichols, Matthew. "Beyond the Rose." *Art in America*, March 1, 2013. http://www.artinamericamagazine.com/news-features/magazine/beyond-the-rose/.

Roth, Moira, ed. "Interview with Leela Elliot." In *Connecting Conversations: Interviews with 28 Bay Area Women Artists*. Oakland, CA: Eucalyptus, 1988.

SFMOMA. "Bruce Conner on Jay DeFeo's *The Rose*." Originally accessed at http://www.sfmoma.org/explore/multimedia/videos/311.

Notes

72   "center": DeFeo, Smithsonian interview, 7.

72   "wedge-like": SFMOMA, "Bruce Conner."

72   "sharpened knives": Davis, "Miss DeFeo's Awesome Painting."

72   organic ... "how flamboyant": DeFeo, Smithsonian interview.

73   "The room itself": SFMOMA, "Bruce Conner."

73   "love affair": Martha Sherrill, "The Story of *The Rose*," in Green and Levy, *Jay DeFeo*, 30.

74   "real moments of happiness": Jay DeFeo Trust, Letter to her mother.

74   "a feeling of aliveness & completeness": Ibid. Jay DeFeo felt she was truly herself in the world of form and paint. At the age of fifteen, Susan Sontag was enthralled by words. In language, she found a form for her inchoate desires: her diaries from adolescence and young adulthood trace the formation of her identity through a series of captures.

"The loneliness is ever and immutably present," Sontag wrote in her diary on May 26, 1948. "Strangely enough I feel a new sharpening of my senses—a mad frustrated longing for absolute honesty—oh, will I never say and do what I really feel!"

Such honesty was at odds with the social scripts of suburbia—"all that surrounds and drowns me," the "morass of mediocrity" in which "the real 'I' is buried, entombed in a mountain of forced feeling." For the young Sontag, finding "the real 'I'" became a never-ending project of self-invention. She constructed for herself an identity out of pieces that captured her; it was the beginning of what would become an edifice in a perpetual state of remodeling.

"There are fleeting (oh so quickly flown) moments when I know as surely as today is Christmas that I am tottering over an illimitable precipice," she wrote on December 25, 1948. Vivaldi's Concerto in B Minor glided over the precocious fifteen-year-old: "I listen with my body and it is my body that aches in response to the passion and pathos. . . . It is the physical 'I' that feels an unbearable pain—and then a dull fretfulness—when the whole world of melody suddenly glistens and comes cascading down."

The sensual, perfectly abstracted experience of music awakens a sense of sheer possibility within Sontag, transporting her to a realm far from parental expectations and social mores: "I must not think of the solar system—of innumerable galaxies spanned by countless light years—of infinities of space."

As the music fades away, the diary moves inward, balanced between hesitance and exhilaration: "What, I ask, drives me to disorder? How can I diagnose myself?" Gradually gaining awareness of herself as a sexual being,

Sontag writes of the "disturbing aspect of my sexual ambitions": "So now I feel that I have lesbian tendencies (how reluctantly I write this)." Such a statement was nothing short of radical in 1940s America, where women were expected to sublimate their sexual desires, and lesbianism was still seen as a pathology.

By the time she'd entered UC Berkeley in early 1949, Sontag was in thrall to desire. "I am in love," she wrote, "with being in love!" Recounting her first sexual experience with a woman, the young writer conveys a moment of pure release: "Everything that was so tight, that hurt so in the pit of my stomach, was vanquished in the straining against her, the weight of her body on top of mine, the caress of her mouth and her hands."

"What am I to do?" she asks, only to answer, in a later, undated annotation, "Enjoy yourself, of course." For Sontag, enjoying herself eventually came to mean unrestrained pursuit of sexual pleasure, with both men and women: "I had never truly comprehended that it was possible to live through your body and not make any of these hideous dichotomies after all!" Bisexuality came to embody for the young writer a rejection of "the jovial claptrap of my classmates and teachers, the maddening bromides I heard at home."

Sontag was captured not merely by her growing awareness of her sexuality, but also by the radical autonomy of the artist. She had discovered an intoxicating freedom in her experiments with language, and in the promise of community among bohemian literati. No longer was she constrained by circumstance, by the "hideous dichotomies" that surrounded her. Sex and even love were suddenly detached from their usual moorings in courtship and marriage, procreation and child rearing. Tossing aside her erstwhile ambitions (graduate school, a career in academia), the teenage Sontag committed herself instead to a highly intellectualized hedonism: "I intend to do everything . . . to have one way of evaluating experience—does it cause me pleasure or pain, and I shall be very cautious about rejecting the painful—I shall anticipate pleasure everywhere and find it, too, for it is everywhere!"

["forced feeling": Journal of Susan Sontag, May 26, 1948, Folder 3, Box 132, Collection 612: Papers of Susan Sontag, UCLA Library Special Collections, Charles E. Young Research Library, Los Angeles, CA.
"cascading down": Susan Sontag, *Reborn: Journals and Notebooks, 1947–1963* (New York: Macmillan, 2009), 10.
"infinities of space": Ibid., 11.
"with being in love": Ibid., 15.
"and her hands": Ibid., 27.
"Enjoy yourself, of course": Ibid., 16.
"maddening bromides": Ibid., 27.
"hideous dichotomies": Susan Sontag, quoted in Carl Rollyson and Lisa Paddock, *Susan Sontag: The Making of an Icon* (New York: Norton, 2000), 16.
"it is everywhere": Sontag, *Reborn*, 28.]
74   "inner core of faith": DeFeo, Smithsonian interview.

DEATH

Sources

Archives

Sexton, Anne. *Anne Sexton: An Inventory of Her Papers.* Series IV. Formerly Closed Materials, 1948, 1954–1974: Correspondence: Orne, Martin, 1959–1964. Box 43, Folder 1. Harry Ransom Center, Austin, TX.

———. *Anne Sexton: An Inventory of Her Papers.* Series IV. Formerly Closed Materials, 1948, 1954–1974: Miscellaneous: Journals, 1961–1963. Box 43. Harry Ransom Center, Austin, TX.

———. *Audiotapes and Papers of Anne Sexton 1956–1988.* Schlesinger Library on the History of Women in America, Cambridge, MA.

Works Referenced

Berman, Jeffrey. *Surviving Literary Suicide.* Boston: University of Massachusetts Press, 1999.

Colburn, Steven E., ed. *Anne Sexton: Telling the Tale.* Ann Arbor: University of Michigan Press, 1988.

George, Diana Hume. *Oedipus Anne: The Poetry of Anne Sexton.* Urbana: University of Illinois Press, 1987.

McClatchy, J. D., ed. *Anne Sexton: The Artist and Her Critics.* Bloomington: Indiana University Press, 1978.

Middlebrook, Diane. *Anne Sexton: A Biography.* New York: Vintage, 1992.

Middlebrook, Diane, and Diana Hume George, eds. *Selected Poems of Anne Sexton.* New York: First Mariner, 2000.

Salvio, Paula M. *Anne Sexton: Teacher of Weird Abundance.* New York: State University of New York Press, 2007.

Sexton, Anne. *The Complete Poems.* New York: Mariner, 1999.

Sexton, Linda Gray, and Lois Ames, eds. *Anne Sexton: A Self-Portrait in Letters.* Boston: Houghton Mifflin, 1977.

Skorczewski, Dawn M. *An Accident of Hope: The Therapy Tapes of Anne Sexton.* New York: Routledge, 2012.

Wagner-Martin, Linda. *Critical Essays on Anne Sexton.* Boston: Hull, 1989.

Notes

74 "Suicide is addicting . . . someone forces us": Middlebrook, *Anne Sexton: A Biography,* 199–200.
75 "life is lovely": Sexton and Ames, *Anne Sexton: A Self-Portrait,* 251.
75 "Since you ask": Ibid., 189.
75 "never ask why build": Ibid., 190.
76 "mouth-hole . . . bad prison": Ibid., 192–93.
76 safety of her children: Middlebrook, *Anne Sexton: A Biography,* 31–32, 39–43.
77 "a caged tiger": Ibid., 36.
77 "not gonna kill myself in the doctor's": Sexton Audiotapes, Nov. 5, 1963.
77 "someone else is me": Sexton, Audiotapes and Papers, April 25, 1964.
77 "Spring. Warm. Leaves": Sexton, Audiotapes and Papers, March 7, 1963.
77 "This heavy thing . . . we're in silence": Sexton, Audiotapes and Papers, April 25, 1964.

A THREAT

Sources

Interview by the Author
"Charlotte" contributed directly to this chapter.
Works Referenced
See scientific articles cited in the notes.

Notes

81   **alerts us to danger:** Y. Bar-Haim, D. Lamy, L. Pergamin, M. J. Bakermans-Kranenburg, and M. H. van IJzendoorn, "Threat-Related Attentional Bias in Anxious and Nonanxious Individuals: A Meta-Analytic Study," *Psychological Bulletin* 133, no. 1 (2007): 1–24, doi:10.1037/0033-2909.133.1.1; K. Mogg, A. Mathews, and M. Eysenck, "Attentional Bias to Threat in Clinical Anxiety States," *Cognition and Emotion* 6, no. 2 (1992): 149–59, doi:10.1080/02699939208411064; C. MacLeod, A. Mathews, and P. Tata, "Attentional Bias in Emotional Disorders," *Journal of Abnormal Psychology* 95, no. 1 (1986): 15–20.

81   **feelings of vulnerability:** H. J. Richards, V. Benson, N. Donnelly, and J. A. Hadwin, "Exploring the Function of Selective Attention and Hypervigilance for Threat in Anxiety," *Clinical Psychology Review* 34, no. 1 (2014): 1–13, doi:10.1016/j.cpr.2013.10.006; M. Kimble, M. Boxwala, W. Bean, K. Maletsky, J. Halper, K. Spollen, and K. Fleming, "The Impact of Hypervigilance: Evidence for a Forward Feedback Loop," *Journal of Anxiety Disorders* 28, no. 2 (2014): 241–45, doi:10.1016/j.janxdis.2013.12.006.

82   **arranged by height:** M. Thobaben, "Obsessive-Compulsive Disorder (OCD): Symptoms and Interventions," *Home Health Care Management and Practice* 24, no. 4 (2012): 211–13, doi:10.1177/1084822312441364.

82   **ever harder to resist:** D. J. Stein, "Neurobiology of the Obsessive–Compulsive Spectrum Disorders," *Biological Psychiatry* 47, no. 4 (2000): 296–304, doi:10.1016/S0006-3223(99)00271-1.

TWO ADDICTS

Sources

Interview by the Author
"Ned" (November 2014).
Works Referenced
Aykroyd, Dan, and John Landis. *The Blues Brothers*. Directed by John Landis. Los Angeles: Universal Studios, 1980.
Belushi, Judith Jacklin. *Samurai Widow*. New York: Carroll and Graf, 1990.
Crowley, Aleister. *The Diary of a Drug Fiend*. San Francisco, CA: Weiser, 2010.
Final 24: John Belushi: His Final Hours. Directed by Michelle Metivier. Discovery Channel, 2006.
Guse, Joe. *The Tragic Clowns: An Analysis of the Short Lives of John Belushi, Lenny Bruce, and Chris Farley*. Lexington, KY: Aardvark, 2007.
Pisano, Judith Belushi, and Tanner Colby. *Belushi: A Biography*. New York: Rugged Man, 2005.
Ramis, Harold, Douglas Kenney, and Chris Miller. *Animal House*. Directed by John Landis. Los Angeles: Universal Studios, 1978.

Smith, Cathy. *Chasing the Dragon.* Toronto: Seal, 1984.

Woodward, Bob. *Wired: The Short Life and Fast Times of John Belushi.* New York: Simon and Schuster, 1984.

Notes

83   **"When he was clean":** Pisano and Colby, *Belushi*, 254.

84   **John Landis . . . puddle of urine . . . "my movie":** Woodward, *Wired*, 19–20.

84   **"start drinking":** *Animal House.*

84   **hired a bodyguard:** Pisano and Colby, *Belushi*, 202; Woodward, *Wired*, 64.

84   **"so, so ashamed":** Woodward, *Wired*, 20.

84   **"afraid of myself . . . respect your decision":** Belushi, *Samurai Widow*, 248–49.

85   **"total fear":** Woodward, *Wired*, 193.

85   **go to Los Angeles:** Pisano and Colby, *Belushi*, 255.

86   **"permanent satisfaction":** Ibid., 244.

86   **early experiences with drugs:** Belushi, *Samurai Widow*, 159, 175.

86   **"got to have it":** Woodward, *Wired*, 21.

CONTROL

Sources

Interview by the Author
"Frances" contributed directly to this chapter.

Notes

94   **eating disorders:** In addition to food cues, cues relating to weight and shape also become highly salient to people with eating disorders. Patients with eating disorders pay greater attention to and have greater memory of negative words associated with weight or shape (e.g., "fat") than positive ones. For example, they focus more on parts of their body they consider "ugly" while paying more attention to the "beautiful" body parts of other people. This selective attention leads to heightened body dissatisfaction, which can lead to "body checking," e.g., weighing oneself obsessively, measuring, or looking in the mirror often. If the patient is satisfied with the results of the check, the overall behavior is negatively reinforced; if he or she isn't, it can lead to purging, excessive exercising, or restricting.

94   **selective and undue attention:** T. Brockmeyer, C. Hahn, C. Reetz, U. Schmidt, and H. C. Friederich, "Approach Bias Modification in Food Craving: A Proof-of-Concept Study," *European Eating Disorders Review* 23, no. 5 (2015): 352–60; V. Cardi, R. Di Matteo, F. Corfield, and J. Treasure, "Social Reward and Rejection Sensitivity in Eating Disorders: An Investigation of Attentional Bias and Early Experiences," *World Journal of Biological Psychiatry* 14, no. 8 (2013): 622–33; V. Cardi, M. Esposito, A. Clarke, S. Schifano, and J. Treasure, "The Impact of Induced Positive Mood on Symptomatic Behaviour in Eating Disorders: An Experimental, Ab/Ba Crossover Design Testing a Multimodal Presentation During a Test-Meal," *Appetite* 87 (2015): 192–98; S. R. Chamberlain, K. Mogg, B. P. Bradley, A. Koch, C. M. Dodds, W. X. Tao, K. Maltby, et al., "Effects of Mu Opioid Receptor Antagonism on Cognition in Obese Binge-Eating Individuals," *Psychopharmacology* 224, no. 4 (2012): 501–9; I. S. Dipl-Psych, B. Renwick, H. de Jong, M. Kenyon, H. Sharpe, C. Jacobi, and U. Schmidt, "Threat-Related Attentional Bias in Anorexia Nervosa," *International Journal*

of Eating Disorders 47, no. 2 (2014): 168–73; G. J. Faunce, "Eating Disorders and Attentional Bias: A Review," *Eating Disorders* 10, no. 2 (2002): 125–39; J. Hewig, S. Cooper, R. H. Trippe, H. Hecht, T. Straube, and W. H. Miltner, "Drive for Thinness and Attention Toward Specific Body Parts in a Nonclinical Sample," *Psychosomatic Medicine* 70, no. 6 (2008): 729–36; S. Horndasch, O. Kratz, A. Holczinger, H. Heinrich, F. Honig, E. Noth, and G. H. Moll, " 'Looks Do Matter'—Visual Attentional Biases in Adolescent Girls with Eating Disorders Viewing Body Images," *Psychiatry Research* 198, no. 2 (2012): 321–23; J. L. Placanica, G. J. Faunce, and R. F. Soames Job, "The Effect of Fasting on Attentional Biases for Food and Body Shape/Weight Words in High and Low Eating Disorder Inventory Scorers," *International Journal of Eating Disorders* 32, no. 1 (2002): 79–90; B. Renwick, I. C. Campbell, and U. Schmidt, "Review of Attentional Bias Modification: A Brain-Directed Treatment for Eating Disorders," *European Eating Disorders Review* 21, no. 6 (2013): 464–74; F. Schmitz, E. Naumann, S. Biehl, and J. Svaldi, "Gating of Attention Towards Food Stimuli in Binge Eating Disorder," *Appetite* 95 (2015): 368–74; F. Schmitz, E. Naumann, M. Trentowska, and J. Svaldi, "Attentional Bias for Food Cues in Binge Eating Disorder," *Appetite* 80 (2014): 70–80; R. Shafran, M. Lee, Z. Cooper, R. L. Palmer, and C. G. Fairburn, "Attentional Bias in Eating Disorders," *International Journal of Eating Disorders* 40, no. 4 (2007): 369–80; R. Shafran, M. Lee, Z. Cooper, R. L. Palmer, and C. G. Fairburn, "Effect of Psychological Treatment on Attentional Bias in Eating Disorders," *International Journal of Eating Disorders* 41, no. 4 (2008): 348–54; E. Smeets, A. Jansen, and A. Roefs, "Bias for the (Un)Attractive Self: On the Role of Attention in Causing Body (Dis)Satisfaction," *Health Psychology* 30, no. 3 (2011): 360–67; E. Smeets, A. Roefs, E. van Furth, and A. Jansen, "Attentional Bias for Body and Food in Eating Disorders: Increased Distraction, Speeded Detection, or Both?," *Behaviour Research and Therapy* 46, no. 2 (2008): 229–38; I. Wolz, A. B. Fagundo, J. Treasure, and F. Fernandez-Aranda, "The Processing of Food Stimuli in Abnormal Eating: A Systematic Review of Electrophysiology," *European Eating Disorders Review* 23, no. 4 (2015): 251–61.

SADNESS

Sources

Interview by the Author
"Wes" (2011, 2012, 2013, and 2014).
Works Referenced
See scientific articles cited in notes.

Notes

97   **both processes are clearly involved:** D. Watson, L. A. Clark, and G. Carey, "Positive and Negative Affectivity and Their Relation to Anxiety and Depressive Disorders," *Journal of Abnormal Psychology* 93, no. 3 (1988): 346–53; I. H. Gotlib, E. Krasnoperova, D. N. Yue, and J. Joormann, "Attentional Biases for Negative Interpersonal Stimuli in Clinical Depression," *Journal of Abnormal Psychology* 113, no. 1 (Feb. 2004): 121–35, doi:10.1037/0021-843X.113.1.121; Ian H. Gotlib and Jutta Joormann, "Cognition and Depression: Current Status and Future Directions," *Annual Review of Clinical Psychology* 6, no. 1

(March 2010): 285–312, doi:10.1146/annurev.clinpsy.121208.131305; Thomas Armstrong and Bunmi O. Olatunji, "Eye Tracking of Attention in the Affective Disorders: A Meta-Analytic Review and Synthesis," *Clinical Psychology Review* 32, no. 8 (Dec. 2012): 704–23, doi:10.1016/j.cpr.2012.09.004.

97 **person's brain:** M. Liotti, H. S. Mayberg, S. K. Brannan, S. McGinnis, P. Jerabek, and P. T. Fox, "Differential Limbic–Cortical Correlates of Sadness and Anxiety in Healthy Subjects: Implications for Affective Disorders," *Biological Psychiatry* 48, no. 1 (July 1, 2000): 30–42; Mary L. Phillips, Wayne C. Drevets, Scott L. Rauch, and Richard Lane, "Neurobiology of Emotion Perception II: Implications for Major Psychiatric Disorders," *Biological Psychiatry* 54, no. 5 (2003): 515–28, doi:S0006322303001719 [pii]; Seth G. Disner, C. G. Beevers, Emily A. P. Haigh, and Aaron T. Beck, "Neural Mechanisms of the Cognitive Model of Depression," *Nature Reviews Neuroscience* 12, no. 8 (July 6, 2011): 467–77, doi:10.1038/nrn3027.

97 **locked in memory:** A. Mathews and C. MacLeod, "Cognitive Vulnerability to Emotional Disorders," *Annual Review of Clinical Psychology* 1 (2005): 167–95, doi:10.1146/annurev.clinpsy.1.102803.143916; Ian H. Gotlib and Jutta Joormann, "Cognition and Depression: Current Status and Future Directions," *Annual Review of Clinical Psychology* 6, no. 1 (March 2010): 285–312, doi:10.1146/annurev.clinpsy.121208.131305; B. P. Bradley, K. Mogg, and N. Millar, "Implicit Memory Bias in Clinical and Non-Clinical Depression," *Behaviour Research and Therapy* 34, no. 11 (Nov. 1996): 865–79; Philip C. Watkins, Karen Vache, Steven P. Verney, and Andrew Mathews, "Unconscious Mood-Congruent Memory Bias in Depression," *Journal of Abnormal Psychology* 105, no. 1 (1996): 34–41, doi:10.1037/0021-843X.105.1.34.

97 **viewed in a negative light:** C. G. Beevers, Tony T. Wells, Alissa J. Ellis, and Kathryn Fischer, "Identification of Emotionally Ambiguous Interpersonal Stimuli Among Dysphoric and Nondysphoric Individuals," *Cognitive Therapy and Research* 33, no. 3 (2009): 283–90, doi:10.1007/s10608-008-9198-6; R. C. Gur, R. J. Erwin, R. E. Gur, A. S. Zwil, C. Heimberg, and H. C. Kraemer, "Facial Emotion Discrimination: II. Behavioral Findings in Depression," *Psychiatry Research* 42, no. 3 (June 1992): 241–51; A. L. Bouhuys, E. Geerts, and M. C. Gordijn, "Depressed Patients' Perceptions of Facial Emotions in Depressed and Remitted States Are Associated with Relapse: A Longitudinal Study," *Journal of Nervous and Mental Disease* 187, no. 10 (Oct. 1999): 595–602.

98 **compulsory return to specific thoughts:** S. Nolen-Hoeksema, "The Role of Rumination in Depressive Disorders and Mixed Anxiety/Depressive Symptoms," *Journal of Abnormal Psychology* 109 (2000): 504–11; Susan Nolen-Hoeksema, Blair E. Wisco, and Sonja Lyubomirsky, "Rethinking Rumination," *Perspectives on Psychological Science: Journal of the Association for Psychological Science* 3, no. 5 (Sept. 2008): 400–24, doi:10.1111/j.1745-6924.2008.00088.x; S. Nolen-Hoeksema, J. Morrow, and B. L. Fredrickson, "Response Styles and the Duration of Episodes of Depressed Mood," *Journal of Abnormal Psychology* 102 (1993): 20–28.

98 **capture associated with depression:** Susan Nolen-Hoeksema, Blair E. Wisco, and Sonja Lyubomirsky, "Rethinking Rumination," *Perspectives on Psychological Science: A Journal of the Association for Psychological Science* 3, no. 5 (Sept. 2008): 400–24, doi:10.1111/j.1745-6924.2008.00088.x.

98 **Characteristic features:** J. T. Schelde, "Major Depression: Behavioral Markers

of Depression and Recovery," *Journal of Nervous and Mental Disease* 186, no. 3 (March 1998): 133–40; T. F. Heatherton and R. F. Baumeister, "Binge Eating as Escape from Self-Awareness," *Psychological Bulletin*, 110, no. 1 (1991): 86–108. Susan J. Paxton and Justine Diggens, "Avoidance Coping, Binge Eating, and Depression: An Examination of the Escape Theory of Binge Eating." *International Journal of Eating Disorders* 22, no. 1 (July 1997): 83–87, doi:10.1002/(SICI)1098-108X(199707)22:1<83::AID-EAT11>3.0.CO;2-J; C. B. Montano, "Recognition and Treatment of Depression in a Primary Care Setting," *Journal of Clinical Psychiatry* 55 (Dec. 1994): 18–34, discussion 35–37.

98   **feeds on itself:** G. Hasler, W. C. Drevets, H. K. Manji, and D. S. Charney, "Discovering Endophenotypes for Major Depression," *Neuropsychopharmacology* 29 (2004): 1765–81, doi:10.1038/sj.npp.1300506 1300506 [pii]; Seth G. Disner, C. G. Beevers, Emily A. P. Haigh, and Aaron T. Beck, "Neural Mechanisms of the Cognitive Model of Depression," *Nature Reviews Neuroscience* 12, no. 8 (July 6, 2011): 467–77, doi:10.1038/nrn3027.

98   **more likely to become depressed:** This increased sensitivity to stress is exacerbated by the fact that depression leads to the acquisition of additional stressors. For example, stress can result in decreased energy and an urge to isolate, which interferes with work responsibilities, possibly leading to financial troubles, thereby giving rise to more stress. D. J. Newport and C. B. Nemeroff, "Stress and Depression: From Vulnerability to Treatment." *European Neuropsychopharmacology* 10 (2000): 164–65, doi:http://dx.doi.org/10.1016/S0924-977X(00)80070-X; Kenneth S. Kendler, Laura M. Karkowski, and Carol A. Prescott, "Causal Relationship Between Stressful Life Events and the Onset of Major Depression," *American Journal of Psychiatry* 156, no. 6 (June 1999): 837–41, doi:10.1176/ajp.156.6.837; S. M. Monroe and A. D. Simons, "Diathesis-Stress Theories in the Context of Life Stress Research: Implications for the Depressive Disorders," *Psychological Bulletin* 110, no. 3 (Nov. 1991): 406–25.

GRANDEUR

Sources

Works Referenced

Hamilton, Ian. *Robert Lowell: A Biography*. New York: Vintage, 1982.

Hamilton, Saskia, ed. *The Letters of Robert Lowell*. New York: Farrar, Straus and Giroux, 2005.

Hart, Henry. *Robert Lowell and the Sublime*. Syracuse, NY: Syracuse University Press, 1986.

Lowell, Robert. *Life Studies*. New York: Farrar, Straus and Cudahy, 1959.

Mariani, Paul. *Lost Puritan: A Life of Robert Lowell*. New York: Norton, 1994.

Meyers, Jeffrey. *Manic Power: Robert Lowell and His Circle*. New York: Arbor House, 1987.

Ratliff, Ben. *Coltrane: The Story of a Sound*. New York: Farrar, Straus and Giroux, 2007.

Notes

99   **"pathological enthusiasm":** Hart, *Robert Lowell and the Sublime*, 22.

99   **"against devils and homosexuals":** Ibid., 107.

99   **"the middle of the highway":** Ian Hamilton, *Robert Lowell: A Biography*, 162.

99 "on the rampage": Ibid., 161.
99 "You must be Robert Lowell": Ibid.
99 "Bristling and manic": From "Commander Lowell," in Lowell, *Life Studies*.
100 "I have been eccentric . . . life is sober": Saskia Hamilton, *The Letters of Robert Lowell*, 3–4.
100 "Christ, Satan, Ahab": Hart, *Robert Lowell and the Sublime*, 118.
100 "prosaic, everyday things": Ratliff, *Coltrane*, 139.
101 "firmly engaged": Ian Hamilton, *Robert Lowell: A Biography*, 30.
101 "deeply and mysteriously . . . press our case": Ibid., 150.
101 "filled his bathtub": Saskia Hamilton, *The Letters of Robert Lowell*, 136–37.
101 "rather tremendous experiences": Ibid., 136.
102 "doctors are learning": Ibid., 137.
102 "another Yaddo": Ibid., 136.
102 "off his rocker": Ibid., 312.

ABANDONMENT

Sources

Works Referenced

Alexander, Paul. *Rough Magic: A Biography of Sylvia Plath*. Cambridge, MA: Da Capo, 1999.

Bundtzen, Lynda K. *The Other Ariel*. Amherst: University of Massachusetts Press, 2001.

Cooper, Brian. "Sylvia Plath and the Depression Continuum." *Journal of the Royal Society of Medicine* 96, no. 6 (June 2003): 296–301. http://www.ncbi.nlm.nih .gov/pmc/articles/PMC539515/.

Hayman, Ronald. *The Death and Life of Sylvia Plath*. New York: Birch Lane, 1991.

Malcolm, Janet. *The Silent Woman: Sylvia Plath and Ted Hughes*. New York: Knopf, 1994.

McCullough, Fran, and Ted Hughes, eds. *The Journals of Sylvia Plath*. New York: Dial, 1982.

Plath, Aurelia Schober, ed. *Letters Home by Sylvia Plath: Correspondence 1950–1963*. New York: Harper and Row, 1975.

Plath, Sylvia. "Edge" and "The Rabbit Catcher." In *The Collected Poems*. New York: Harper Perennial, 1992.

———. *The Unabridged Journals of Sylvia Plath*. New York: Anchor, 2000.

Rollyson, Carl. *American Isis: The Life and Art of Sylvia Plath*. New York: Picador, 2013.

Sigmund, Elizabeth, and Gail Crowther. *Sylvia Plath in Devon: A Year's Turning*. Havertown, PA: Fonthill Media, 2014.

Stevenson, Anne. *Bitter Fame: A Life of Sylvia Plath*. Boston: Houghton Mifflin, 1989.

Notes

102 "Frustrated? Yes": McCullough and Hughes, *Journals of Sylvia Plath*, 23–24.
103 her father died: Her father was Otto Emil Plath (1885–1940), who immigrated to America from Grabow in Germany in 1900. He worked as a professor of biology and German at Boston University.
103 relationship with her mother: Plath's mother was Aurelia Schober Plath (1906–1994). Plath's husband was Edward James (Ted) Hughes (1930–1998).

103 **"huge, sad hole"**: Aurelia Schober Plath, *Letters Home*, 288–89; Cooper, "Sylvia Plath and the Depression Continuum."

103 **"Images of [Hughes's] faithlessness"**: Sylvia Plath, *Unabridged Journals*, 447.

103 **David and Assia Wevill**: David Wevill (b. 1935) was an aspiring poet. His wife, Assia Wevill (1927–1969), worked for an advertising agency in London.

104 **market town**: The town was called North Tawton.

104 **second child**: Nicholas Farrar Hughes (1962–2009).

104 **"magically run by two electric currents"**: Sylvia Plath, *Unabridged Journals*, 395.

104 **sexual charge**: For an account of this visit, see Stevenson, *Bitter Fame*.

104 **"felt a still busyness"**: Sylvia Plath, "The Rabbit Catcher," in *The Collected Poems*, 194.

105 **"everything in life"**: Alexander, *Rough Magic*, 283.

105 **briefest of conversations**: The description of this event features in all biographies. Alexander, *Rough Magic*, has a detailed account and an important source: Aurelia Plath.

105 **"your whole heart"**: Alexander, *Rough Magic*, 284.

105 **public humiliation**: Sylvia Plath to Aurelia Plath, Oct. 21, 1962, unpublished letter.

105 **"babies & a novel"**: Sylvia Plath to Aurelia Plath, Oct. 21, 1962, unpublished letter.

105 **her fiercest competition**: Sylvia Plath to Aurelia Plath, Oct. 23, 1962, unpublished letter.

106 **a poem a day**: These poems would eventually form her collection *Ariel*, published posthumously in 1965.

106 **"domesticity had choked me"**: Aurelia Schober Plath, *Letters Home*, 463–65, 467–69.

106 **one editor after another**: An account of one such reaction was a meeting between poet and essayist Al Alvarez and Karl Miller, editor of *The New Statesman*, in Rollyson, *American Isis: The Life and Art of Sylvia Plath*, 211.

106 **described them as "alarming"**: Plath herself used this word to describe her new poems, in a letter written Nov. 16, 1962, to Peter Davison at *The Atlantic*.

106 **legal action for slander**: This detail is taken from Ted Hughes's journal fragments held in the British Library.

106 **"pathologically depressed"**: Alexander, *Rough Magic*, 325.

106 **"Barren" Assia**: The first of many times Sylvia Plath uses this phrase is in a letter to Aurelia Schober Plath, Tuesday, Oct. 16, 1962. Letter held at Lilly Library, Indiana University.

107 **"The woman is perfected"**: "Edge," in *The Collected Poems*, 224.

GOING MAD

Sources

Works Referenced

Bell, Anne Olivier, ed. *The Diary of Virginia Woolf: Vol. 5, 1936–1941*. San Diego, CA: Harcourt, 1984.

Bell, Quentin. *Virginia Woolf: A Biography*. New York: Harvest, 1974.

Bond, Alma Halbert. *Who Killed Virginia Woolf? A Psychobiography*. San Jose, CA: iUniverse, 1989.

Briggs, Julia. *Virginia Woolf: An Inner Life*. London: Penguin, 2005.

Caramagno, Thomas C. *The Flight of the Mind: Virginia Woolf's Art and Manic-Depressive Illness*. Berkeley: University of California Press, 1992.

Coates, Irene. *Who's Afraid of Leonard Woolf? A Case for the Sanity of Virginia Woolf*. New York: Soho, 1998.

Crinquand, Sylvie, ed. *Last Letters*. Newcastle, UK: Cambridge Scholar Printing, 2008.

Dally, Peter. *The Marriage of Heaven and Hell: Manic Depression and the Life of Virginia Woolf*. New York: St. Martin's, 1999.

Dalsimer, Katherine. *Virginia Woolf: Becoming a Writer*. New Haven, CT: Yale University Press, 2001.

DeSalvo, Louise. *Virginia Woolf: The Impact of Childhood Sexual Abuse on Her Life and Work*. Boston: Beacon, 1989.

Gordon, Lyndall. *Virginia Woolf: A Writer's Life*. New York: Norton, 1984.

Lee, Hermione. *Virginia Woolf*. New York: Knopf, 1997.

Marder, Herbert. *The Measure of Life: Virginia Woolf's Last Years*. Ithaca, NY: Cornell University Press, 2000.

Mulvihill, Maureen E. "Dancing on Hot Bricks: Virginia Woolf in 1941." *Rapportage* 12 (2009), 52–64.

Nicolson, Nigel, and Joanne Trautmann, eds. *The Letters of Virginia Woolf: Vol. 6, 1936–1941*. New York: Harcourt, 1980.

Poole, Roger. *The Unknown Virginia Woolf*. 3rd ed. Atlantic Highlands, NJ: Humanities Press, 1990.

Roe, Sue, and Susan Sellers, eds. *The Cambridge Companion to Virginia Woolf*. Cambridge, UK: Cambridge University Press, 2000.

Rosenthal, Michael. *Virginia Woolf*. New York: Columbia University Press, 1979.

Schulkind, Jeanne, ed. *Moments of Being*. 2nd ed. San Diego: Harcourt, 1976.

Spater, George. *A Marriage of True Minds: An Intimate Portrait of Leonard and Virginia Woolf*. New York: Harcourt, 1979.

Stape, John Henry, ed. *Virginia Woolf: Interviews and Recollections*. Iowa City: Iowa University Press, 1995.

Trombley, Stephen. *All That Summer She Was Mad: Virginia Woolf—Female Victim of Male Medicine*. New York: Continuum, 1981.

Woolf, Leonard. *The Journey Not the Arrival Matters: An Autobiography of the Years 1939 to 1969*. New York: Harcourt, 1969.

———, ed. *Virginia Woolf: A Writer's Diary*. San Diego: Harcourt, 1954.

Notes

107 **"From now on"**: Quentin Bell, *Virginia Woolf: A Biography*, 44.

108 **up one's drive for a visit**: Stape, *Virginia Woolf: Interviews and Recollections*, 110.

108 **"cocoon of quiescence"**: Leonard Woolf, *The Journey*, 79.

108 **"odd why sleeplessness"**: Caramagno, *The Flight of the Mind*, 41.

109 **"minute pinpricks"**: Dalsimer, *Virginia Woolf: Becoming a Writer*, 184.

109 **"Sank into a chair . . . pain in my life"**: Ibid., 186, 187.

109 **"live in my brain"**: Ibid., 192.

110 **"madness is terrific"**: Ibid., 189; Dally, *The Marriage of Heaven and Hell*, 108.

110 **"Never has a time been"**: Lee, *Virginia Woolf*, 180.

110 **specific plans for suicide**: Virginia rejected Leonard's suicide plans. On May 15, 1940, she wrote in her diary, "I don't want the garage to see the end of me. I've a

wish for 10 years more, and to write my book which as usual darts into my brain" [Anne Olivier Bell, *The Diary of Virginia Woolf*, 285].

111 **"no future":** Nicolson and Trautmann, *The Letters of Virginia Woolf: Vol. 6*, 475.

111 **"the lowest depths":** Spater, *A Marriage of True Minds*, 183; Nicolson and Trautmann, *The Letters of Virginia Woolf: Vol. 6*, xvi.

111 **biographical works were "failures":** Lee, *Virginia Woolf*, 61, 745.

111 **acknowledged her fears:** Quentin Bell, *Virginia Woolf: A Biography*, 225.

111 **"The interview was difficult":** Ibid.; Dally, *The Marriage of Heaven and Hell*, 182.

111 **"I am certain now":** Dally, *The Marriage of Heaven and Hell*, 182; Quentin Bell, *Virginia Woolf: A Biography*, 226; Crinquand, *Last Letters*, 54, 55.

ACCUMULATION OF BURDENS

Sources

Interview by the Author
Hemingway, Valerie (2011).

Archives
Ernest Hemingway Collection. Other Materials: Medical Records, 1960–61, undated. Dr. Howard P. Rome letter to Mary Hemingway, Nov. 1, 1961. National Archives, John F. Kennedy Presidential Library and Museum, Boston.

Works Referenced

Castillo-Puche, Jose Luis. *Hemingway in Spain*. Translated by Helen R. Lane. New York: Doubleday, 1974.

Hardy, Richard E., and John G. Cull. *Hemingway: A Psychological Portrait*. Sherman Oaks, CA: Banner, 1977.

Hays, Peter L. *The Critical Reception of Hemingway's "The Sun Also Rises."* Rochester, NY: Camden House, 2011.

Hemingway, Ernest. *To Have and Have Not*. New York: Scribner, 2003.

Hemingway, John. *Strange Tribe: A Family Memoir*. Guilford, CT: Lyons Press, 2007.

Hemingway, Mary Welsh. *How It Was*. New York: Knopf, 1976.

Hemingway, Valerie. *Running with the Bulls: My Years with the Hemingways*. New York: Ballantine, 2004.

Kert, Bernice. *The Hemingway Women*. New York: Norton, 1983.

Moorehead, Caroline. *Gellhorn: A Twentieth-Century Life*. New York: Holt, 2003.

———, ed. *Selected Letters of Martha Gellhorn*. New York: Holt, 2006.

Nagel, James, ed. *Ernest Hemingway: The Oak Park Legacy*. Tuscaloosa: University of Alabama Press, 1998.

Reynolds, Michael. *Hemingway in the 1930s*. New York: Norton, 1998.

———. *Hemingway: The Paris Years*. New York: Norton, 1999.

———. *The Young Hemingway*. New York: Norton, 1998.

Spanier, Sandra, and Robert W. Trogdon, eds. *The Letters of Ernest Hemingway: Vol. 1, 1907–1922*. Cambridge: Cambridge University Press, 2011.

Yardley, Jonathan. "A Writer's Companion." *Washington Post*, Nov. 11, 2004. http://www.washingtonpost.com/wp-dyn/articles/A41351-2004Nov10.html.

Notes

113 **"When I met Hemingway . . . period":** Valerie Hemingway, in discussion with the author, Nov. 2011. All subsequent quotations refer to this interview.

A UNIFIED THEORY

Sources

Interview by the Author
Hoffman, Ralph (2014).
Works Referenced
See scientific articles cited in the notes.

Notes

117 **sleep deprivation makes us particularly prone:** See E. T. Kahn-Greene, D. B. Killgore, G. Kamimori, T. J. Balkin, and W. D. S. Killgore, "The Effects of Sleep Deprivation on Symptoms of Psychopathology in Healthy Adults," *Sleep Medicine* 8 (2007): 215–21.

117 **schizophrenic:** Empirical data suggest that many symptoms of schizophrenia (SCZ) arise due to the aberrant assignment of salience to stimuli in the environment that the healthy brain would perceive as irrelevant. For example, in one study, Roiser et al. (2009) asked participants with SCZ and controls to learn associations between the dimensions of a stimulus and the probability of a reward in order to earn money in a reaction-time game. In the task, the stimulus varied along two visual dimensions, color and form, but only one of these dimensions was associated with the probability of reward. The authors found that patients with SCZ exhibited less "adaptive salience" (were less likely to learn the association between the reward-associated dimension and the probability of reward) than controls. Patients with delusions were more likely to attribute aberrant salience to the dimension not associated with rewards, and this pattern of aberrant salience attribution was also correlated with negative symptoms. In another study, Holt et al. (2006) presented unpleasant, pleasant, and neutral words to patients with schizophrenia (one group with delusions and one without) and to a healthy control group. The authors found that the SCZ patients with delusions were more likely to classify words as unpleasant, and took significantly longer to correctly classify neutral words. The authors interpreted these findings as indicative of misattribution of salience to neutral stimuli, specifically among delusional patients with SCZ.

Several theories exist to explain this putative relationship between aberrant salience assignment and symptoms of SCZ. Fletcher and Frith (2009) propose a Bayesian model of attention and salience attribution in SCZ. In this model, disruptions of prediction-error signaling related to dopamine neurotransmission result in the abnormal signaling of a violated expectation. Violation of expectation then signals a greater allocation of attention and attribution of salience to the occurrence. The individual may perceive the occurrence to signal a change in the environment, and as such, the need to update his or her existing beliefs. In this way, noisy prediction-error signaling may lead to patients' attributions of salience and meaning to incidental stimuli, and thus adoption of delusional beliefs based on seemingly unimportant or uncertain information. In two theory papers, Kapur and colleagues (Kapur et al., 2005; Kapur, 2003) propose a similar model whereby a hyperdopaminergic state leads to the aberrant assignment of salience to elements of one's experience. The authors further suggest that hallucinations are a reflection of this experience of aberrantly salient internal representations, and delusions represent the individual's cognitive effort to make sense of these seemingly salient experiences. Building on this framework, Kapur and colleagues propose that the mechanism of action of antipsychotic medication is to "dampen the salience" of these experiences, thereby ameliorating symptoms.

Electrophysiological studies have begun to shed light on the neural basis of abnormal salience assignment in SCZ, particularly in the context of early perceptual "gating" contributors to the assignment of salience. One consistent finding of such studies has been a blunted "mismatch negativity" (MMN), an event-related potential (ERP) that signals the detection of change in the environment, which has been considered a prediction-error signal (see Todd, Michie, Schall, Ward, and Catts, 2012, for review). This deficit has been interpreted as reflective of impairment in prediction-error generation, or the attribution of salience to novel stimuli. Electrophysiological studies have also focused on abnormalities in early perceptual and sensory "gating" that may give way to biased allocation of attention in SCZ. For example, in one study, patients with SCZ were found to exhibit less differential electrophysiological responding to salient versus nonsalient stimuli than healthy controls in several stages of stimulus processing, which the authors interpreted as a failure to selectively "gate in" salient information (Brenner et al., 2009). In another ERP study, patients with schizophrenia and healthy controls were asked to listen to a variety of tones and to press a button in response to a specific tone when it occurred (Michie, Fox, Ward, Catts, and McConaghy, 1990). The authors found reduced reactivity to salient tones at multiple points, such as a blunted P300 component typically associated with decision making, in response to target tones. These results were interpreted as reflective of multiple attentional deficits associated with failure to plan and execute a selective listening strategy.

Neuroimaging studies have found abnormalities in the structure and function of regions associated with attention and salience attribution in patients with SCZ, and even found specific associations between these neuroimaging measures and distorted perceptions of reality. For example, Palaniyappan et al. (2011) reported volumetric reductions in the anterior cingulate and anterior insula in patients with SCZ, and found that these volumetric reductions in the right hemisphere were significantly correlated with the severity of patients' reality distortion. This finding was interpreted to suggest that a deficit of gray matter in regions implicated in attribution is an important mechanism associated with hallucinations and delusions in SCZ. Another study used a monetary incentive delay task to explore reward processing in patients with SCZ (Walter, Kammerer, Frasch, Spitzer, and Abler, 2009). Results revealed that while healthy participants showed the highest level of activity in the right ventrolateral prefrontal cortex in response to the most salient outcomes (omission or receipt of reward versus a neutral outcome), SCZ participants did not show this pattern of differential, salience-based activation. Another study later replicated this finding, and found a pattern of linear increase in anterior cingulate cortex activity with increasing reward in healthy participants, which did not exist in the SCZ group (Walter et al., 2010). Another study examined functional connectivity between the bilateral insula and anterior cingulate cortex, regions comprising a "salience network" thought to be important for recruiting relevant brain regions for the processing of sensory information (White et al., 2010). In that study, participants with SCZ and a healthy control group were presented with tactile stimulation of the right index finger during fMRI scanning. Results revealed reduced connectivity within this salience network among the SCZ, compared to the control group, providing evidence for one potential mechanistic contributor to aberrant assignment of salience in SCZ. P. C. Fletcher and C. D. Frith, "Perceiving Is Believing: A Bayesian Approach to Explaining the Positive Symptoms of Schizophrenia," *Nature Reviews Neuroscience* 10, no. 1 (2009): 48–58; V. B. Gradin, G. Waiter, A. O'Connor, L. Romaniuk, C. Stickle, K. Matthews, et al., "Salience Network–Midbrain Disconnectivity and Blunted Reward Signals in Schizophrenia," *Psychiatry Research: Neuroimaging* 211, no. 2 (2013): 104–11; B. Hahn, B. M. Robinson, S. T. Kaiser, A. N. Harvey, V. M. Beck, C. J. Leonard, et al., "Failure of Schizophrenia Patients to Overcome Salient Distractors During Working Memory Encoding," *Biological Psychiatry* 68, no. 7, (2010): 603–9; A. Heinz and F. Schlagenhauf, "Dopaminergic Dysfunction in Schizophrenia: Salience Attribution Revisited," *Schizophrenia Bulletin* 36, no. 3 (2010): 472–85; D. J. Holt, D. Titone, L. S. Long, D. C. Goff, C. Cather, S. L. Rauch, et al., "The Misattribution of Salience in Delusional Patients with Schizophrenia," *Schizophrenia Research* 83, no. 2–3 (2006): 247–56; J. Jensen and S. Kapur, "Salience and Psychosis: Moving from Theory to Practise," *Psychological Medicine* 39, no. 2 (2009): 197–98; S. Kapur, "Psychosis as a State of Aberrant Salience: A Framework Linking Biology, Phenomenology, and Pharmacology in Schizophrenia," *American Journal of Psychiatry* 160, no. 1 (2003): 13–23; S. Kapur, R. Mizrahi, and M. Li, "From Dopamine to Salience to Psychosis: Linking Biology, Pharmacology and Phenomenology of Psychosis," *Schizophrenia Research* 79, no. 1 (2005): 59–68; L. Palaniyappan and P. F. Liddle, "Does the Salience Network Play a Cardinal Role in Psychosis? An Emerging Hypothesis of Insular Dysfunction," *Journal of Psychiatry and Neuroscience* 37, no. 1 (2012): 17–27; L. Palaniyappan, P. Mallikarjun, V. Joseph, T. P. White, and P. F. Liddle, "Reality Distortion Is Related to the Structure of the Salience Network in Schizophrenia," *Psychological Medicine* 41, no. 8 (2011): 1701–8; J. P. Roiser, K. E. Stephan, H. E. M. den Ouden, T. R. E. Barnes, K. J. Friston, and E. M. Joyce, "Do Patients with Schizophrenia Exhibit Aberrant Salience?," *Psychological Medicine* 39, no. 2 (2009): 199–209; H. Walter, S. Heckers, J. Kassubek, S. Erk, K. Frasch, and B. Abler, "Further Evidence for Aberrant Prefrontal Salience Coding in Schizophrenia," *Frontiers in Behavioral Neuroscience* 3 (2010): 62; T. P. White, V. Joseph, S. T. Francis, and P. F. Liddle, "Aberrant Salience Network (Bilateral Insula and Anterior Cingulate Cortex) Connectivity During Information Processing in Schizophrenia," *Schizophrenia Research* 123, no. 2–3 (2010): 105–15.

117 **it is salient:** Psychosis has been conceptualized as a state of aberrant attribution of salience, driven largely by errant impartation of significance to various stimuli (or patterns, as described in Dr. Hoffman's traffic light example) observed in daily life. See S. Kapur, "Psychosis as a State of Aberrant Salience: A Framework Linking Biology, Phenomenology, and Pharmacology in Schizophrenia," *American Journal of Psychiatry* 160, no. 1 (2003): 13–23.

117 **"an addiction to a certain affective state":** Many people with bipolar disorder describe manic states as quite enjoyable. In *An Unquiet Mind: A Memoir of Moods and Madness*, Kay Redfield Jamison, a professor of psychiatry at Johns

Hopkins University, describes her experience of mania: "When you're high it's tremendous. The ideas and feelings are fast and frequent like shooting stars, and you follow them until you find better and brighter ones. Shyness goes, the right words and gestures are suddenly there, the power to captivate others a felt certainty. There are interests found in uninteresting people. Sensuality is pervasive and the desire to seduce and be seduced irresistible. Feelings of ease, intensity, power, well-being, financial omnipotence, and euphoria pervade one's marrow." Perhaps unsurprisingly, some individuals with bipolar disorder resist treatment, desiring to experience such euphoric states in the future. Gruber and Persons (2010) provide recommendations for clinicians seeking to help their patients with bipolar disorder overcome treatment refusal. J. Gruber and J. B. Persons, "Unquiet Treatment: Handling Treatment Refusal in Bipolar Disorder," *Journal of Cognitive Psychotherapy* 24 (2010): 16–25; K. R. Jamison, *An Unquiet Mind* (New York: Vintage Books, 1996).

117   **"sense of their own specialness":** After a particular success or the attainment of a goal, individuals with bipolar disorder may be more likely to generalize from these outcomes, believing that they are reflections of their positive traits. For example, in one recent study, researchers administered the Hypomanic Personality Scale (Eckblad and Chapman, 1986) and a measure of positive generalization, or the tendency to generalize from a single positive outcome to one's broader sense of self (Eisner, Johnson, and Carver, 2008). Results indicated that those at higher risk for mania were significantly more likely to make positive generalizations, and suggest that this type of responding to success was uniquely associated with mania risk. M. Eckblad and L. J. Chapman, "Development and Validation of a Scale for Hypomanic Personality," *Journal of Abnormal Psychology* 95, no. 3 (1986): 214–22; L. R. Eisner, S. L. Johnson, and C. S. Carver, "Cognitive Responses to Failure and Success Relate Uniquely to Bipolar Depression Versus Mania," *Journal of Abnormal Psychology* 117 (2008): 154.

118   **some rare gift:** Grandiose delusions are the most common type of psychotic symptom during mania. See E. Dunayevich and P. E. Keck, "Prevalence and Description of Psychotic Features in Bipolar Mania," *Current Psychiatry Reports* 2 (2000): 286–90.

118   **antipsychotics, target the neurological circuitry of salience:** See S. Kapur, "How Antipsychotics Become Anti-'Psychotic'—From Dopamine to Salience to Psychosis," *Trends in Pharmacological Sciences* 25 (2004): 402–6.

118   **turn to drugs and alcohol or food:** J. D. Swendsen and K. R. Merikangas, "The Comorbidity of Depression and Substance Use Disorders," *Clinical Psychology Review* 20, no. 2 (2000): 173–89, doi:10.1016/S0272-7358(99)00026-4; T. F. Heatherton and R. F. Baumeister, "Binge Eating as Escape from Self-Awareness," *Psychological Bulletin* 110, no. 1 (1991): 86–108.

118   **learned helplessness:** M. E. P. Seligman, *Helplessness: On Depression, Development, and Death* (San Francisco: W. H. Freeman, 1975); M. E. P. Seligman, "Learned Helplessness," *Annual Review of Medicine* 23 (1972): 407–12.

CHAPTER 4: WHEN CAPTURE TURNS ON THE SELF

DAVID FOSTER WALLACE

Sources

Interviews by the Author
See the listing in chapter 1.

Works Referenced

Esposito, Scott. "Who Was David Foster Wallace?—Wallace's Masterpiece." *The Quarterly Conversation*, June 6, 2011. http://quarterlyconversation.com/david-foster-wallace-infinite-jest.

Granada House. "An Ex-Resident's Story." http://www.granadahouse.org/people/letters_from_our_alum.html.

Kakutani, Michiko. Review of *The Broom of the System*, by David Foster Wallace. *New York Times*, Dec. 27, 1986.

Karr, Mary. *The Liars' Club: A Memoir*. New York: Penguin, 1995.

Maslin, Janet. "In Layered Fiction and Wry Notes to Mom, a Cosmic Genius Distilled." *New York Times*, Dec. 17, 2014. http://www.nytimes.com/2014/12/18/books/the-david-foster-wallace-reader-a-compilation.html?_r=0.

Miller, Laura. "David Foster Wallace." *Salon*, March 9, 1996. http://www.salon.com/1996/03/09/wallace_5/.

Sheppard, R. Z. "Mad Maximalism." *Time* 147, no. 8 (Feb. 19, 1996).

Wallace, David Foster. *The David Foster Wallace Reader*. New York: Little, Brown, 2014.

———. "Good People." *The New Yorker*, Feb. 5, 2007. http://www.newyorker.com/magazine/2007/02/05/good-people.

Yablonowitz. "Struggling to Make Sense." *Yablonowitz* (blog discontinued), *Salon*, Sept. 14, 2008.

Notes

121 he'd lost his tribe: Amy Wallace.
122 "ashen gray": James and Sally Wallace.
122 "personal hell": Charlie McLagan interview.
122 "I love this place": Mark Costello.
123 "the attention of professors": Fred Brooke.
124 straight A-pluses: D. T. Max, *Every Love Story Is a Ghost Story: A Life of David Foster Wallace* (New York: Viking, 2012), 28, 39.
124 "no wealth": Fred Brooke.
124 "banzai workload": Stacey Schmeidel, "Brief Interview with a Five Draft Man," *Amherst* magazine, Spring 1999. Republished at https://www.amherst.edu/aboutamherst/magazine/extra/node/66410.
125 "got a cavity": Amy Wallace.
126 "unquiet mind": Daniel Javit interview.
126 "hypervigilance": Dave Colmar; Mark Costello.
127 "But it's all . . . a big one": Larry McCaffery, "A Conversation with David Foster Wallace," *Review of Contemporary Fiction* 13, no. 2 (Summer 1993).
127 "unsolvable in some ways": Daniel Javit.
128 "terrorized himself": Ibid.
129 *Sabrina*: Corey Washington interview.
130 "unpleasant neuroses": Ibid.

131 **"the click in literature"**: McCaffery, "A Conversation with David Foster Wallace."

131 **"The Bad Thing is you"**: Wallace, *The David Foster Wallace Reader*, 12.

132 **Lenore . . . "what can be said"**: Wallace, *The Broom of the System* (New York: Penguin, 2004), 119.

132 **"looking at solipsism"**: McCaffery, "A Conversation with David Foster Wallace."

132 **Wallace's great regard for Wittgenstein**: David Wallace and Ludwig Wittgenstein were not merely kindred philosophic spirits. They also faced similar inner struggles as they endeavored to meet their own exacting demands.

Like Wallace, Wittgenstein explored the freedoms, and confines, of language, an intellectual enterprise that ran parallel to a keenly felt isolation. Both men were captured by a perfectionism that alternately urged them toward fulfilling lofty ambitions and crippled them with doubt. Wittgenstein described his unhappiness in terms of a circling search around an uncertain anchor. "I am wandering about with great restlessness, but around what point of equilibrium I do not know," the philosopher wrote in his diary. "I am at present swinging without knowing my centre of gravity."

Wittgenstein anxiously sought to pivot into a new position from which he could better view the emotional and philosophical questions that plagued him. The goal gradually became less a search for answers than a recasting of the questions themselves, a change not in his manner of living but in his perspective on that life. In the *Philosophical Investigations*, the magnum opus of his later career, he proclaims that "the axis of our reference must be rotated, but about the fixed point of our real need." As the philosopher Richard Eldridge has put it, the philosopher was calling for a "turning around of the soul."

Uncompromising and charismatic in equal measure, Wittgenstein taught a slew of young followers to question the discipline of philosophy as they knew it, and with unpitying rigor he applied the same skepticism to his own work and even his life. Over the course of that life, he would reject, in turn, his grand Viennese upbringing, the religion and possibly the sexual orientation he was born into, his intellectual mentors, his early philosophical work, his academic life in Cambridge, and eventually the legitimacy and goals of philosophy itself. All this in pursuit of an emotional and intellectual freedom that he never quite attained—an effort, as he famously put it, to "let the fly out of the fly bottle."

The philosopher's later work represents a turn away from the analytic tradition and toward the mystical—and, at the same time, a sustained response to his private suffering. Indeed, it is no coincidence that Wittgenstein's most abiding philosophical instinct was to acknowledge the limits of philosophical language. "A picture held us captive," he wrote. "And we could not get outside it, for it lay in our language and language seemed to repeat it to us inexorably."

Ludwig Wittgenstein was born in Austria in 1889, the youngest son of one of Europe's wealthiest families. His father, Karl Wittgenstein, was a self-made steel magnate, and Ludwig's childhood home in Vienna was fantastically opulent. It was also exceedingly unhappy. Karl was a severe and inflexible patriarch who held his children to stringent standards and expected the boys to follow in his professional footsteps.

The children were kept out of school and for most of their lives educated at home—a home that was at the center of a rich Viennese cultural scene. Schoenberg and Brahms visited the house for concerts; a Rodin sculpture stood in the grand entrance hall. Despite the frequent musical interludes, Karl's severity cast a chill over the household, one that his subservient wife did little to warm. It also seems likely that depression ran in the family. Certainly two and probably three of Ludwig's four brothers eventually killed themselves. The other survivor, Paul, Ludwig's closest brother in both age and affection, overcame similar mental afflictions, and the loss of one arm in World War I, to become a famous concert pianist. The boys settled into their relative temperaments early: practicing alone at the piano, Paul once leaped up to yell at Ludwig in another room, "I cannot play when you are in the house, as I feel your skepticism seeping towards me from under the door!"

From childhood, Wittgenstein suffered from what his sister once described as "almost pathological distress in any surroundings that were uncongenial to him." Fania Pascal, a friend in Cambridge, remembered him as "a person who in some manner was always given to despair." As Wittgenstein himself put it, "Often I feel that there is something in me like a lump which, were it to melt, would let me cry or I would then find the right words (or perhaps even a melody). But this something (is it the heart?) in my case feels like leather and cannot melt. Or is it just that I am too cowardly to let the temperature rise sufficiently?"

Certain that he was "soiling everything with [his] vanity," the young philosopher became obsessed with his precariousness: "You hang trembling with all you have above the abyss. It is horrible that such a thing can be." Wittgenstein couldn't escape the gloom of his own shadow, and his attempts to cure himself of anxiety were largely in vain. "It is difficult as it were to keep our heads up," he wrote. "We feel as if we had to repair a torn spider web with our fingers."

Yet no matter how vigorously he sawed at his flaws, paring away vast swaths of his life and its indulgences, a nagging doubt haunted the young Ludwig. In a letter to his close friend Paul Engelmann, Wittgenstein explained, "I simply had to chop off a limb or two, those that remain are the healthier for it." To another friend he wrote, "What I recognize is actually: how terribly unhappy a human being can become. Again and again I find myself dwelling on base thoughts, yes, on the basest thoughts."

At one point, Bertrand Russell, Wittgenstein's mentor at Cambridge, began to fear for his protégé's life:

> Wittgenstein is on the verge of a nervous breakdown, not far removed from suicide, feeling himself a miserable creature, full of sin. Whatever he says he apologizes for having said. He has fits of dizziness and can't work— the Dr. says it is all nerves. He wanted to be treated morally, but I persisted in treating him physically—I told him to ride, to have biscuits by his bedside to eat when he lies awake—to have better meals and so on. I suppose genius always goes with excitable nerves—it is a very uncomfortable possession. He makes me terribly anxious and I hate seeing his misery—it is so real, and I know it all so well.

When World War I broke out, Wittgenstein tried to escape his endless self-doubt by volunteering as a gunner for the Austro-Hungarian army. If he had hoped that the company of his fellow soldiers would prove more agreeable than the society of philosophers, he was soon proven mistaken. While he was decorated for his bravery in combat, Wittgenstein's coded diaries from that period are filled with suicidal thoughts and self-loathing, and biting disdain for his comrades in arms.

At the same time, the diaries reveal a dramatic turn in his intellectual life. During a furlough in Poland, Wittgenstein picked up a copy of Tolstoy's *The Gospels in Brief*. Tolstoy's approach to the Gospels—his rejection of religious institutions and his emphasis on the spiritual freedom of the individual—inspired in the philosopher a near-conversion experience. "I am on the path to great discovery," he wrote. "But will I reach it?" Tolstoy's words began to reverberate in the echo chamber of his mind: "Man is powerless in the flesh but free through the spirit." Suddenly uplifted, Wittgenstein pleaded for salvation from a nameless force: "May the spirit be in me!" He began to carry *The Gospels in Brief* everywhere he went, "like a talisman," convinced that the book bestowed on him an "indescribable" blessing.

The philosopher's ultimate goal became "a state of indifference to the difficulties of *external* life." When he felt himself nearing such a state, he noted down further reflections on logic. But such flashes of clarity were outnumbered by spells of despair: "This is what 'sin' is, the unreasoning life, a false view of life. From time to time I become an animal. Then I can think of nothing but eating, drinking, and sleeping. . . . And then I suffer like an animal, too, without the possibility of internal salvation. I am then at the mercy of my appetites and aversions."

In June 1916, Wittgenstein's jottings became more explicitly spiritual in tenor: "What do I know about God and the purpose of life? . . . I cannot bend the happenings of the world to my will: I am completely powerless." Belief in God promised to clarify his long-standing "questions about the meaning of life": "The world is given me, i.e. my will enters into the world completely from outside as into something that is already there." Religious faith promised to free us from our own meekness, the inevitable disappointments of the will when it chafed against circumstance.

Wittgenstein's God filled the void that had long tormented the philosopher. "A man who is happy must have no fear," he wrote. "Not even in the face of death." The coded entries from his diary are even more revealing:

> *28 Mar 16: . . . and would have to take my life. I suffered the agonies of Hell. And yet the image of life was so attractive to me that I wanted to live again. I will not poison myself until I really want to poison myself.*
>
> *6 Apr 16: Life is an. . .*
>
> *7 Apr 16: ordeal, from which one gains relief only occasionally in order to remain susceptible to new torments. A horrible assortment of torments. An exhausting march, coughing through the night, the company of drunkards, of base and stupid people. Do good, and be glad of your virtue. Am ill and have a bad life. God help me. I am a poor, unhappy man.*

Life on the front unfolded in an endless series of petty degradations. As the passage of time slowed, then ground to halt, Wittgenstein felt himself challenged not only by the deprivations of war but also by his failure to find meaning in pain.

Wittgenstein's wartime experiences may not have clarified his years of suffering, but they did leave a deep imprint on his evolving philosophy. Far from the petty squabbles of the academy and from the stifling formality of his child-hood home, he reimagined his life as a Romantic quest, suffused with spiritual meaning:

> 5 May 16: At the observation post I am like the prince in the enchanted castle. Now during the day everything is quiet, but in the night it will certainly be frightful! Will I endure it???? Tonight will show. God stand by me!!
> 6 May 16: In constant danger of death. Through the grace of God the night went well. From time to time I become afraid. That is the fault of a false view of life! Understand the men! Whenever you want to hate them, try to understand them instead. Live in inner peace! But how does one arrive at inner peace? ONLY by living so as to please God! Only in this way is it possible to bear life.

Wittgenstein's conversion experience did not render meaningless his years of philosophical toil. Rather, the stakes of philosophy became indistinguishable from the moral imperatives of religion: "Only from the consciousness of the *uniqueness of my life* arises religion—science and art."

What seems a dramatic, perhaps even inexplicable, transformation in Witt-genstein's thought has important precedents in Western philosophy. Schopen-hauer, whom Wittgenstein read as a boy, described vividly the experience of complete immersion in the world, when the perceiving subject becomes indis-tinguishable from the object of its perception. For the poets and philosophers of the Romantic era, this kind of immersion was the sine qua non of aesthetic ex-perience: the sublime. Schopenhauer's "supporter of the *world* and of all objec-tive existence" became in Wittgenstein's characteristically gnomic prose a theological revelation: "I am my world."

For such an experience of complete immersion to occur, Wittgenstein wrote, the object of perception "needed to take on for the individual an importance." To illustrate his point, he described the experience of looking at an ordinary stove. As a mere "thing among things," the stove is as insignificant as any other object. When it becomes a world unto itself, however, it radiates with untold meaning. If I allow the stove to become, albeit temporarily, my entire world, and everything else is bleached "colorless by contrast with it," then I will become so absorbed by the object of my attention that I no longer recognize myself as an independent being. The usual boundary separating subject and object will begin to blur. Indeed, "it is equally possible to take the bare present image as the worthless momentary picture in the whole temporal world" and to perceive it as "the true world among shadows." It was this delicate truth that Wittgenstein spent the rest of his career trying to illuminate.

This was not simply an academic exercise. Wittgenstein argued that when the whole of consciousness is absorbed by a single object, we become aware of our utter contingency and even precariousness. What is so need not be: it was this feeling that Wittgenstein sought to convey, and around which his increasingly idiosyncratic faith revolved. Wittgenstein soon realized that aesthetic contem-plation, like spiritual ecstasy, could evoke in the beholder this very sense of his own contingency. Indeed, "the work of art is the object seen sub specie aeterni-tatis," just as "the good life is the world seen sub specie aeternitatis." "This," Wittgenstein wrote, "is the connexion between art and ethics."

Despite his eloquent public disquisitions on the limits of philosophy, in private Wittgenstein remained racked by guilt throughout his life. His diary records a conscience that "plagues me & won't let me work." Reading Kierkegaard on suffering and faith only heightened his anguish.

> I don't want to suffer; that is what unsettles me. I don't want to let go of any conveniences or of any pleasure. (I would not fast, for example, or even restrain myself in my eating.) But I also don't want to oppose anyone & involve myself in discord. At least not as long as the matter is not placed right before my eyes. But even then I fear that I might dodge it. In addition an ineradicable immodesty dwells in me. In all my pitifulness I still always want to compare myself to the most significant persons. It is as if I could find solace only in the recognition of my pitifulness.

Wittgenstein's coded diaries bespeak an intense longing for salvation. Yet a stubborn "immodesty" seemed to stand in the way of this total selflessness, or complete immersion in the fact of the divine.

> Few things are as difficult for me as modesty. Now I am noticing this again as I read in Kierkegaard. Nothing is as difficult for me as to feel inferior; even though it is only a matter of seeing reality as it is. Would I be able to sacrifice my writing for God?

This pattern of self-recrimination became increasingly obsessive and self-referential. Wittgenstein admonished himself for often seeking to gain favor with his audience through "a somewhat comic turn" in his lectures, "to entertain them so that they willingly hear me out." In conversation, he depended on the opinion of others: "A good word from someone or a friendly smile has a lasting effect on me, pleasantly encouraging & assuring, & an unpleasant, that is, unfriendly word has an equally long effect, depressing." Everything he did, "these entries included," was "tinted by vanity." "The best I can do," he wrote ruefully, is "to separate, to isolate the vanity & do what's right in spite of it, even though it is always watching. I cannot chase it away." But the desire to rid himself of vanity became in itself a pathology, a mark of sin: "When I say I would like to discard vanity, it is questionable whether my wanting this isn't yet again only a sort of vanity."

As his mind darkened, the philosopher felt himself edging toward madness: "A storm is blowing and I cannot collect my thoughts." His philosophical work was "lacking in seriousness & love of truth." His lectures were mere stumblings in the dark: "I have . . . cheated often by pretending to already understand something while I was still hoping that it would become clear to me." Yet even as Wittgenstein found fault in all his endeavors, he remained somehow cognizant that his endless accusations had little, if any, basis in fact: "But if I now think of my sins & it is only a hypothesis that I have performed these acts, why do I regret them as if any doubt about them was impossible?" Unassuaged, Wittgenstein resigned himself to a life of endless gloom, hoping that misery would "somehow cleanse" him of fault but unable to convince himself that his suffering would ever cease.

> Let me confess this: After a difficult day for me I kneeled during dinner today & prayed & suddenly said, kneeling & looking up above: "There is no one here." That made me feel at ease as if I had been enlightened in an important matter.

Trying to quiet his mind, Wittgenstein turned to Kierkegaard's Johannes as a spiritual model. "What must I do so that It becomes bearable as it is?" Kierkegaard had defined "purity of heart" as the capacity "to will [only] one thing," yet this one thing proved elusive for Wittgenstein. To convey the single-minded dedication required of the truly pure, Kierkegaard describes purposeful action in terms of locomotion: "He becomes solitary, and then he undertakes the movement." This slow-motion, perfectly distilled model of intentionality demanded that one "concentrate the whole substance of his life and the meaning of actuality into one single desire." Only then is true clarity of thought possible: "In the next place, the knight will have the power to concentrate the conclusion of all his thinking into one act of consciousness."

Having committed himself to this state of perfect self-awareness, Wittgenstein soon found himself ensnared in an endless dialogue with his rational faculties, beset by the nagging doubt of the metaphysician: "Any fight in this is only a fight against myself," he declared. "The harder I beat, the harder I get beaten. But it is my heart that would have to submit, not simply my hand." Placing himself just outside the circle of salvation, he bemoaned his lot: "Were I a believer, that is, would I intrepidly do what my inner voice asks me to do, this suffering would be over."

In the 1930s, Wittgenstein began to turn away from his earlier understanding of philosophy as digging down to the real; no longer was the structure of language dictated by reality. Before the war, Wittgenstein had believed in a logical relationship between philosophical propositions and the world. He argued that propositions—that is, thoughts expressed in logical language—can give us an accurate picture of the world, much as an architect's model can accurately represent a house. In other words, the only possible relationship between language and the objects it describes is referential. If language is to be both meaningful and accurate, words such as "dog" and "flower" must function as mental pictures, or representations of dogs and flowers.

Gradually Wittgenstein's interest shifted to the images that hold us captive, the forces that tempt us, the way things appear to us. Few of the scholars who have tried to understand Wittgenstein's later philosophy have realized that he was making sense of his own suffering through his work. In a lecture on moral philosophy, the eminent philosopher struggled to convey the impossibility of his task. The logician had abandoned the certainties of logic: "Our words will only express facts, as a teacup will only hold a teacup full of water, [even] if I were to pour out a gallon over it."

Held captive by an imperfect language, we can see the world only one way. The solution, then, was simple: to accept things as they are. In other words, once we have "exhausted the justifications" for suffering, we reach "bedrock," and our "spade is turned." In such a moment the challenge is not merely dwelling upon it. "The difficulty here is: to stop." The "solution to the problem of life in space and time," then, lies "outside space and time." In the absence of any scientific solutions, the solution could lie only in the vanishing of the question, "when [the problem] no longer concerns us" and we go on living.

One particular thought, however, continued to torment him. "I have been through it several times before," he once wrote to a wartime friend. "It is the state of *not being able to get over a particular fact*." On Nov. 19, 1936, Wittgenstein wrote in his diary, "About 12 days ago I wrote to Hansel a confession of my lie concerning my ancestry. Since that time I have been thinking again and again how I can & should make a full confession to everyone I know. I hope & fear! Today I feel a bit sick, chilled. I thought: 'Does God want to put an end to me before I could do the difficult thing?' May it turn out well!"

> *Weary & disinclined to work or really incapable . . . After having now made that one confession it is as if I couldn't support the whole edifice of lies anymore, as if it had to collapse entirely. If only it had entirely collapsed already!*

The next week, on November 24, he wrote, "Today I mailed the letter with my confession to Mining. Even though the confession is candid, I am still lacking the seriousness that is appropriate to the situation."

> *My Dear Friend Hansel:*
> *I lied to you & several others back then during the Italian internment when I said that I was descended one quarter from Jews and three quarters from Arians, even though it is just the other way round. This cowardly lie has burdened me for a long time & like many other lies I also told this one to others. Until today I have not found the strength to confess it.*

Late in life, Wittgenstein came to see all mental suffering as analogous to "the suffering of an ascetic who stood raising a heavy ball, amid groans." We set him free simply by telling him to "drop it."

Why, then, had he not dropped it earlier? He had accommodated himself to a faulty system, a way of seeing the world that was intended to provide solace but in fact caused pain. Indeed, the solution to all philosophical and existential problems could be compared to "a gift in a fairy tale: In the magic castle it appears enchanted, and if you look at it outside in daylight it is nothing but an ordinary bit of iron."

["centre of gravity": Ludwig Wittgenstein, quoted in Michael Nedo, Guy Moreton, and Alec Finlay, *Ludwig Wittgenstein: There Where You Are Not* (London: Black Dog, 2005), 51.

"our real need": Ludwig Wittgenstein, *Philosophical Investigations*, trans. G. E. M. Ascombe, P. M. S. Hacker, and Joachim Schulte (Oxford: Blackwell, 2009), sec. 108.

"turning around of the soul": Richard Eldridge, "Rotating the Axis of Our Investigation," in John Gibson and Wolfgang Huemer, eds., *The Literary Wittgenstein* (London: Routledge, 2004), 212.

"repeat it to us inexorably": Wittgenstein, *Philosophical Investigations*, sec. 115.

"under the door": Anthony Gottlieb, "A Nervous Splendor," *The New Yorker*, April 6, 2009, 70.

"uncongenial to him": Hermine Wittgenstein, quoted in James C. Klagge, *Wittgenstein in Exile* (Cambridge, MA: MIT Press, 2010), 179, note 24.

"given to despair": Fania Pascal, quoted in F. A. Flowers, ed., *Portraits of Wittgenstein*, vol. 2 (Bristol, UK: Thoemmes, 1999), 235.

"rise sufficiently": Ludwig Wittgenstein, quoted in James C. Klagge, *Wittgenstein: Biography and Philosophy* (Cambridge, UK: Cambridge University Press, 2001), 111.

"such a thing can be": Ludwig Wittgenstein, diary entry (Feb. 22, 1937), in James C. Klagge and Alfred Nordmann, eds., *Ludwig Wittgenstein: Public and Private Occasions* (Lanham, MD: Rowman and Littlefield, 2003), 209–11.

"with our fingers": Wittgenstein, *Philosophical Investigations*, sec. 106.

"healthier for it": Ludwig Wittgenstein, quoted in Rush Rhees, ed., *Recollections of Wittgenstein* (Oxford: Oxford University Press, 1984), 32.

"basest thoughts": Ludwig Wittgenstein, diary entry (Feb. 17, 1937), in Klagge and Nordmann, *Public and Private*, 183.

"I know it all so well": Bertrand Russell, letter to Ottoline Morrell (Oct. 31, 1912), in Brian McGuinness, *Wittgenstein: A Life*, vol. 1 (Oakland: University of California Press, 1988), 154.

"But will I reach it?": Ludwig Wittgenstein, diary entry (Sept. 5, 1914), in Ray Monk, *Ludwig Wittgenstein: The Duty of Genius* (New York: Free Press, 1990), 117.

"May the spirit be in me!": Ludwig Wittgenstein, diary entry (Sept. 13, 1914), in McGuinness, *Wittgenstein: A Life*, 221.

"indescribable" blessing: Ludwig Wittgenstein, letter to Ludwig von Ficker (July 24, 1915), in Martin J. B. Stokhof, *World and Life as One: Ethics and Ontology in Wittgenstein's Early Thought* (Redwood City, CA: Stanford University Press, 2002), 258, note 60.

"*external* life": Ludwig Wittgenstein, diary entry (Oct. 11, 1914), in *Geheime Tagebücher, 1914–1916*, ed. Wilhelm Baum (Vienna: Turia and Kant, 1992). Unpublished translation by John Bishop.

"appetites and aversions": Ludwig Wittgenstein, diary entry, in Monk, *Ludwig Wittgenstein: The Duty of Genius*, 146.

"something that is already there": Ludwig Wittgenstein, *Notebooks, 1914–1916*, ed. G. E. M. Anscombe and G. H. von Wright (Chicago: University of Chicago Press, 1984), 72–73.

"in the face of death": Wittgenstein, *Notebooks*, 74–75.

"poor, unhappy man": Wittgenstein, diary entries (March 28–April 7, 1916), in *Geheime Tagebücher, 1914–1916*.

"possible to bear life": Wittgenstein, diary entries (May 5–6, 1916), in *Geheime Tagebücher, 1914–1916*.

"science and art": Wittgenstein, *Notebooks*, 79.

"I am my world": Ludwig Wittgenstein, *Tractatus Logico-Philosophicus*, trans. D. F. Pears and B. F. McGuinness (London: Routledge, 2001), 5:621.

trying to illuminate: Wittgenstein, *Notebooks*, 63.

"art and ethics": Ibid.

"won't let me work": Wittgenstein, diary entry (Feb. 13, 1937), in Klagge and Nordmann, *Public and Private*, 175.

"my pitifulness": Ibid.

"hear me out": Wittgenstein, diary entry (May 2, 1930), in Klagge and Nordmann, *Public and Private*, 21.

"chase it away": Ibid., 23.

"sort of vanity": Wittgenstein, diary entry, in Klagge and Nordmann, *Public and Private*, 139.

"collect my thoughts": Wittgenstein, diary entry (Nov. 30, 1936), in Klagge and Nordmann, *Public and Private*, 155.

"clear to me": Wittgenstein, diary entry (Nov. 23, 1936), in Klagge and Nordmann, *Public and Private*, 152.

"doubt about them was impossible": Wittgenstein, diary entry (Nov. 15, 1936), in Klagge and Nordmann, *Public and Private*, 141.

would ever cease: Wittgenstein, diary entry (Feb. 21, 1937), in Klagge and Nordmann, *Public and Private*, 203.

"an important matter": Wittgenstein, diary entry (Feb. 19, 1937), in Klagge and Nordmann, *Public and Private*, 193.

"bearable as it is": Ibid., 191.

"act of consciousness": Clare Carlisle, *Kierkegaard's Philosophy of Becoming: Movements and Positions* (Albany: State University of New York Press, 2006), 95–96, 147.

"suffering would be over": Wittgenstein, diary entry (Feb. 19, 1937), in Klagge and Nordmann, *Public and Private*, 191.

dogs and flowers: David Foster Wallace, in conversation with Larry McCaffrey, *The Review of Contemporary Fiction* 13, no. 2 (Summer 1993): n.p.

"a gallon over it": Ludwig Wittgenstein, *Philosophical Occasions, 1912–1951*, ed. James C. Klagge and Alfred Nordmann (Indianapolis: Hackett, 1993), 40.

"spade is turned": Wittgenstein, *Philosophical Investigations*, sec. 217.

"to stop": Ludwig Wittgenstein, *Zettel*, ed. G. E. M. Anscombe and G. H. von Wright (Oakland: University of California Press, 2007), sec. 314.

"outside space and time": Ludwig Wittgenstein, *Tractatus Logico-Philosophicus*, 6.4312.

go on living: See Hans Sluga, "Whose House Is That? Wittgenstein on the Self," in *Cambridge Companion to Wittgenstein*, ed. Hans Sluga and David G. Stern (Cambridge, UK: Cambridge University Press, 1996), 342–43.

"*a particular fact*": Ludwig Wittgenstein, letter to Paul Engelmann (June 21, 1920), in Brian McGuinness, *Wittgenstein: A Life*, vol. 1, 293.

"May it turn out well": Wittgenstein, diary entry (Nov. 19, 1936), in Klagge and Nordmann, *Public and Private*, 151.

"collapsed already": Ibid.

"appropriate to the situation": Wittgenstein, diary entry (Nov. 24, 1936), in Klagge and Nordmann, *Public and Private*, 153.

"confess it": Wittgenstein, letter to Ludwig Hansel (Nov. 7, 1936), in Klagge and Nordmann, *Public and Private*, 281.

"drop it": Wittgenstein, *Philosophical Occasions*, 175.

caused pain: Ludwig Wittgenstein, *The Big Typescript: TS 213*, trans. C. Grant Luckhardt and Maximilian E. Aue (Oxford: Blackwell, 2005), 307.

"bit of iron": Ludwig Wittgenstein, *Culture and Value*, trans. Peter Winch (Chicago: University of Chicago Press, 1984), 11.]

132 *Philosophical Investigations* . . . "all in here together": McCaffery, "A Conversation with David Foster Wallace."

133 Nardil . . . most of his adult life: James and Sally Wallace; Mark Costello.

134 "madman or a troublemaker": Letter from Wallace to Nadell.

135 "energetic refusal to compromise": Kakutani, review of *The Broom of the System*, by David Foster Wallace, *New York Times*, Dec. 27, 1986.

136 "just drugs to just alcohol": Granada House, "An Ex-Resident's Story."

137 "Everybody, but everybody": Wallace, *Infinite Jest*, 349.

137 "enslaving Substance": Ibid., 200.

138 "An Ex-Resident's Story": Granada House, "An Ex-Resident's Story."

138 **"Mr. Rogers with tattoos"**: Wallace, *Infinite Jest*, 357.

138 **"American type of sadness"**: Laura Miller, "David Foster Wallace"; Yablono-witz, "Struggling to Make Sense."

140 **professor at Emerson**: Max, *Every Story Is a Love Story*, 148.

140 **Karr's marriage**: Ibid., 163.

141 **"narcissistically-deprived"**: Dale Peterson interview; Max, *Every Story Is a Love Story*, 170.

141 **a small child . . . all for her**: Wallace, *Brief Interviews with Hideous Men*, 286.

142 **asked Karr to marry**: Max, *Every Story Is a Love Story*, 170.

142 **"genius"**: Maslin, "In Layered Fiction and Wry Notes to Mom, a Cosmic Genius Distilled."

142 **"masterpiece"**: Esposito, "Who Was David Foster Wallace?—Wallace's Master-piece."

142 **"virtuoso display"**: Sheppard, "Mad Maximalism."

143 **"restless mind"**: Kakutani, review of *Infinite Jest*.

143 **"wasteful and wrong"**: DeLillo correspondence. The letter is dated May 20 with no year, but Wallace indicates in the letter that he is thirty-five at the time.

143 **"Kate Gompert knew"**: Yablonowitz, "Struggling to Make Sense."

144 **"terrible, lingering death"**: Ralph Spaulding interview.

145 **"totally hosed"**: Wallace, *This Is Water*, 53–55.

145 **"What if he was just afraid"**: David Foster Wallace, "Good People," *The New Yorker*, Feb. 5, 2007.

145 **"moments when he would be down"**: Ralph Spaulding interview.

146 **high-romance, low-intimacy**: Max, *Every Story Is a Love Story*, 233.

146 **"I'm a piece of shit"**: Ralph Spaulding interview.

147 **"shift directions . . . promised Karen"**: Ibid.

148 **"universal pain"**: Wallace, *Infinite Jest*, 696.

CHAPTER 5: WHEN CAPTURE LEADS TO VIOLENCE

STRIKING OUT

Sources

Interview by the Author

Blair, James (April 2014).

Works Referenced

Chase, Alston. *Harvard and the Unabomber: The Education of an American Terrorist*. New York: Norton, 2003.

Clarke, James W. *Defining Danger: American Assassins and the New Domestic Terrorists*. New Brunswick, NJ: Transaction, 2012.

Johnson, Sally C. Psychiatric Competency Report of Dr. Sally C. Johnson, Sept. 11, 1998. *United States v. Theodore John Kaczynski*, 232 F.3d 1034 (9th Cir. 2001). CR. NO. S-96-259 GEB.

Kintz, Theresa. "Interview with Ted Kaczynski." The Anarchist Library. http://the anarchistlibrary.org/library/theresa-kintz-interview-with-ted-kaczynski. Retrieved Dec. 11, 2009, from http://www.insurgentdesire.org.uk.

Waits, Chris, and Dave Shors. *Unabomber: The Secret Life of Ted Kaczynski*. Helena, MT: *Independent Record* and *Montana Magazine*, 1999.

Notes

153 **July 24, 1978:** Waits and Shors, *Unabomber*, 275.
154 **"not really politically oriented":** Kintz, "Interview with Ted Kaczynski."
154 **"civilization is a monstrous octopus":** Waits and Shors, *Unabomber*, 274.
154 **very angry:** Johnson, Psychiatric Competency Report.
154 **"strongly conditioned":** Chase, *Harvard and the Unabomber*, 292.
155 **"kill that psychiatrist":** Waits and Shors, *Unabomber*, 270.
155 **"duzzent gnaw":** Ibid., 272.

THE ASSASSINATION OF ROBERT KENNEDY

Sources

Works Referenced

Ayton, Mel. *The Forgotten Changes: Sirhan Sirhan and the Assassination of Robert F. Kennedy.* Washington, DC: Potomac Books, 2007.

Federal Bureau of Investigation. *Robert F. Kennedy Assassination: The FBI Files.* Minneapolis: Filiquarian, 2007.

Jansen, Godfrey H. *Why Robert Kennedy Was Killed: The Story of Two Victims.* New York: Third Press, 1970.

Kaiser, Robert Blair. *R.F.K. Must Die! Chasing the Mystery of the Robert Kennedy Assassination.* New York: Overlook Press, 2008.

Lawrence, David. "Paradoxical Bob." *Independent Star-News*, May 26, 1968. http://www.newspapers.com/newspage/31810108/.

O'Sullivan, Shane. *Who Killed Bobby? The Unsolved Murder of Robert F. Kennedy.* New York: Union Square Press, 2008.

People v. Sirhan Bishara Sirhan, 7 Cal. 3d 710 (1972). Reporter's Transcript on Appeal. Crim. No. 14026. [S.C. CA June 16, 1972], 4969. https://www.maryferrell.org/mffweb/archive/docset/getList.do?docSetId=1659.

Notes

157 **CBS aired a documentary:** Ayton, *The Forgotten Changes*; Jansen, *Why Robert Kennedy Was Killed*, 193.
158 **"I had loved Robert Kennedy":** Jansen, *Why Robert Kennedy Was Killed*, 193–94.
158 **"I saw that television program":** Ibid., 191–96.
158 **"Jews kicked us out":** *People v. Sirhan Bishara Sirhan*, 4832.
158 **"RFK must die":** Kaiser, *R.F.K. Must Die!*, 383.
158 **"beginning of the Israeli assault":** *People v. Sirhan Bishara Sirhan*, 4972, 4973.
159 **"supporting the cause of Israel":** Lawrence, "Paradoxical Bob."
159 **furious and disoriented:** Albert Bandura, a social psychologist, argues that most people impose "self-sanctions" that regulate certain behaviors, including killing. Through learning and social norms, certain behaviors provoke a negative emotional response. Fear, anxiety, or negative affect develop at the thought or actions associated with the behavior. A way to overcome those "self-sanctions" is to restructure the moral value of killing—that killing is justified in the name of some greater cause—so that the act is free from self-censuring restraints.
159 **"instead of seeing my own face":** *People v. Sirhan*, 4977.
159 **"looked like a saint to me":** Ayton, *The Forgotten Changes*, 74.

159 **"A fire started burning"**: Federal Bureau of Investigation, *Robert F. Kennedy Assassination.*

160 **"boiled him up again"**: Jansen, *Why Robert Kennedy Was Killed*, 201.

160 **"Kennedy must fall"**: Ibid., 195.

THE COLUMBINE SCHOOL SHOOTINGS

Sources

Works Referenced

Campher, Rosemary, ed. *Violence in Children: Understanding and Helping Those Who Harm.* London: Karnac, 2008.

Cullen, Dave. *Columbine.* New York: Hachette, 2010.

Harwood, Valerie, and Julie Allan. *Psychopathology at School: Theorizing Mental Disorders in Education.* New York: Routledge, 2014.

Kelly, Katharine D., and Mark Totten. *When Children Kill: A Social-Psychological Study of Youth Homicide.* Peterborough, ON: Broadview, 2002.

Langman, Peter. *Why Kids Kill: Inside the Minds of School Shooters.* New York: Palgrave Macmillan, 2010.

Larkin, Ralph W. *Comprehending Columbine.* Philadelphia: Temple University Press, 2007.

Monahan, John, and Henry J. Steadman. *Violence and Mental Disorder: Developments in Risk Assessment.* Chicago: University of Chicago Press, 1996.

Scherz, Jared M., and Donna Scherz. *Catastrophic School Violence: A New Approach to Prevention.* New York: Rowman and Littlefield, 2014.

Shepard, C. "A Columbine Site." http://www.acolumbinesite.com.

Notes

161 **"a statement"**: Cullen, *Columbine*, 329.

161 **self-awareness**: Larkin, *Comprehending Columbine*, 137.

161 **"better be fuckin good"**: Harris journal, Shepard, "A Columbine Site."

162 **"OFFICIALLY lower . . . welt"**: Ibid.

163 **"original-copycats"**: Ibid.

163 **"MY fault!"**: Ibid.

163 **"napalm on sides of skyscrapers"**: Ibid.

163 **"war's war"**: Cullen, *Columbine*, 327.

164 **"nuke the world"**: Ibid.

164 **"girls and looks and such"**: Ibid.

164 **"weird looking Eric"**: Ibid.

THE MURDER OF JOHN LENNON

Sources

Work Referenced

Jones, Jack. *Let Me Take You Down: Inside the Mind of Mark David Chapman, the Man Who Killed John Lennon* New York: Villard Books, 1992.

Notes

165 **Jack Jones**: Jack Jones, *Let Me Take You Down.* (I am indebted to Mr. Jones for the extensive interviews with Mark David Chapman compiled in his book.)

165 "to kill the phony": Ibid., 45.
165 impromptu concerts in his garage: James R. Gaines, "Mark Chapman: The Man Who Shot Lennon," *People*, Feb. 23, 1987, http://www.people.com/people/archive/article/0,,20095701,00.html.
165 easily disarmed: Jones, *Let Me Take You Down*, 107.
165 "more popular than God": Ibid., 120.
165 Fort Chaffee, Arkansas: Ibid., 126.
166 started a band: Ibid.
166 Gerald Ford: Ibid.
166 "enough good": Ibid., 125.
166 Jessica Blankenship: Ibid., 127.
166 "I became a nobody": Ibid., 141.
166 "celebrate death": Ibid., 139.
166 suicide: Ibid., 141.
167 changing his name: Ibid., 201.
168 "around his music": Ibid., 187.
168 "the tornado": Ibid., 185.
168 *"Kill John Lennon"*: Ibid., 190.
168 "no power on earth": Ibid., 188.
168 waited for the police to arrive: James R. Gaines, "Mark Chapman: The Man Who Shot Lennon."

THE MURDERS AT SANDY HOOK ELEMENTARY SCHOOL

Sources

Works Referenced

Altimari, Dave. "Summary of Sandy Hook Report to Be Released." PoliceOne.com. Nov. 16, 2013. http://www.policeone.com/active-shooter/articles/6596773-Summary-of-Sandy-Hook-report-to-be-released/.

Caldwell, Maggie. "Sandy Hook Crime Report: Adam Lanza Obsessed with Mass Murder and Dance Dance Revolution." *Mother Jones,* Nov. 26, 2013. http://www.motherjones.com/mojo/2013/11/what-we-learned-sandy-hook-crime-report.

Crevier, Nancy K. "Child Advocate's Office Reports on Adam Lanza's Troubled Life." *Newtown Bee*, Dec. 28, 2015. http://www.newtownbee.com/news/0001/11/30/child-advocate-s-office-reports-adam-lanza-s-troub/242115.

Happé, F., and U. Frith, "The Weak Coherence Account: Detail-Focused Cognitive Style in Autism Spectrum Disorders." *Journal of Autism and Developmental Disorders* 36 (2006); 5–25.

Lysiak, Matthew. *Newtown: An American Tragedy*. New York: Gallery Books, 2013.

McShane, Larry. "Newtown Killer Adam Lanza Wrote Tales About 'Hide and Go Die' Game, Graphic Violence." *New York Daily News,* Dec. 28, 2013. http://www.nydailynews.com/news/national/newtown-killer-wrote-tales-graphic-violence-documents-article-1.1560456.

Office of the Child Advocate. "Shooting at Sandy Hook Elementary School: Report of the Office of the Child Advocate." Nov. 21, 2014. http://www.ct.gov/oca/lib/oca/sandyhook11212014.pdf.

Office of the State's Attorney, Judicial District of Danbury. "Report of the State's Attorney for the Judicial District of Danbury on the Shootings at Sandy Hook Elementary School." Nov. 25, 2013.

Notes

169   *DanceDanceRevolution*: Caldwell; Office of the Child Advocate, 98.
169   **Bushmaster . . . Savage Mark II:** Caldwell.
170   **"significant developmental challenges":** Office of the Child Advocate, 6.
170   **Asperger's . . . theory:** Happé and Frith, 5–25.
171   **"Hide and Go Die":** McShane.
171   **"not the sort of creation":** Office of the Child Advocate, 29.
171   **"did not think highly of himself":** Office of the State's Attorney.
172   **"increasingly intense":** Office of the Child Advocate, 36.
172   **"living in a box":** Ibid., 34.
175   **"The aesthetic of pistols":** Ibid., 105.

A THEORY OF HUMAN CAPITAL

Sources

Works Referenced

Denver Post, "Live Blog: The Aurora Theater Shooting Trial," http://live.denverpost
.com/Event/LIVE_BLOG_The_Aurora_Theater_Shooting_Trial.
Oulis, P., et al., "Differential Diagnosis of Obsessive-Compulsive Symptoms from
Delusions in Schizophrenia: A Phenomenological Approach." *World Journal of
Psychiatry* 3, no. 3 (2013): 50–56.
All quotes in this section are taken from the *Denver Post* blog.

Notes

176   **"broken brain":** I observed much of the trial of James Holmes. In addition,
many of the quotes are drawn heavily from detailed blog posts by the *Denver
Post*, generally written by Jordan Steffen or Larry Ryckman, and edited by Eric
J. Lubbers. The blog, based on live reporting from the courtroom on the nine-
week trial, synthesized all the testimony and statements by defense and prose-
cution attorneys.

Videotapes of the entire trial, which also provided crucial information, are archived online (ABC 7 News Denver, "Daily Archive," http://www.thedenver
channel.com/aurora-movie-theater-shooting/daily-trial-video-archive). The
writing of James Holmes appears in his "computational notebook" and was entered into evidence during the trial as People's Exhibit no. 341.

The observations of University of Colorado psychiatrist Dr. Lynne Fenton were
drawn primarily from her trial testimony, as summarized in news blogs created by
local Denver television channel 9. The distinction between obsessions and delusions is described in P. Oulis et al., "Differential Diagnosis of Obsessive-Compulsive
Symptoms from Delusions in Schizophrenia." The e-mail conversations between
James Holmes and his girlfriend, Gargi Datta, occurred on March 25, 2012, and
were entered into evidence during the trial as People's Exhibit no. 649.

Access to the full set of documents entered as evidence in the trial, including jury instructions, e-mail exchanges, exhibits, and theater diagrams, is also
available online. [detailed blog posts: *Denver Post Live*, "Live Blog: The Aurora Theater Shooting
Trial," http://live.denverpost.com/Event/LIVE_BLOG_The_Aurora_Theater_
Shooting_Trial.]

archived online: ABC 7 News Denver, "Daily Archive," http://www.thedenver channel.com/aurora-movie-theater-shooting/daily-trial-video-archive.

People's Exhibit no. 341: James Holmes, personal notebook, entered into evidence as Exhibit 341 in trial of James Holmes that began on April 27, 2015, http://extras.denverpost.com/trial/docs/notebook.pdf.

Denver television channel 9: Allison Sylte, "Day 32," *Aurora Theater Trial* (blog), http://www.9news.com/story/news/local/aurora-theater-trial/2015/06/16/aurora-theater-trial-updates/28803939/.

People's Exhibit no. 649: James Holmes and Gargi Datta, e-mail correspondence, entered into evidence as Exhibit no. 649 in trial of James Holmes that began on April 27, 2015, http://extras.denverpost.com/trial/docs/649.pdf.

available online: *Denver Post*, "Aurora Theater Shooting Trial: Documents," http://extras.denverpost.com/trial/docs.html.]

178 **"what almost everyone else believes":** Oulis, *World Journal of Psychiatry*, 52.

CHAPTER 6: CAPTURE AND IDEOLOGY

Sources

Interview by the Author
Kateb, George (2014).

Works Referenced

Arendt, Hannah. *The Life of the Mind*. New York: Harcourt, Brace, 1978.

———. *The Origins of Totalitarianism*. New York: Harcourt, Brace, 1951.

Kateb, George. *The Inner Ocean: Individualism and Democratic Culture*. Ithaca, NY: Cornell University Press, 1994.

Kateb, George, and Hannah Arendt. *Politics, Conscience, Evil*. Totowa, NJ: Roman and Allanheld, 1983.

Notes

185 **theorist of political evil:** For an analysis of Arendt's political thought and intellectual legacy, see Kateb and Arendt, *Politics, Conscience, Evil*. See also Kateb, *The Inner Ocean*.

185 **"they get us to act":** George Kateb, in discussion with the author, January 2014. All subsequent quotations refer to this interview.

185 **great tragedies of modern history:** For a discussion of the role of ideology in twentieth-century totalitarian movements, see Arendt, *The Origins of Totalitarianism*. See also Arendt, *The Life of the Mind*.

THE AMERICA I HAVE SEEN

Sources

Works Referenced

Bergesen, Albert J., ed. *The Sayyid Qutb Reader: Selected Writings on Politics, Religion, and Society*. New York: Taylor and Francis, 2008.

Bonner, Michael. *Jihad in Islamic History: Doctrines and Practice*. Princeton, NJ: Princeton University Press, 2006.

Calvert, John. *Sayyid Qutb and the Origins of Radical Islamism*. New York: Columbia University Press, 2010.

DeLong-Bas, Natana J. *Wahhabi Islam: From Revival and Reform to Global Jihad.* New York: Oxford University Press, 2004.

Qutb, Sayyid. "The America I Have Seen: In the Scale of Human Values." In *America in an Arab Mirror: Images of America in Arabic Travel Literature, 1688 to 9/11 and Beyond,* ed. Kamal Abdel-Malek. 1951; repr. London: Palgrave Macmillan, 2011.

———. *In the Shade of the Qur'an.* London: MWH, 1979.

———. *Milestones.* Oneonta, NY: Islamic Publications International, 2006.

———. *Social Justice in Islam.* Oneonta, NY: Islamic Publications International, 2000.

Rodenbeck, Max. "The Father of Violent Islamism." *New York Review of Books,* May 9, 2013.

Ruthven, Malise. *Fundamentalism: The Search for Meaning.* London: Oxford University Press, 2005.

Storm, Morton. *Agent Storm: My Life Inside Al Qaeda and the CIA.* New York: Atlantic Monthly Press, 2014.

Toth, James. *Sayyid Qutb: The Life and Legacy of a Radical Islamic Intellectual.* New York: Oxford University Press, 2013.

Weston, Mark. *Prophets and Princes: Saudi Arabia from Muhammad to the Present.* New York: Wiley, 2000.

Notes

187 **"foot does not take part"**: Qutb, "The America I Have Seen," 14.

187 **"Beat him to a pulp"**: Ibid.

187 **"without heart or conscience"**: Qutb, quoted in Rodenbeck, "The Father of Violent Islamism."

188 **"in their language, 'fun'"**: Ibid., 19.

188 **"Each church races to advertise itself"**: Ibid., 20.

189 **"arts and the mass media"**: Qutb, *In the Shade of the Qur'an,* quoted in Weston, *Prophets and Princes,* 366.

190 **"obligatory on Islam"**: DeLong-Bas, *Wahhabi Islam,* 260.

190 **"what is to be done"**: Ruthven, *Fundamentalism,* 90–91.

190 **"devoid of those vital values"**: Qutb, *Milestones,* quoted in Bergesen, *The Sayyid Qutb Reader,* 35.

190 **"coming to restore this religion"**: Qutb, *Social Justice in Islam,* quoted in Calvert, *Sayyid Qutb and the Origins of Radical Islamism,* 131.

191 **"die or be slain"**: Calvert, *Sayyid Qutb and the Origins of Radical Islamism,* 263.

191 **"those who are God-fearing"**: Qutb, *Milestones,* quoted in Bergesen, *The Sayyid Qutb Reader,* 144.

191 **heavenly paradise that awaited**: Calvert, *Sayyid Qutb and the Origins of Radical Islamism,* 247.

191 **"they have his provision"**: Ibid.

THE OBLIGATION OF OUR TIME

Sources

Works Referenced

Egerton, Brooks. "Imam's E-Mails to Fort Hood Suspect Hasan Tame Compared to Online Rhetoric." *Dallas Morning News,* Nov. 29, 2009.

Federal Bureau of Investigation, Records Management, FOIPA: Subject, Requester, Unclassified, Parts 1–36.

Janes, Dominic, and Alex Houen. *Martyrdom and Terrorism: Pre-Modern to Contemporary Perspectives.* Oxford: Oxford University Press, 2014.

Meleagrou-Hitchens, Alexander. *As American as Apple Pie: How Anwar al-Awlaki Became the Face of Western Jihad.* London: International Centre for the Study of Radicalisation and Political Violence, 2011.

Newton, Paula. "Purported al-Awlaki Message Calls for Jihad Against the U.S." CNN, March 17, 2010. http://www.cnn.com/2010/WORLD/europe/03/17/al.awlaki.message/.

Shane, Scott, and Souad Mekhennet. "Imam's Path from Condemning Terror to Preaching Jihad." *New York Times*, May 8, 2010. (I want to credit Scott Shane and Souad Mekhennet for their incisive reporting, helping me to understand Anwar al-Awlaki.)

SITE Intelligence Group. "AQAP Releases Interview with Anwar al-Awlaki." https://news.siteintelgroup.com/Multimedia/awlaki52310.html.

———. "Awlaki Justifies Deaths of Millions." https://news.siteintelgroup.com/Articles-Analysis/awlaki-justifies-deaths-of-millions.html.

Storm, Morton. *Agent Storm: My Life Inside Al Qaeda and the CIA.* New York: Atlantic Monthly Press, 2014.

Notes

192 **"Sayyid was with me in my cell speaking"**: Storm, *Agent Storm*, 146; Shane and Mekhennet, "Imam's Path from Condemning Terror to Preaching Jihad."

192 **"saw spies everywhere"**: Storm, *Agent Storm*, 146.

192 **"*Jahiliyya*"**: Meleagrou-Hitchens, *As American as Apple Pie*, 40.

192 **"by the tip of the sword"**: Egerton, "Imam's E-Mails to Fort Hood Suspect Hasan Tame Compared to Online Rhetoric."

192 **"They are the party of Satan"**: SITE, "AQAP Releases Interview with Anwar al-Awlaki."

193 **a small mosque**: Shane and Mekhennet, "Imam's Path from Condemning Terror to Preaching Jihad."

193 **conspiracy theorist**: Ibid.

193 **"every *haram*"**: Ibid.

193 **lesser prophets of Islam**: Ibid.

194 **"not to destroy" . . . known to the FBI**: Ibid.

194 **"the radical sound reasonable"**: Storm, *Agent Storm*, 98.

194 **"9/11 was justified"**: Ibid., 99.

194 **"they voted . . . financing this war"**: SITE, "Awlaki Justifies Deaths of Millions."

194 **"nation of evil"**: Newton, "Purported al-Awlaki Message Calls for Jihad Against the U.S."

195 **"The ballot has failed us"**: Janes and Houen, *Martyrdom and Terrorism*, 234.

I'M GOING TRAVELING

Sources

Works Referenced

Berger, J. M. "Anwar Awlaki E-Mail Exchange with Fort Hood Shooter Nidal Hasan." *Intel Wire*, July 19, 2012. http://news.intelwire.com/2012/07/the-following-e-mails-between-maj.html.

Fernandez, Manny. "Fort Hood Gunman Told His Superiors of Concerns." *New York Times*, Aug. 20, 2013. http://www.nytimes.com/2013/08/21/us/fort-hood -gunman-nidal-malik-hasan.html.

Gjelten, Tom, Daniel Zwerdling, and Scott Neuman. "Answers Sought on Fort Hood Suspect's Link to Imam." NPR, Nov. 10, 2009. http://www.npr.org/templates/ story/story.php?storyId=120266334.

Lieberman, Joseph, and Susan Collins. *A Ticking Time Bomb: Counterterrorism Lessons from the U.S. Government's Failure to Prevent the Fort Hood Attack*. Special Report by the U.S. Senate Committee on Homeland Security and Governmental Affairs, Washington, DC, Feb. 3, 2011. http://www.hsgac.senate .gov//imo/media/doc/Fort_Hood/FortHoodReport.pdf.

McKinley, James C. "Fort Hood Gunman Gave Signals Before His Rampage." *New York Times*, Nov. 8, 2009. http://www.nytimes.com/2009/11/09/us/09reconstruct .html.

New York Times. "E-Mails from Maj. Nidal Malik Hasan." Aug. 20, 2013. http:// www.nytimes.com/interactive/2013/08/21/us/21hasan-emails-document.html.

Smith, Cindy, and Imityaz Delawala. "Cousin of Fort Hood Shooter Speaks Out Against Violent Extremism." ABC News, Sept. 4, 2011. http://abcnews.go .com/Politics/ft-hood-shooters-cousin-speaks-violent-extremism/ story?id=14445896.

Spencer, Robert. "U.S.-Born Islamic Cleric: Nidal Hasan 'Did the Right Thing.'" *Jihad Watch*, Nov. 9, 2009. http://www.jihadwatch.org/2009/11/us-born -islamic-cleric-nidal-hasan-did-the-right-thing.

Washington Post. "Biography of Nidal Hasan, Suspect in Shooting at Fort Hood." Nov. 7, 2009. http://www.washingtonpost.com/wp-dyn/content/article/2009/11/06/ AR2009110601978.html.

———. "Hasan on Islam." Nov. 10, 2011. http://www.washingtonpost.com/wp-dyn/ content/gallery/2009/11/10/GA2009111000920.html?sid=ST2010101600356.

Weinstein, Mikey. "Hasan and the Proselytization Factor." Military Religious Freedom Foundation, Nov. 10, 2009. http://www.militaryreligiousfreedom.org/ press-releases/washpost_hasan.html.

Notes

195 **one-way e-mail:** Eighteen e-mails sent from Dec. 17, 2008, through June 16, 2009.

195 **"consider them shaheeds":** E-mail 1, from Hasan to Awlaki, Dec. 17, 2008.

195 **"eye for a sister":** E-mail 10, from Awlaki to Hasan, Feb. 22, 2009.

195 **"looking for a wife":** E-mail 9, from Hasan to Awlaki, Feb. 19, 2009.

195 **"activist and leader":** E-mail 7, from Hasan to Awlaki, Feb. 16, 2009.

195 **"potential repercussions":** E-mail 9, from Hasan to Awlaki, Feb. 19, 2009.

195 **"striving for Jannat Firdaus":** E-mail 11, from Hasan to Awlaki, Feb. 22, 2009.

196 **"the right thing":** Spencer, "U.S.-Born Islamic Cleric."

196 **as an enlisted soldier:** McKinley, "Fort Hood Gunman Gave Signals Before His Rampage"; *Washington Post*, "Biography of Nidal Hasan."

196 **childhood had been largely secular:** Smith and Delawala, "Cousin of Fort Hood Shooter Speaks Out Against Violent Extremism."

196 **possibility of leaving the service:** Weinstein, "Hasan and the Proselytization Factor"; McKinley, "Fort Hood Gunman Gave Signals Before His Rampage."

196 **"fairly benign":** Gjelten, Zwerdling, and Neuman, "Answers Sought on Fort Hood Suspect's Link to Imam."

197 **gave a presentation:** *Washington Post*, "Hasan on Islam."

198 **"War on Islam":** McKinley, "Fort Hood Gunman Gave Signals Before His Rampage."

198 **"ticking time bomb":** Lieberman and Collins, *A Ticking Time Bomb*.

199 **troubling war stories:** Fernandez, "Fort Hood Gunman Told His Superiors of Concerns"; *New York Times*, "E-Mails from Maj. Nidal Malik Hasan."

199 **"going traveling":** McKinley, "Fort Hood Gunman Gave Signals Before His Rampage."

@SLAVEOFALLAH

Sources

Websites and Blogs Consulted

Ask.fm

Colorado Muslim Society. https://coloradomuslimsociety.org/.

Diary of a Muhajirah. Tumblr. Originally accessed at http://diary-of-a-muhajirah
.tumblr.com/post/103704073209/as-salaamualaykum-warahamat-allah-wabara
katuh#notes.

InviteToIslam.org. Tumblr. Originally accessed at http://invitetoislam.tumblr.com/
post/74631082758/drop-the-nationalistic-flags-raise-the-banners.

Works Referenced

Arapahoe County Sheriff's Office. Offense Report. Case Number 2014-00035607.
Oct. 21, 2014. https://www.documentcloud.org/documents/1345794-colorado-police-
report.html.

Awlaki, Anwar al-. "44 Ways of Promoting Jihad." http://www.kavkazcenter.com/
eng/content/2009/02/16/10561.shtml.

———. "Qualities of Great Women." YouTube. https://www.youtube.com/watch?
v=Tu5Ko2TaRos.

"Carrier of Sins." Tumblr. http://carrierofsins.tumblr.com/post/101913533157/
assalamu-alaykum-warahmatullahi-wabarakatuh-where#notes.

Hall, Ellie. "Inside the Online World of Three American Teens Who Allegedly
Wanted to Join ISIS." *BuzzFeed News*, Oct. 27, 2014. http://www.buzzfeed
.com/ellievhall/inside-the-online-world-of-three-teens-who-allegedly-wanted#
.cjmYNRKKv.

Katz, Rita. "From Teenage Colorado Girls to Islamic State Recruits: A Case Study in
Radicalization via Social Media." *Insite Blog*, Nov. 11, 2014. http://news.siteintel
group.com/blog/index.php/entry/309-from-teenage-colorado-girls-to-islamic
-state-recruits-a-case-study-in-radicalization-via-social-media.

Kuruvilla, Carole. "Colorado Soccer Team Dons Headscarves After Muslim Teammate Is Banned from Field." *New York Daily News*, March 18, 2014. http://
www.nydailynews.com/news/national/colorado-soccer-team-dons-headscarves
-muslim-teammate-banned-field-article-1.1725987.

Paul, Jesse. "Colorado Father Was Anguished When Daughter Fled to Join Islamic
State." *Denver Post*, Oct. 27, 2014. http://www.denverpost.com/news/
ci_26811442/colorado-father-was-anguished-when-daughter-fled-join.

SITE Intelligence Group. "Girl Talk: Calling Western Women to Syria." *Insite Blog*, July 30, 2014. https://news.siteintelgroup.com/blog/index.php/entry/218-girl-talk-lives-of-%20western-women-who-have-migrated-to-syria,%20as%20well %20as%20many%20other%20sources.

U.S. News & World Report. "Overland High School: Student Body." http://www .usnews.com/education/best-high-schools/colorado/districts/cherry-creek-school -district/overland-high-school-4021/student-body.

Notes

199 **the life of Asiya:** al-Awlaki, "Qualities of Great Women."

199 **"Jihad must be practiced":** Ibid.

199 **Teenage sisters:** Names used in this text are pseudonyms. Because the girls are minors as of this writing, their full names have not been released to the public.

200 **they would be expected:** Found in Ask.fm; quoted in Katz, "From Teenage Colorado Girls to Islamic State Recruits."

200 **indication that they were unhappy:** Arapahoe County Sheriff's Office, Offense Report.

200 **Colorado Muslim Society:** http://coloradomuslimsociety.org.

200 **African American majority:** *U.S. News & World Report*, "Overland High School: Student Body."

200 **barred a Muslim teammate:** Kuruvilla, "Colorado Soccer Team Dons Headscarves After Muslim Teammate Is Banned from Field."

200 **she posted primarily . . . "and stay on twitter":** Ask.fm; quoted in Katz, "From Teenage Colorado Girls to Islamic State Recruits."

201 **"drop the nationalist flags":** Invite to Islam.org.

201 **subculture of women:** SITE Intelligence Group, "Girl Talk."

201 **"drop your whatsapp":** *Diary of a Muhajirah* blog.

201 **"makes things a lot easier":** Ask.fm, quoted in Katz, "From Teenage Colorado Girls to Islamic State Recruits."

201 **Arabic pseudonyms:** Hall, "Inside the Online World of Three American Teens Who Allegedly Wanted to Join ISIS."

201 **"Lower your gaze":** Ibid.

202 **tweeted in early June:** Ibid.

202 **"What are you doing":** Ibid.

202 **feigning illness:** Ibid.

202 **left her house:** Ibid.

202 **board a plane:** Ibid.

202 **"made a mistake":** Paul, "Colorado Father Was Anguished When Daughter Fled to Join Islamic State."

202 **"Where are you from":** "Carrier of Sins."

CHAPTER 7: CAPTURE AND SPIRITUALITY

Sources

Works Referenced

Otto, Rudolf. *The Idea of the Holy*. Translated by John W. Harvey. London: Oxford University Press, 1950.

Wallace, David Foster. *This Is Water: Some Thoughts, Delivered on a Significant Occasion, About Living a Compassionate Life.* New York: Little, Brown, 2009.

Notes

206 "sweeping like a gentle tide": Otto, *Idea of the Holy*, 12.

206 **Rudolf Otto:** It was in a Moroccan synagogue in 1911 that Rudolf Otto first experienced spiritual transcendence. Only one synagogue of its kind still exists in the country today.

Otto recorded the visit in his diary: "On the Sabbath. A small, dimly lighted room, not 10 meters long, hardly 5 meters wide. Subdued light floats in from above. The walls, fitted with brown wainscoting, are censed by smoke from 30 hanging oil-lamps."

Otto's guide, Chayyim el Malek, had led him through "the labyrinthine streets of the ghetto and [up] two narrow, gloomy staircases" to this hidden temple. "Already in the dark, incredibly filthy vestibule," Otto recounted, "we hear the 'blessings' of the prayers and the Scripture readings, those half-sung, half-spoken nasal chants that the synagogue bequeathed to both the church and the mosque."

The Lutheran theologian was soon taken by the intoned melody as he began "to distinguish certain, regular modulations and cadences, which follow one another like leitmotifs. At first the ear tries to separate and understand the words in vain, and soon one wants to quit trying. Then suddenly the tangle of voices resolves itself and . . . a solemn fear overcomes one's limbs."

The congregation chanted in unison:

*Qadôš qadôš qadôš 'elohîm adonay ṣebaôt
Male'û haššamayim wehaareṣ kebôdô!*

For Otto, the force of these words, the Kadusha, cut across barriers of time and place. "I have heard the *Sanctus, sanctus, sanctus* of the Cardinals in St. Peter's, the *Swiat, swiat, swiat* in the Cathedral of the Kremlin, and the *Hagios, hagios, hagios* of the patriarch in Jerusalem," he later wrote. "In whatever language these words are spoken, the most sublime words that human lips have ever uttered, they always seize one in the deepest ground of the soul, arousing and stirring with a mighty shudder the mystery of the otherworldly that sleeps therein."

A hundred years before Rudolf Otto wrote those words, the eighteenth-century German philosopher and theologian Friedrich Schleiermacher argued that religious experience was, at its core, opposed to intellection: it sprang from a feeling of utter dependence, the loss of a recognizable self. Whereas the rational self sought to broaden its boundaries through observation and analysis, to project itself out into the world, the spiritual self experienced faith as a form of ecstatic obliteration. Schleiermacher's thesis would have a lasting influence on Rudolf Otto, as it would on many others who studied religion and human psychology over the course of the following century.

Rather than focusing on the validity of a given religious tradition, Schleiermacher approached all religious traditions as fundamentally similar. More important than doctrine or ritual was the felt experience of faith: "It is the consciousness that the whole of our spontaneous activity comes from a Source outside us." Schleiermacher's radical views on religion were consistent with his upbringing. In 1749 his grandfather, known for his unorthodox views on theology, was charged with sorcery and witchcraft. His father was an Enlightenment pastor who immersed himself in the works of Kant, Spinoza, and Plato.

According to Schleiermacher, all human beings have a natural capacity for religion. "The more ardent the thirst and the more persistent the drive to grasp the infinite," he wrote, "the more manifoldly will the mind itself be seized by it." In his early works, Schleiermacher wrote that "the original intuition . . . from which all these views derive determines the character of its feeling." In other words, the initial impulse or "intuition" that sparks religious faith gives shape to subsequent spiritual experiences.

The defining characteristic of all such experiences was absolute dependence on an Other. "The moment before consciousness divides into thought and action"—for Schleiermacher, the "secret moment"—marked the point at which the human mind was receptive to the divine. This is the moment before the mind forces perceptions into the mold of thought and is instead rapt by sheer sensory experience: "That first mysterious moment that occurs in every sensory perception, before intuition and feeling have separated, where sense and its objects have, as it were, flowed into one another and become one."

Such moments were, by their very definition, fleeting: "I know how indescribable it is and how quickly it passes away. . . . Would that I could and might express it, at least indicate it without having to desecrate it! It is as fleeting and transparent as the first scent with which the dew gently caresses the waking flowers, as modest and delicate as a maiden's kiss, as holy and fruitful as a nuptial embrace; indeed, not like these, but it is itself all of these." When the mind is in such a state of sheer receptivity, the familiar boundaries separating the self from the objects of its perception dissolve: "A manifestation, an event develops quickly and magically into an image of the universe. Even as the beloved and ever-sought-for form fashions itself, my soul flees toward it; I embrace it, not as a shadow, but as the holy essence itself." The believer lies in "the bosom of the infinite world," becoming its very "soul," feeling "all its powers and its infinite life" as his own.

For Schleiermacher, then, the numinous was not simply a matter of everyday perception. Rather, it referred to a particular psychological, or even aesthetic, disposition. In the language of cognitive neuroscience, one might describe Schleiermacher's "natal hour" as "prereflective consciousness," an experience of the self before the mind walls it off from the surrounding world: "This feeling, of which you are frequently scarcely aware, can in other cases grow to such intensity that you forget both the object and yourselves because of it; your whole nervous system can be so permeated by it that for a long time that sensation alone dominates and resounds and resists the effect of other impressions."

For Schleiermacher, however, religion was not alone in fostering a sense of a power greater than the self. Speaking to a group of student-soldiers who were about to help liberate Prussia from French forces, Schleiermacher extolled the transcendent potential of political struggle. With the students' firearms piled near the altar or resting against the walls of the church, Schleiermacher ascended to the pulpit: "There, in this holy place, and at this solemn hour, stood the physically so small and insignificant man, his noble countenance beaming with intellect, and his clear, sonorous, penetrating voice ringing through the overflowing church. . . . And when, at last, with the full fire of enthusiasm, he addressed the noble youths already equipped for battle, and next, turning to their mothers, the greater number of whom were present, he concluded with the words, 'Blessed is the womb that has borne such a son! blessed the breast that has nourished such a babe!' a thrill of deep emotion ran through the assembly, and amid loud sobs and weeping, Schleiermacher pronounced the closing Amen."

["hanging oil-lamps": Rudolf Otto, *Autobiographical and Social Essays*, ed. Gregory D. Alles (Berlin: Walter de Greyter, 1996), 80–81. All subsequent quotations from Otto refer to this source.

"a Source outside us": Friedrich Schleiermacher, *The Christian Faith*, ed. H. R. Mackintosh and J. S. Stewart (Edinburgh: Clark, 1999), 16.

"seized by it": Friedrich Schleiermacher, *On Religion: Speeches to Its Cultured Despisers*, ed. Richard Crouter (Cambridge, UK: Cambridge University Press, 1996), 29.

"character of its feeling": Ibid., 119.

"and become one": Ibid., 31.

"its infinite life": Ibid., 31–32.

"other impressions": Ibid., 29.

"closing Amen": Bishop Eilert, quoted in Friedrich Schleiermacher, *Selected Sermons of Schleiermacher*, ed. W. Robertson Nicoll, trans. Mary F. Wilson (London: Hodder and Stoughton, 1890), 29.]

206 "eat you alive": Wallace, *This Is Water*.

CAPTURE BY THE DIVINE

Sources

Works Referenced

Borden, Sarah. *Edith Stein*. New York: Continuum, 2003.

Carmelite Nuns of Baltimore. "Letter of Saint Edith Stein to Pope Pius XI in 1933." Feb. 23, 2003. http://www.baltimorecarmel.org/saints/Stein/letter%20to%20 pope.htm.

Graef, Hilda C. *The Scholar and the Cross: The Life and Work of Edith Stein*. Westminster, MD: Newman, 1955.

Herbstrith, Waltraud. *Edith Stein: A Biography*. San Francisco, CA: Ignatius, 1992.

Mosley, Joanne. *Edith Stein: Modern Saint and Martyr*. Mahwah, NJ: HiddenSpring, 2006.

Oben, Freda Mary. *The Life and Thought of St. Edith Stein*. New York: Alba House, 2001.

Oesterreicher, John M. *Walls Are Crumbling: Seven Jewish Philosophers Discover Christ*. London: Hollis and Carter, 1953.

St. Edith Stein Parish. "Biography of Edith Stein Continued." http://stedithstein parish.info/biography%20cont.html.

Stein, Edith. *The Hidden Life: Essays, Meditations and Spiritual Texts*. Washington, DC: ICS, 2014.

———. *Life in a Jewish Family 1891–1916: An Autobiography*. Washington, DC: ICS, 1999.

———. *The Life of a Philosopher and Carmelite*. Washington, DC: ICS, 2005.

———. *Philosophy of Psychology and the Humanities*. Translated by Mary Catherine Baseheart and Marianna Sawicki. Washington, DC: ICS, 2000.

Sullivan, John, ed. *Edith Stein: Essential Writings*. New York: Alba Orbis Books, 2002.

———. *Holiness Befits Your House: Canonization of Edith Stein*. Washington, DC: ICS, 2000.

Notes

208   **tragic death:** Graef, *The Scholar and the Cross*, 24.
208   **Edmund Husserl:** Ibid., 22–24.
208   **"unbelief collapsed":** Herbstrith, *Edith Stein: A Biography*, 56.
209   **"especially dear":** Stein, *Life in a Jewish Family*, 72.
209   **"resting in God":** Mosley, *Edith Stein: Modern Saint and Martyr*, 12.
209   **"duty to act":** Stein, *Philosophy of Psychology and the Humanities*, 85.
209   **Saint Teresa:** Herbstrith, *Edith Stein: A Biography*, 65.
209   **"This is truth":** Oesterreicher, *Walls Are Crumbling*, 297; Stein, *The Life of a Philosopher and Carmelite*, 63.
210   *Life in a Jewish Family:* Stein, *Life in a Jewish Family*, 7.
210   **"mock any sense":** Carmelite Nuns of Baltimore, "Letter of Saint Edith Stein to Pope Pius XI in 1933."
210   **"divine things only . . . 'beyond himself'":** Sullivan, *Holiness Befits Your House*, 15.
211   **"eternal Love":** Stein, *The Hidden Life,* 151; St. Edith Stein Parish, "Biography."

PAYING ATTENTION

Sources

Works Referenced

Du Plessix Gray, Francine. *Simone Weil*. New York: Viking, 2001.
Weil, Simone. *Waiting for God*. Translated by Emma Craufurd. New York: Putnam's Sons, 1951.

Notes

211   **spiritual capture:** For Paul Tillich, one of the greatest theologians of the early twentieth century, grace was not a matter of good deeds or just deserts. Indeed, Tillich insisted that God demands of us "no religious or moral or intellectual presupposition"—"nothing," that is, "but acceptance." In a handwritten note, now in the archives of the Harvard Divinity School, Tillich offered a startling interpretation of his faith: "[Christianity] offers . . . a way out, a way in which the existentialist analysis is accepted, in which materialism, idealism, and supernaturalism are overcome, in which the self is liberated."

Despite the seeming confidence of this proclamation, Tillich endured a life marked by profound pain, and his philosophy dwelled more often on imprisonment than on liberation. His first spiritual crisis came with the death of his mother, when he was seventeen years old. A poem he penned shortly thereafter foreshadowed his lifelong philosophical interests, what he would come to call "being and nonbeing":

> *Am I then I? who tells me that I am!*
> *Who tells me what I am, what I shall become?*
> *What is the world's and what life's meaning?*

What is being and passing away on earth?
O abyss without ground, dark depth of madness!
Would that I had never gazed upon you and were sleeping like a child!

For Tillich, the loss of his mother was not simply the loss of a loved one; in a single moment, the very "ground" of his existence had dropped away. Later in life his friend Rollo May, the American psychologist, described the abyss into which the adolescent had been suddenly cast: "At [his mother's] death he felt the whole world disappear from under his feet. In all its concrete vividness he experienced the reality of nothingness.... His orientation ... was gone; there was no longer any up or down."

Years later, Tillich's experience as a military chaplain in World War I would have a similarly profound and destabilizing effect on his understanding of human psychology. Having witnessed some of the bloodiest battles of the war, including Verdun and the Somme, he recorded the sheer chaos in a letter to his father: "Hell rages around us." At the Battle of the Marne, the sight of so many wounded and the experience of tending to the dying and the dead "absolutely transformed me." All told, Tillich was hospitalized three times during the war, for "nervous breakdowns."

Even after establishing himself as a prominent theologian and philosopher, Tillich remained haunted by these images of death: "True experience has its roots in suffering, and happiness is a blossom which opens itself up only now and then." Over the course of the war, Tillich had become unrecognizable to himself; long after the fighting had ended, he struggled to revive his interest in intellectual and spiritual pursuits, in love and marriage. "I have constantly the most immediate and very strong feeling that I am no longer alive," he wrote. "Therefore I don't take life seriously. To find someone, to become joyful, to recognize God, all these things are things of life. But life itself is not dependable ground.... I preach almost exclusively 'the end.'"

While in his writings Tillich was the most logically rigorous of philosophers, in person he was rarely free of anxiety. "Paulus lived in fear," explained his wife, Hannah. "His nervous body was tense; his desires, many. His fingers would fiddle with a pebble from the beach, a silver coin, or a paper clip. He breathed unevenly and sighed heavily, an ever-guilt-ridden Christian in distress." Tillich would later describe his anxiety as "the threat of non-being," a threat that "belongs to existence itself." His philosophy dwelled on the central paradox of all human endeavor: "We are not always aware of our having to die, but in the light of the experience of our having to die, our whole life is experienced differently." He believed that any balm for psychic distress, whether spiritual or psychotherapeutic, had to address this fundamental paradox of human psychology.

For Tillich, the threat of nonbeing encompassed not only physical death but the loss of one's identity, of whatever had once seemed to define us. To an audience of pastoral counselors, he described our looming destiny as a simple fact: "Anxiety is the awareness of finitude. Man comes from nothing and goes to nothing. He always lives in the conscious or unconscious anxiety of having to die. Non-being is present in every moment of his being." Awareness of our finitude was necessarily accompanied by feelings of alienation: "Everyone participates in the estranged character of existence," Tillich explained. "All men are estranged from what they essentially are. It is their tragic predicament to be guilty of this estrangement, although it is universal and inescapable." This "bondage to estrangement" results from our inability to accept our finitude, our inevitably fallen state: "Nothing is more difficult than to say 'yes' to oneself, especially if we see ourselves in the mirror of what we essentially are and should be."

In his sermons, Tillich probed this very challenge. "And now," he intoned, "let us look down into ourselves to discover there the struggle." Coaxing his audience along a path of understanding, he asked, "Who has not, at some time, been lonely in the midst of a social event?" Paradoxically, the sense of our "separation from the rest of life" is sharpest when we are surrounded by others. Only then do we realize how "fundamentally strange we are to each other, how estranged life is from life." It was these moments of estrangement that captured Tillich's imagination: "Each one of us draws back into himself. We cannot penetrate the hidden centre of another individual; nor can that individual pass beyond the shroud that covers our own being. Even the greatest love cannot break through the walls of the self. Who has not experienced that disillusionment of all great love?"

The self was fundamentally alone in the world. Simply leaving behind that solitude was not an option: "If one were to hurl away his self in complete self-surrender, he would become a nothing, without form or strength, a self without self, merely an object of contempt and abuse." Rather, we are saddled with the inherent paradoxes, even cruelties, of selfhood. "How often," Tillich continued, "we commit certain acts in perfect consciousness, yet with the shocking sense that we are being controlled by an alien power." Solace and salvation do not depend simply on choosing to act justly; we cannot hope to overcome our sinful and irrational nature.

For the suffering soul, then, the only hope lay in being "struck by grace." Grace, for Tillich, referred not merely to "moral self-control," but also to the experience of radical reconciliation with a power greater than oneself. This unnamable power "strikes us when we walk through the dark valley of a meaningless and empty life": "It strikes us when, year after year, the longed-for perfection of life does not appear, when the old compulsions reign within us as they have for decades, when despair destroys all joy and courage. Sometimes at that moment a wave of light breaks into our darkness, and it is as though a voice were saying: *You are accepted.*"

["but acceptance": Paul Tillich, *The Shaking of the Foundations* (New York: Scribner, 1953), 161–62.

"the self is liberated": Papers of Paul Tillich, Lecture Notes (collection bMS 649), Andover Theological Library, Harvard Divinity School, Cambridge, MA.

"Am I then . . . a child": Paul Tillich, quoted in Rollo May, *Paulus: Tillich as Spiritual Teacher* (Dallas: Saybrook, 1988), 41. Originally published as *Paulus: Reminiscences of Friendship* (New York: Harper and Row, 1973).

"up or down": Paul Tillich, quoted in Wilhelm and Marion Pauck, *Paul Tillich: His Life and Thought* (Eugene, OR: Wipf and Stock, 2015), 49.

"rages around us": May, *Paulus*, 40–41.

"transformed me": Paul Tillich, quoted ibid., 18.

"now and then": Paul Tillich, quoted in Pauck and Pauck, *Paul Tillich*, 43.

"'the end'": Ibid., 51.

"Christian in distress": Hannah Tillich, *From Time to Time* (New York: Stein and Day, 1974), 24.

"existence itself": Paul Tillich, *The Courage to Be* (New Haven, CT: Yale University Press, 1952), 48.

"experienced differently": Ibid., 56.

"essentially are and should be": Paul Tillich, "The Theology of Pastoral Care," *Pastoral Psychology* 10, no. 97 (October 1959): 23.

"discover there the struggle": Paul Tillich, *The Shaking of the Foundations*, 157.

"social event": Ibid.

"of all great love": Ibid.

"alien power": Paul Tillich, quoted in Stefan S. Jäger, *Glaube und Religiöse Rede bei Tillich und im Shin-Buddhismus: Eine religionshermeneutische Studie* (Berlin: Walter de Gruyter, 2011), 558.

"*You are accepted*": Paul Tillich, quoted in Pauck and Pauck, *Paul Tillich*, 92–93.]

212 "mediocrity of my natural faculties": Weil, *Waiting for God*, 23.

212 "longs for truth": Ibid.

212 "the affliction of others": Ibid., 25.

212 "the data": Ibid., 22.

212 "born inside": Ibid., 24.

212 "to add dogma": Ibid.

213 "wives of the fishermen": Ibid., 25, 26.

213 "each sound hurt me": Ibid., 26.

213 In a letter Weil wrote: Ibid., xxv.

213 reciting the Lord's Prayer: Ibid., 29.

214 "very first words": Ibid.

214 "infinitely more real": Ibid.

214 "sheer stupidity": Ibid., 60.

214 "Attention consists of suspending our thought": Ibid., 62.

214 "very rare and difficult thing": Ibid., 64.

CAPTURED BY A MESSAGE

Sources

Works Referenced

Dixie, Quinton, and Peter Eisenstadt, *Visions of a Better World: Howard Thurman's Pilgrimage to India and the Origins of African American Nonviolence*. Boston: Beacon, 2011.

Thurman, Howard. *Jesus and the Disinherited*. Boston: Beacon, 1976.

———. *With Head and Heart*. Orlando, FL: Harcourt Brace, 1979.

Notes

215 **Miriam Slade:** Dixie and Eisenstadt, *Visions of a Better World*, 97.

215 **"delighted":** Ibid., 97–98.

215 **Mahatma's questions:** Ibid., 100.

215 **"vitality":** Ibid., 107.

216 **"untouchability":** All dialogue in this chapter is taken from Howard Thurman's autobiography, *With Head and Heart*, 133–34.

216 **" 'Harijan,' . . . 'Child of God' ":** Ibid., 134.

216 **justified slavery:** Thurman, *Jesus and the Disinherited*, 19.

216 **"not niggers":** Thurman, *With Head and Heart*, 20–21.

217 **"religion of Jesus":** Thurman, *Jesus and the Disinherited*, 7.

217 **"backs against the wall"**: Ibid., 98.
217 **"Good News for the Underprivileged"**: Dixie and Eisenstadt, *Visions of a Better World*, 187; Thurman, *Jesus and the Disinherited*, 7.
218 **"social inferiority"**: Thurman, *Jesus and the Disinherited*, 39.
218 **"moral responsibility"**: Ibid., 75.
219 **"the unadulterated message"**: Dixie and Eisenstadt, *Visions of a Better World*, 112; Thurman, *With Head and Heart*, 134.
219 **"thank God"**: Thurman, *Jesus and the Disinherited*, 102.

The Revelation of Nature

Sources

Works Referenced

Addison, Joseph. *The Works of Joseph Addison*, vol. 3. New York: Harper and Brothers, 1837.
Coburn, Kathleen, ed. *The Notebooks of Samuel Taylor Coleridge*. New York: Pantheon, 1957.
Day, Henry J. M. *Lucan and the Sublime: Power, Representation, and Aesthetic Experience*. New York: Cambridge University Press, 2013.
Longinus. *On the Sublime*. Translated by H. L. Havell. London: Macmillan, 1890.
Perry, Seamus. *Coleridge and the Uses of Division*. Oxford: Oxford University Press, 1999.
Wordsworth, William. "Tintern Abbey," in *Lyrical Ballads*. London: J. and A. Arch, 1798.

Notes

219 **"Five years have passed . . . objects of all thought"**: Wordsworth, "Tintern Abbey," 300.
220 **"deep feeling"**: Perry, *Coleridge and the Uses of Division*, 125.
220 **"Percipient & the Perceived"**: As the *I* recedes, subject and object, perceiver and perceived, are fused in one of Wordsworth's "spots of time," a moment that crystallizes out of the flux of time, engulfing the entirety of our pasts and futures. "To view the world *sub specie aeternitatis*," Wittgenstein would write some hundred years later, "is to view it as a whole—a limited whole." Indeed, this limited whole was Wordsworth's private consolation for the disappointments of earthly existence. As both Wordsworth and Wittgenstein knew well, the numinous need not depend on institutionalized religion, or even on God. Aesthetic experience, too, can temporarily halt the flux of time, blurring the boundary between the self and the surrounding world. Indeed, Wittgenstein described the work of art as "the object seen *sub specie aeternitatis*": art both represents the world and at the same time transfigures it, elevating it above the contingencies of space and time.
　["limited whole": Ludwig Wittgenstein, quoted in James R. Atkinson, *The Mystical in Wittgenstein's Early Writings* (London: Routledge, 2009), 89.
　"*sub specie aeternitatis*": Ludwig Wittgenstein, diary entry (July 10, 1916), quoted in J. Mark Lazenby, *The Early Wittgenstein on Religion* (London: Bloomsbury, 2006), 50.]
221 **"idea of pain"**: Day, *Lucan and the Sublime*, 49.
222 **"agreeable kind of horror"**: Addison, *The Works of Joseph Addison*, 374.

CHAPTER 8: CAPTURE AND CHANGE

Sources

Work Referenced
Taber, Harry Persons, and Elbert Hubbard. *The Philistine: A Periodical of Protest* 24 (1906).

Notes

223 **form of capture:** Hannah Robinson at HarperCollins brought me this quotation from the nineteenth-century writer Elbert Hubbard that well expresses the quandary: "Fanaticism is a disease of the mind, just as alcoholism is a disease of the body, and the rational cure for both is the diminishing dose. That is, you are weaned from one thing by the substitution of something less harmful."

MARTIN LUTHER'S *ANFECHTUNGEN*

Sources

Interview by the Author
Reichelt, Silvio (2012).
Works Referenced
Brecht, Martin. *Martin Luther: His Road to Reformation, 1483–1521.* Stuttgart: Fortress, 1981.
Marius, Richard. *Martin Luther: The Christian Between God and Death.* Cambridge, MA: Harvard University Press, 1999.
Marty, Martin E. *Martin Luther: A Life.* New York: Penguin Books, 2004.
McKim, Donald K., ed. *The Cambridge Companion to Martin Luther.* Cambridge, UK: Cambridge University Press, 2003.
Meurer, Moritz. *Life of Martin Luther: Related from Original Authorities.* New York: H. Ludwig, 1848.
Mullett, Michael A. *Martin Luther.* New York: Routledge, 2015.
Osborn, Ian. *Can Christianity Cure Obsessive-Compulsive Disorder?* Grand Rapids, MI: Brazos, 2008.
Wilson, Derek. *Out of the Storm: The Life and Legacy of Martin Luther.* New York: St. Martin's, 2007.

Notes

224 *anfechtung:* Osborn, *Can Christianity Cure Obsessive-Compulsive Disorder?*, 53.
225 **"deception and delusion":** Meurer, *Life of Martin Luther,* 22.
226 **"I am not":** Mullett, *Martin Luther,* 65.
226 **"I am driven":** Wilson, *Out of the Storm,* 95.
226 **"round and round":** Ibid., 56.
227 **"so terribly angry":** Brecht, *Martin Luther,* 80.
228 **Eckhart . . . St. John:** Wilson, *Out of the Storm,* 77, 198.
228 **"nothing sounds sweeter":** Ibid., 61.
229 *"The just lives by faith":* McKim, *The Cambridge Companion to Martin Luther,* 90.

MEANINGFUL ASSOCIATION

Sources

Interview by the Author
Styron, Rose (2013, 2014).

Works Referenced

Styron, Alexandra. *Reading My Father*. New York: Simon and Schuster, 2011.
Styron, William. *Darkness Visible: A Memoir of Madness*. New York: Random House, 1990.

Notes

229 "an irreversible decision": William Styron, *Darkness Visible*, 64.
230 disposing of his journal, visiting his lawyer to draw up his final will: Ibid., 65.
230 "inadequate apologies": Ibid.
230 "sixty when the illness struck": Ibid., 38.
230 "hulking milestone of mortality": Ibid., 78.
231 "general feeling of worthlessness": Ibid., 5.
231 "depression-free": Rose Styron.
231 "unproductive and fallow . . . little beasts": Alexandra Styron, *Reading My Father*, 255.
231 "not even a visitor totally unannounced": William Styron, *Darkness Visible*, 79.
232 "soaring passage from the Brahms": Ibid., 37.
233 "admitted to the hospital": Ibid.

MOMENTS OF CLARITY

Sources

Works Referenced

"Bill Wilson's Story." Transcripts. http://silkworth.net/gsowatch/1938/manu38/manu38.htm.
B., Dick. *Henrietta B. Seiberling: Ohio's Lady with a Cause*. Kihei, HI: Paradise Research, 2006.
———. "The Oxford Group and Alcoholics Anonymous: Part One." 2003. Accessed Aug. 3, 2015. http://silkworth.net/aahistory/oxford_group_connection1.html.
Bishop, Michler F. *Managing Addictions: Cognitive, Emotive, and Behavioral Techniques*. Northvale, NJ: Aronson, 2001.
Cheever, Susan. *My Name Is Bill: Bill Wilson—His Life and the Creation of Alcoholics Anonymous*. New York: Washington Square, 2004.
Denzin, Norman K. *The Alcoholic Self*. Newbury Park, CA: SAGE, 1987.
———. *The Alcoholic Society: Addiction & Recovery of the Self*. New Brunswick, NJ: Transaction, 2007.
———. *The Recovering Alcoholic*. Newbury Park, CA: SAGE, 1987.
DiClemente, Carlo C. *Addiction and Change: How Addictions Develop and Addicted People Recover*. New York: Guilford, 2003.
Gorski, Terence T. *Understanding the Twelve Steps: An Interpretation and Guide for Recovering People*. New York: Simon and Schuster, 1989.
Hartigan, Francis. *Bill W.: A Biography of Alcoholics Anonymous Cofounder Bill Wilson*. New York: Thomas Dunne, 2000.

Khantzian, Edward J. *Treating Addiction as a Human Process*. New York: Rowman and Littlefield, 1999.

Kurtz, Ernest. *Not God: A History of Alcoholics Anonymous*. San Francisco: Harper and Row, 1991.

Markel, Howard, MD. "An Alcoholic's Savior: God, Belladonna, or Both?" April 19, 2010. http://www.nytimes.com/2010/04/20/health/20drunk.html.

Marlatt, G. Alan, and Dennis M. Donovan, eds. *Relapse Prevention: Maintenance Strategies in the Treatment of Addictive Behaviors*. New York: Guilford, 2005.

Marlatt, G. Alan, and Judith R. Gordon, eds. *Relapse Prevention: Maintenance Strategies in the Treatment of Addictive Behaviors*. New York: Guilford, 1985.

Parker, Jan, and Diana L. Guest. *The Clinician's Guide to 12-Step Programs: How, When, and Why to Refer a Client*. Westport, CT: Auburn House, 1999.

Seppala, Marvin D. *Clinician's Guide to the Twelve Step Principles*. New York: McGraw Hill, 2001.

Washton, Arnold, and Donna Bundy. *Willpower's Not Enough: Recovering from Addictions of Every Kind*. New York: Harper and Row, 1989.

Z., Phillip. *A Skeptic's Guide to the 12 Steps*. Center City, MN: Hazelden, 1990.

Notes

233 **"power of God":** Dick B., "The Oxford Group"; and Dick B., *Henrietta B. Seiberling*, 303.

234 **"survey of their defects":** "Bill Wilson's Story."

234 **"awful compulsion":** Ibid.

234 **"terrible balking":** Ibid.

234 **"obstinacy was crushed . . . His world":** Kurtz, *Not God: A History of Alcoholics Anonymous*, 19–20.

235 **"deflation at depth":** "Bill Wilson's Story."

235 **There is no doubt:** K. S. Walitzer, K. H. Dermen, and C. Barrick, "Facilitating Involvement in Alcoholics Anonymous During Out-Patient Treatment: A Randomized Clinical Trial," *Addiction* 104, no. 3 (March 2009): 391–401, doi: 10.1111/j.1360-0443.2008.02467.x; M. D. Litt, R. M. Kadden, E. Kabela-Cormier, and N. M. Petry, "Changing Network Support for Drinking: Network Support Project 2-Year Follow-Up," *Journal of Consulting and Clinical Psychology* 77, no. 2 (April 2009): 229–42, doi: 10.1037/a0015252; Project MATCH Research Group, "Matching Alcoholism Treatments to Client Heterogeneity: Project MATCH Posttreatment Drinking Outcomes," *Journal of Studies on Alcohol and Drugs* 58, no. 1 (Jan. 1997): 7–29; R. I. Longabaugh, P. W. Wirtz, A. Zweben, and R. L. Stout, "Network Support for Drinking, Alcoholics Anonymous and Long-Term Matching Effects," *Addiction* 93, no. 9, (Sept. 1998): 1313–33.

235 **this moment itself is freedom:** How do alcoholics achieve recovery? What does "recovery" even mean in the context of alcoholism? I asked Dr. Arnold Ludwig, a professor of psychiatry, who has spent decades studying what goes on in the alcoholic's mind.

"I think what you see with the alcoholic and the drug addict who have been able to break free, and the person who smokes tobacco, is not too dissimilar from what you see with people who are born again after some kind of religious revelation, some kind of insight; something strikes them," he told me.

He relayed his own experience of trying to stop smoking a pipe, first describing his failed attempts to overcome the strength of his conditioned response to keep smoking: "I would vow that I would quit, and then the next morning, I'd pick up the pipe and smoke it. This went on for years, and I tried all kinds of ways of stopping, like keeping all the empty cans around to remind myself how much I was smoking and how ridiculous it was. And yet it didn't help.

"Then one day—and I can put a date on it, February 20, 1980—there was this kind of a defining moment. I had a cold, a bad sore throat, and when I tried to smoke, I couldn't—it was too painful, so I stopped for a few days, and the cold started getting better. And my throat felt better. I was going to reach out for my pipe and stuff it with tobacco again and smoke it, and as I lit up, all of a sudden it hit me: this is absolutely stupid. It just categorically hit me. From that moment on, I never experienced another craving. It was like the door to my mind shut, and my whole attitude about smoking and all the things that I associated with it changed entirely."

I asked more about how these sudden shifts in attitude, these moments of awareness, help people overcome their addictions. "I'm not sure they're fully understood," he said. "But unless that kind of thing occurs, it's going to be a struggle, over and over and over again, as it was for my wife to give up smoking."

I mentioned what I thought happened with cigarette smoking:

"We didn't change the reinforcing properties of nicotine. We didn't change the makeup of cigarettes—what we did was change societal norms. There was a critical perceptual shift with regard to how we view cigarettes. For decades, cigarettes were associated with glamour, sexiness, or other positive attributes, and we shifted this perception. The conception changed from 'This is a product that I want' to 'I don't want it; it's a deadly, disgusting, addictive product.'"

Dr. Ludwig and I agreed that something more than a volitional shift had occurred, that there had been a change in the affective valence of the cigarette. From a public health perspective, this perceptual shift had been crucial.

A CREATIVE LIFE

Sources

Interview by the Author
Ware, Chris (2013, 2014).
Works Referenced

Ball, David M., and Martha B. Kuhlman, eds. *The Comics of Chris Ware: Drawing Is a Way of Thinking.* Jackson: University Press of Mississippi, 2010.

Raeburn, Daniel. *Chris Ware.* New Haven, CT: Yale University Press, 2004.

Ware, Chris. *The Acme Novelty Date-Book: Sketches and Diary Pages in Facsimile, Vol. 1, 1986–1995.* Montreal: Drawn and Quarterly, 2013.

———. *The Acme Novelty Date-Book: Sketches and Diary Pages in Facsimile, Vol. 2, 1995–2002.* Montreal: Drawn and Quarterly, 2007.

———. *Building Stories.* New York: Pantheon Books, 2012.

DISTRACTING THE BLACK DOG

Sources

Works Referenced

Attenborough, Wilfred. *Churchill and the "Black Dog" of Depression: Reassessing the Biographical Evidence of Psychological Disorder.* London: Palgrave Macmillan, 2014.

Churchill, Sir Winston. *Painting as a Pastime.* London: Unicorn, 2013.

Manchester, William. *The Last Lion: Winston Spencer Churchill, Visions of Glory, 1874–1932.* New York: Bantam Books, 1983.

Manchester, William, and Paul Reid. *The Last Lion: Winston Spencer Churchill, Defender of the Realm, 1940–1965.* New York: Bantam Books, 2012.

Notes

245   **"black dog returns"**: Attenborough, *Churchill and the "Black Dog" of Depression*, 72.

247   **"illuminated"**: Churchill, *Painting as a Pastime*, 7.

247   **"convulsive grasp"**: Ibid., 9.

247   **"Time stands respectfully aside"**: Ibid., 85–86.

BELIEF

Sources

Interview by the Author

Rivers, Rachel (2014).

GOOD-BYE TO ALL THAT

Sources

Works Referenced

Graves, Robert. *The Centenary Selected Poems.* Manchester, UK: Carcanet Poetry, 1995.

———. *Good-Bye to All That.* New York: Anchor Books, 1998.

Notes

250   **"My breaking point"**: Graves, *Good-Bye to All That*, 198.

250   **"Shells . . . bursting on my bed"**: Ibid., 287.

251   **"undisguised history"**: Ibid., 321.

252   **put miles between**: "I don't remember well," writes Branko Lustig. "When I read some books and they describe the scenes, then I remember, yes, this happened to me. But I don't want to remember. I never wanted to remember, and I think this is the reason why I never had nightmares."

Branko Lustig lived for two years as an adolescent in the death camps at Auschwitz and Bergen-Belsen. He survived to become a film producer whose credits include the Oscar-winning *Schindler's List*, and he has been honored for his many efforts to keep the memory of the Holocaust alive. When he returned to Auschwitz in 2011 to celebrate the bar mitzvah that he missed during his imprisonment, the trip was memorialized in a *New York Times* video "op-doc," where he talks about how forgetting traumatic experiences has served him well.

Branko Lustig offers no hint of how he accomplished this fine balance of effica-
cious forgetting and purposeful remembering; we can only guess at the mix of
personality and circumstance that armored him as a child.
["op-doc": Topaz Adizes, "Branko: Return to Auschwitz," *New York Times*
online ed., April 14, 2013, http://www.nytimes.com/2013/04/15/opinion/branko
-return-to-auschwitz.html.]
252   **night-stumbling:** Graves, *Centenary Selected Poems*, 91.

RECONCILIATION AND FORGIVENESS

Sources

Interview by the Author
Shezi, Thandi (2015).
Works Referenced
Posel, Deborah, and Graema Simpson, eds. *Commissioning the Past: Understanding
    South Africa's Truth and Reconciliation Commission.* Johannesburg: Wits Uni-
    versity Press, 2002.
Truth and Reconciliation Commission: Human Rights Violations, Women's Hearing.
    July 28, 1997. http://www.justice.gov.za/trc/special%5Cwomen/shezi.htm.
Urbsaitis, Bryan Mark. "Reconciliation Contested: Understandings of Reconcilia-
    tion by NGOs in South Africa." PhD diss., Steinhardt School of Education,
    New York University, New York, 2007.

Notes

253   **"Do not kill her in front of me":** Posel and Simpson, *Commissioning the Past,*
      118.
254   **"control of my soul":** Ibid., 121.
254   **common commitment:** Adam von Trott zu Solz adored Germany as most par-
      ents love their children. A lawyer and diplomat, Trott played a key role in the
      anti-Nazi resistance and the July 20 plot to assassinate Adolf Hitler. During his
      life, and in the decades following his death, however, Trott was often misrepre-
      sented as a Nazi collaborator. In fact, he fixated on the idea of saving Germany
      both from the Nazi regime and from destruction and dismemberment at the
      hands of its opponents. He was willing to sacrifice personal relationships and
      his own safety in order to serve that goal.
         Returning to Germany from Sweden in 1944, just days before the assassina-
      tion attempt, Trott wrote a letter to his wife, Clarita, that captures the sheer
      intensity of his commitment. If Trott's cohort successfully carried out the assas-
      sination, he explained, they would save his fatherland from inevitable self-
      destruction under Nazism.
         Looking out the airplane window, he describes the villages and hamlets, the
      pale ocher of the roofs, the white stucco walls, a solitary church spire poking up
      over each village, and rows of trees and hedges bordering the green and hay-
      colored fields. He was looking down on more than his home; Germany—its
      culture, history, art, music, and philosophy—was the origin of his identity. He
      and his country were so intimately joined that, were Hitler to prevail, he knew
      his soul would wither.
         Trott felt the selfishness inherent in his collaboration with the other con-
      spirators: he was, after all, risking his life and imperiling his family. Yet when he

joined the conspiracy, his identity seemed to melt into something much larger and more important than he. He identified so completely with his mission that he felt his life fuse with it. Whatever the outcome, when the mission was complete, it was as if his life, his *true life*, would be over. A Swedish friend, Fru Almstrom, wrote to Trott's wife about her husband's last visit, just days before this flight home: "I remember one evening, when he was exhausted mentally and physically and I asked him to go back to the hotel and sleep. He looked at me and said: 'Why should I sleep when there is so much to do? . . . And besides, old people do not need so much sleep.' Whereupon I . . . answered, 'But you are only 35.' 'No,' Adam said, 'I am at least 60 and I will never be younger—I think I have done what I was supposed to do in my life, whatever was asked of me—and I am ready to die.'" It was as if the pattern of his entire life was completing itself with this mission.

Now, as he flew home only days before the planned assassination, the drone of the passenger plane's engines set him dreaming of the past, his youth, his time in England as a Rhodes Scholar obsessed with political philosophy, and his later years of travel, as he secretly tried, and failed, to draw England and America into an effort to thwart Hitler. He remembered his father, busy in his post as minister of culture for the kingdom of Prussia during World War I. During those years, his mother would spend the summers with her children at the family's estate at Imshausen. As a child, Adam was entranced by the property's beauty; he remembered asking his father with wonder, "Is this all *ours*?" The prescient answer: "It has been entrusted to us." He was now shouldering that trust in the most serious of ways, taking responsibility for the fate of Germany. Recalling those visits now, looking out over familiar landscapes, he felt love for those woods, the ponds and sloping fields, the smell of the grass and summer rain around their home.

He once wrote: "We can make the ethical life exist only to the extent that we develop it in our own person, out of ourselves. . . . In a chaotic world, this is the best defense: order in one's inner life." It soothed his fears whenever he felt the least doubt about the violent act he had committed himself to. Killing might be wrong, but in this case, it promised to restore to the outer world around him the sense of order and meaning he had found in his inner life. He later remembered the moment that he reentered German airspace: "I was filled again by a sense of deep love and joy that I have been thrust into the place where I am, and in such difficult times, to fight with others for our homeland. I do not believe that any relationship with any other human being can mean as much to me as this does, and to prove myself in this, and to serve such a cause—this is my first duty. To recognize one's own unique task, this liberates a man and gives him firmness in life; this allows him a clear choice amid the manifold and conflicting values and principles which confront any citizen of the modern world."

This passion simplified Trott's entire life into a single purpose; all other considerations gradually dissolved. A year earlier, he had met Claus Schenk von Stauffenberg; the two soon became intimate friends. Along with his coconspirators, Trott had worked tirelessly on a plan, with Stauffenberg as the designated assassin. Trott's wife later wrote, "It is quite possible that for Adam this friendship was the fulfillment of something which he had been seeking all his life, in spite of the fact that he had so many friends to whom he was united by affection, respect, and by tasks and responsibilities shared in common." Shortly after Trott's return from Sweden, on July 15, 1944, Stauffenberg attempted to set off a bomb in close proximity to Hitler, but failed. The next day, Trott met with Stauffenberg at the man's apartment, along with other members of the conspiracy, to discuss what options remained. It may have been this meeting that doomed him by alerting the Nazis to his complicity in the plot. Four days later, his hopes had dwindled, and those who met him reported that he had looked pale, desperate, and defeated.

Something within him had collapsed. He seemed to have given up on life, as if he'd already prepared himself to die. He'd discovered a deep sense of purpose in his mission; when the conspiracy failed, he lost *himself*. Though it isn't clear how the Gestapo so quickly tracked down members of the conspiracy after the assassination attempt, Trott seemed to know his fate. He made no attempt to save himself from arrest when the Nazis came for him. In a final letter to his wife from prison, just days before his execution, he grieved for his lost opportunity: "What hurts me most is that I can no longer serve our country."

[serve that goal: For a discussion of von Trott's life and historical legacy, see Henric L. Wuermeling, *Doppelspiel: Adam von Trott zu Solz im Widerstand gegen Hitler* (Munich: Deutsche Verlags-Anstalt, 2004); Klemens von Klemperer, ed., *A Noble Combat: The Letters of Shiela Grant Duff and Adam von Trott zu Solz 1932–1939* (Oxford: Oxford University Press, 1988); Giles MacDonogh, *A Good German: A Biography of Adam von Trott zu Solz* (reprint, Woodstock, NY: Overlook, 1993); Andreas Schott, *Adam von Trott zu Solz: Jurist im Widerstand* (Paderborn, Germany: Schoningh, 2001); Christopher Sykes, *Troubled Loyalty: A Biography of Adam von Trott zu Solz* (London: Collins, 1968); Benigna von Krusenstjern, *"Daß es Sinn hat zu sterben—gelebt zu haben": Adam von Trott zu Solz, 1909–1944* (Göttingen, Germany: Wallstein, 2009); Clarita von Trott zu Solz, *Adam von Trott zu Solz* (Berlin: Lukas, 2009); Heinric L. Wuermeling, *Adam von Trott zu Solz: Schlüsselfigur im Widerstand gegen Hitler* (Munich: Pantheon, 2009).

self-destruction under Nazism: Adam von Trott, letter to Clarita von Trott (July 1944), in Christopher Sykes, *Troubled Loyalty: A Biography of Adam von Trott zu Solz* (London: Collins, 1968), 415.]

254 **"violent all the time":** Posel and Simpson, *Commissioning the Past*, 124.
255 **"ten years":** Urbsaitis, "Reconciliation Contested," 131.
257 **"deal with it":** Ibid., 130.

A TOOLKIT BORROWED FROM BUDDHISM

Sources

Interviews by the Author
Kabat-Zinn, Jon (2014).
Thompson, Evan (2014).

Works Referenced
Austin, James H. *Zen and the Brain: Toward an Understanding of Meditation and Consciousness*. Cambridge, MA: MIT Press, 1999.

———. *Zen-Brain Reflections: Reviewing Recent Developments in Meditation and States of Consciousness*. Cambridge, MA: MIT Press, 2006.

Buswell, Robert E., Jr., and Donald S. Lopez Jr. *The Princeton Dictionary of Buddhism*. Princeton, NJ: Princeton University Press, 2014.

Fingarette, Herbert. *The Self in Transformation: Psychoanalysis, Philosophy, and the Life of the Spirit*. New York: Harper and Row, 1963.

Germer, Christopher K., Ronald Siegel, and Paul R. Fulton. *Mindfulness and Psychotherapy*. New York: Guilford, 2005.

Goleman, Daniel. *Destructive Emotions: How Can We Overcome Them? A Scientific Dialogue with the Dalai Lama*. New York: Bantam, 2003.

Kabat-Zinn, J. "Bringing Mindfulness to Medicine": Interview by Karolyn A. Gazella. *Alternative Therapies in Health and Medicine* 11, no. 3 (May–June 2005): 56–64.

———. "Bringing Mindfulness to Medicine: An Interview with Jon Kabat-Zinn, PhD." Interview by Karolyn Gazella. *Advances in Mind-Body Medicine* 21, no. 2 (2005): 22–27.

———. "An Outpatient Program in Behavioral Medicine for Chronic Pain Patients Based on the Practice of Mindfulness Meditation: Theoretical Considerations and Preliminary Results." *General Hospital Psychiatry* 4, no. 1 (April 1982): 33–47.

Kabat-Zinn, J., R. J. Davidson, J. Schumacher, M. Rosenkranz, D. Muller, S. F. Santorelli, F. Urbanowski, A. Harrington, K. Bonus, and J. F. Sheridan. "Alterations in Brain and Immune Function Produced by Mindfulness Meditation." *Psychosomatic Medicine* 64, no. 4 (July–Aug. 2003): 564–70.

Kabat-Zinn, J., A. O. Massion, J. Kristeller, L. G. Peterson, K. E. Fletcher, L. Pbert, W. R. Lenderking, and S. F. Santorelli. "Effectiveness of a Meditation-Based Stress Reduction Program in the Treatment of Anxiety Disorders." *American Journal of Psychiatry* 149, no. 7 (July 1992): 936–43.

Kabat-Zinn, J., E. Wheeler, T. Light, A. Skillings, M. J. Scharf, T. G. Cropley, D. Hosmer, and J. D. Bernhard. "Influence of a Mindfulness Meditation–Based Stress Reduction Intervention on Rates of Skin Clearing in Patients with Moderate to Severe Psoriasis Undergoing Phototherapy (UVB) and Photochemotherapy (PUVA)." *Psychosomatic Medicine* 60, no. 5 (Sept.–Oct. 1998): 625–32.

Kabat-Zinn, J., and R. Davidson, eds. *The Mind's Own Physician*. Oakland, CA: New Harbinger, 2011.

Ludwig, D. S., and J. Kabat-Zinn. "Mindfulness in Medicine." *Journal of the American Medical Association* 300, no. 11 (Sept. 17, 2008): 1350–52.

Meili, T., and J. Kabat-Zinn. "The Power of the Human Heart: A Story of Trauma and Recovery and Its Implications for Rehabilitation and Healing." *Advances in Mind-Body Medicine* 20, 1 (2004): 6–16.

Paulson S., R. Davidson, A. Jha, and J. Kabat-Zinn. "Becoming Conscious: The Science of Mindfulness." *Annals of the New York Academy of Sciences* 1303 (Nov. 2013): 87–104.

Thompson, Evan, ed. "Between Ourselves: Second-Person Issues in the Study of Consciousness." *Journal of Consciousness Studies* 8, no. 5–7. Thorverton, UK: Imprint Academic, 2001.

———. *Colour Vision: A Study in Cognitive Science and the Philosophy of Perception*. Philosophical Issues in Science. London: Routledge, 1995.

———. *Mind in Life: Biology, Phenomenology, and the Sciences of Mind*. Cambridge, MA: Belknap, 2007.

———. *The Problem of Consciousness: New Essays in Phenomenological Philosophy of Mind. Canadian Journal of Philosophy Supplementary Volume.* Calgary: University of Calgary Press, 2003.

———. *Waking, Dreaming, Being: Self and Consciousness in Neuroscience, Meditation, and Philosophy.* New York: Columbia University Press, 2015.

Varela, Francisco J., Evan Thompson, and Eleanor Rosch. *The Embodied Mind: Cognitive Science and Human Experience.* Cambridge, MA: MIT Press, 1991.

Wallace, Alan B. *The Attention Revolution: Unlocking the Power of the Focused Mind.* Somerville, MA: Wisdom, 2006.

Zelazo, Philip David, Morris Moscovitch, and Evan Thompson. *The Cambridge Handbook of Consciousness.* Cambridge, UK: Cambridge University Press, 2007.

Is There Freedom from Capture?

Sources

Interview by the Author
Roeske, Danielle (2013, 2014).

Works Referenced

Basch, Michael Franz. *Doing Brief Psychotherapy.* New York: HarperCollins, 1995.

Basseches, Michael, and Michael F. Mascolo. *Psychotherapy as a Developmental Process.* New York: Routledge, 2010.

Bohart, Arthur C., and Karen Tallman. *How Clients Make Therapy Work: The Process of Active Self-Healing.* Washington, DC: American Psychological Association, 1999.

Castonguay, Louis G., and Larry E. Beutler. *Principles of Therapeutic Change That Work.* Oxford: Oxford University Press, 2006.

Castonguay, Louis G., and Clara E. Hill, eds. *Transformation in Psychotherapy: Corrective Experiences Across Cognitive Behavioral, Humanistic, and Psychodynamic Approaches.* Washington, DC: American Psychological Association, 2012.

Corsini, Raymond J., and Danny Wedding, eds. *Current Psychotherapies.* Belmont, CA: Brooks/Cole, 2008.

Crenshaw, David A. *Evocative Strategies in Child and Adolescent Psychotherapy.* New York: Aronson, 2006.

Cummings, Nicholas A., and Janet L. Cummings. *The Essence of Psychotherapy: Reinventing the Art in the New Era of Data.* San Diego: Academic Press, 2000.

Dryden, Windy. *Developments in Psychotherapy: Historical Perspectives.* London: SAGE, 1996.

Ehrendwald, Jan. *The History of Psychotherapy: From Healing Magic to Encounter.* New York: Aronson, 1976.

Eigen, Michael. *Reshaping the Self: Reflections on Renewal Through Therapy.* London: Karnac Books, 2013.

Gaylin, Willard. *How Psychotherapy Really Works: How It Works When It Works and Why Sometimes It Doesn't.* Chicago: Contemporary Books, 2001.

Gendlin, Eugene T. *Focusing-Oriented Psychotherapy: A Manual of Experiential Method.* New York: Guilford, 1996.

Johnson, Susan M. *The Practice of Emotionally Focused Couple Therapy: Creating Connection.* New York: Brunner-Routledge, 2004.

Jones-Smith, Elsie. *Theories of Counseling and Psychotherapy: An Integrative Approach.* Los Angeles: SAGE, 2012.

Magnavita, Jeffrey J., and Jack C. Anchin. *Unifying Psychotherapy: Principles, Methods, and Evidence from Clinical Science.* New York: Springer, 2014.

Mahoney, Michael J. *Human Change Processes: The Scientific Foundations of Psychotherapy.* New York: Basic, 1991.

Nardone, Giorgio, and Claudette Portelli. *Knowing Through Changing: The Evolution of Brief Strategic Therapy.* Norwalk, CT: Crown House, 2005.

Nelson, Thorana S. *Doing Something Different: Solution-Focused Brief Therapy Practices.* New York: Routledge, 2010.

Norcross, John C., ed. *Psychotherapy Relationships That Work: Evidence-Based Responsiveness.* New York: Oxford University Press, 2011.

O'Donohue, William, and Steven R. Graybar, eds. *Handbook of Contemporary Psychotherapy: Toward an Improved Understanding of Effective Psychotherapy.* Los Angeles: SAGE, 2009.

Rosenfeld, George W. *Beyond Evidence-Based Psychotherapy: Fostering the Eight Sources of Change in Child and Adolescent Treatment.* New York: Taylor and Francis, 2009.

Ryan, Jane, ed. *How Does Psychotherapy Work?* London: Karnac, 2005.

Sommers-Flanagan, John, and Rita Sommers-Flanagan. *Counseling and Psychotherapy Theories in Context and Practice: Skills, Strategies, and Techniques.* Hoboken, NJ: Wiley, 2012.

Sperry, Len. *Highly Effective Therapy: Developing Essential Clinical Competencies in Counseling and Psychotherapy.* New York: Taylor and Francis, 2010.

Symington, Neville. *A Healing Conversation: How Healing Happens.* New York: Karnac, 2006.

Walker, Nigel. *A Short History of Psychotherapy in Theory and Practice.* New York: Taylor and Francis, 1999.

Wallerstein, Robert S. *The Talking Cures.* New Haven, CT: Yale University Press, 1995.

Walter, Bromberg, and Winfred Overholser. *Man Above Humanity: A History of Psychotherapy.* Philadelphia, PA: Lippincott, 1954.

Watzlawick, Paul, John Weakland, and Richard Fisch. *Change: Principles of Problem Formation and Problem Resolution.* New York: Norton, 1974.

Yalom, Irvin D., and Molyn Lescz. *The Theory and Practice of Group Psychotherapy.* New York: Basic, 2005.

Zeig, Jeffrey K., ed. *The Evolution of Psychotherapy.* New York: Brunner/Mazel, 1987.

Notes

262 **release:** Mary Karr, David Foster Wallace's lover for a time, became sober with him. Twenty years later she described her experience: "You assume that when you quit drinking, you're surrendering to that kind of nasty schoolmarm rule-maker. But for me getting sober has been freedom—freedom from anxiety and freedom from . . . my head. What has kept me sober is not that strict rule-following schoolmarm. There's more of a loving presence that you become aware of that is I think everyone's real, actual self—who we really are. Blake said, 'We are put on earth a little space, / That we might learn to bear the beams of love.' And I think, quote-unquote, 'bearing the beams of love' is where the freedom is, actually. Every drunk is an outlaw, and certainly every artist is.

Making amends, to me, is again about freedom. I do that to be free of the past, to not be haunted. That schoolmarm part of me—that hypercritical finger-wagging part of myself that I thought was gonna keep me sober—that was actually what helped me stay drunk. What keeps you sober is love and connection to something bigger than yourself. When I got sober, I thought giving up was saying goodbye to all the fun and all the sparkle, and it turned out to be just the opposite. That's when the sparkle started for me."

["That's when the sparkle started": Nina Puro, "Mary Karr: David Foster Wallace and I Kept Each Other Alive," *Salon*, May 23, 2013. http://www.salon .com/2013/05/23/mary_karr_infinite_jest_was_unkind_partner/.]

ACKNOWLEDGMENTS

There would be no *Capture* without those whose names are in the body of this book. I thank them and add to their ranks those who appear below. As is obvious, there are many people to whom I am indebted, but any errors are mine alone.

Many years, many characters, many days and nights later: I benefited beyond measure from the insights and talents of the writer Nell Casey. Without her and Thomas Dolinger, my thoughts and theories would be in lesser versions still. I am enormously grateful for their patience and craft.

Kelsey Osgood, Zara Houshmand, David Dorsey, Karyn Feiden, Carrie Hagen, and Rebecca Markovits also lent their considerable skills to helping me tell the stories found in *Capture*.

As is usual with my work, Dick Todd has been an editor without peer.

Once again, Chip Kidd, through his jacket design, has understood and encouraged me. The poet Sarah Manguso read *Capture* and got to its essence with her beautiful epigraph. The photographer Peter Turnley did the best anyone could with my author's picture. Drew Altman got to the heart of the book when he suggested the subtitle.

My gurus at Harper, Julie Will and Karen Rinaldi, stuck with me and made me better.

Tanya Boroff and Richard Alwyn Fisher kept me and the manuscript in line by, among other things, transcribing and cite-checking.

Hannah Robinson and John Jusino have been gallant in production, and Brian Perrin is the able marketer, at Harper.

Jenna Dolan proved a worthy copy editor and Kate Mertes, a meticulous indexer. Sheri Gilbert was the scrupulous permissions editor. Tamara Ghattas provided the *Chicago Manual of Style* formatting. Emily Lupton Metrish and Lisa Rowlinson de Ortiz worked on the Library of Congress classifications. Ernie Franic would often arrive in the night to provide tech support.

Emily Loftis, Imad Dphrepaulezz, and Jacob John Gabdaw spent weeks in my basement fact-checking. Jean Cannon and Emily Roehl labored in the archives of the Harry Ransom Center at The University of Texas at Austin. Matthew Scherbel conducted additional research, as did translators Moira G. Weigel and Jeb Bishop. I could not have done without the proofreading skills of Eliza Childs, Mary Bagg, Susan Gamer, Christina Gaugler, Sheila Himmel, Chris Jerome, Kate Johnson, Ella Kusnetz, Elizabeth Macfie, Sheila Oakley, and Nina Questal. Nancy Adler, Mark Costello, Joel Ehrenkranz, Howard Gardner, Katie Hafner, Suzanne Kahn, Marty Morse, Michael Pollan, Lisa Raskin, Richard Walter, and Laura Yorke read versions of *Capture* and provided helpful comments.

The scientists and clinicians Sunny Dutra, Seth Disner, Chris Palmer, Rebecca Pearl, Sara Szczepanski, and Jelena Markovic provided me with their expertise and research help and deserve credit for the science section and references in the book.

A number of literary biographers were generous with their knowledge, including Gail Crowther, coauthor of *Sylvia Plath in Devon*; Dawn Skorczewski, author of *An Accident of Hope: The*

Therapy Tapes of Anne Sexton; and Dan Max, author of *Every Love Story Is a Ghost Story: A Life of David Foster Wallace.*

I was enriched through my conversations with John Allman, Lori Altshuler, Brian Anderson, Christopher Bobonich, Nancy Carlisle, Marvin Chun, Bruce Cuthbert, Arlene Durk, David Durk, Jenny Gaither, Kelly Green, Jonathan Harr, Jake Ifshin, Laurent Itti, John F. Kelly, James Klagge, Anja Koski-Jannes, John Krystal, Leah Levy, Kristian Makron, Luiz Pessoa, Silvio Reichelt, Michael Roeske, David Sander, Natasha Staller, Ronald Tiersky, Rebecca Todd, Martha Umphrey, and Steve Yantis.

At my academic home, the University of California, San Francisco, I was aided by Changzhao Feng, Miranda Chiu, and Donna Ferriero and by Bernard Lo and Lisa Denney on matters related to the Institutional Review Board. Elizabeth Dupuis and Thomas Leonard at the University of California, Berkeley, Library have provided invaluable help.

As always, Tina Andreadis, Meghan Cleary, Heather Drucker, Sharan Jayne, Doug Levy, Jeff Nesbit, Jim O'Hara, and Marci Robinson have aided me in communicating my message to the public.

An exception to my rule of not mentioning here those whose names appear in the book are those who pseudonymously confided in me. I am especially grateful for the candor of Nora, Wes, Jerome, Frances, Ned, Margaret, Jackie, and Charlotte.

Also exceptional in the course of my research, thinking, and writing were Jim, Sally, and Amy Wallace; Mark Costello; and Corey Washington.

My agent Kathy Robbins has always believed I have something worth saying. Aided by Janet Oshiro, Kathy has been a constant advocate of my work.

This project could not have existed without the support of Lynda and Stewart Resnick and Dagmar Dolby.

I hold in admiration the late senators Ted Kennedy and Paul Wellstone, and all those who fought to make the Mental Health Parity Act a reality. Mayada Akil and Meredith Cary are providers of whom the senators would be proud.

The dedication of this book to my son, Ben, honors a great conversation that spanned years on the subjects in *Capture*. My daughter, Elise Snyder, and her husband, Mike, always encourage me, and their daughter, Lena, is a new source of inspiration. My mother, Roz, reads me and questions me, always. My entire adult life, every adventure worth having has been with my wife, Paulette. This one was no exception.

INDEX

Titles of works will generally be found under the author's or artist's name. Page numbers followed by an "n" indicate references to the notes.